THE CATALYTIC CHEMISTRY
OF NITROGEN OXIDES

PUBLISHED SYMPOSIA

Held at the
General Motors Research Laboratories
Warren, Michigan

Friction and Wear, 1959
Robert Davies, *Editor*

Internal Stresses and Fatigue in Metals, 1959
Gerald M. Rassweiler and William L. Grube, *Editors*

Theory of Traffic Flow, 1961
Robert Herman, *Editor*

Rolling Contact Phenomena, 1962
Joseph B. Bidwell, *Editor*

Adhesion and Cohesion, 1962
Philip Weiss, *Editor*

Cavitation in Real Liquids, 1964
Robert Davies, *Editor*

Liquids: Structure, Properties, Solid Interactions, 1965
Thomas J. Hughel, *Editor*

Approximation of Functions, 1965
Henry L. Garabedian, *Editor*

Fluid Mechanics of Internal Flow, 1967
Gino Sovran, *Editor*

Ferroelectricity, 1967
Edward F. Weller, *Editor*

Interface Conversion for Polymer Coatings, 1968
Philip Weiss and G. Dale Cheever, *Editors*

Associative Information Techniques, 1971
Edwin L. Jacks, *Editor*

Chemical Reactions in the Urban Atmosphere, 1971
Charles S. Tuesday, *Editor*

The Physics of Opto-Electronic Materials, 1971
Walter A. Albers, Jr., *Editor*

Emissions From Continuous Combustion Systems, 1972
Walter Cornelius and William G. Agnew, *Editors*

Human Impact Response, Measurement and Simulation, 1973
William F. King and Harold J. Mertz, *Editors*

The Physics of Tire Traction, Theory and Experiment, 1974
Donald F. Hays and Alan L. Browne, *Editors*

The Catalytic Chemistry of Nitrogen Oxides, 1975
Richard L. Klimisch and John G. Larson, *Editors*

THE CATALYTIC CHEMISTRY
OF NITROGEN OXIDES

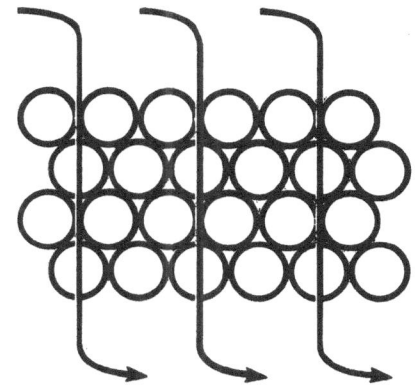

Edited by
RICHARD L. KLIMISCH and **JOHN G. LARSON**

General Motors Research Laboratories

PLENUM PRESS ● **NEW YORK – LONDON** ● **1975**

Library of Congress Cataloging in Publication Data

Symposium on the Catalytic Chemistry of Nitrogen Oxides, Warren, Mich., 1974.
The catalytic chemistry of nitrogen oxides.

Includes index.
1. Nitrogen oxides — Congresses. 2. Catalysis — Congresses. I. Klimisch, Richard L. II. Larson, John G. III. General Motors Corporation. Research Laboratories.
QD181.N1S95 1974   546'.711'2   75-22332
ISBN-13: 978-1-4615-8743-9   e-ISBN-13: 978-1-4615-8741-5
DOI: 10. 1007/978-1-4615-8741-5

Proceedings of the Symposium on The Catalytic Chemistry of Nitrogen Oxides
held at the General Motors Research Laboratories,
Warren, Michigan, October 7-8, 1974

© 1975 Plenum Press, New York
Softcover reprint of the hardcover 1st edition 1975
A Division of Plenum Publishing Corporation
227 West 17th Street, New York, N.Y. 10011

United Kingdom edition published by Plenum Press, London
A Division of Plenum Publishing Company, Ltd.
Davis House (4th Floor), 8 Scrubs Lane, Harlesden, London, NW10 6SE, England

PREFACE

This book contains the papers and discussions from the symposium on "The Catalytic Chemistry of Nitrogen Oxides" held at the General Motors Research Laboratories on October 7-8, 1974. This symposium is the eighteenth in the annual series presented by the Research Laboratories. The topics for these symposiums have covered a broad range. Each topic was selected to be of intense current interest and of significant technical importance.

There is no question that the subject of the 1974 Symposium satisfies these two criteria. The control of automotive nitrogen oxides has been perhaps the most difficult and controversial area of automotive emissions both in terms of what is necessary and in terms of what is technically feasible. This area has been a source of considerable discussion not only in the technical community but also in governments both in the U. S. and abroad.

This meeting brought together scientists working in surface chemistry with engineers working on system design. It also brought together representatives of government, academia and industry. We feel that an important side benefit of the meeting was the improved understanding that was developed between these groups. Participants came from Europe and Japan as well as Canada and the United States.

The technical papers spanned the range from fundamental interactions of NO on surfaces through bench scale kinetic and mechanistic studies and ended with catalytic applications. Although the emphasis was on automotive NO_x removal, stack gas NO_x control was also covered.

The interaction of nitrogen containing compounds with surfaces is certainly one of the most fascinating and important areas of science. The fact that nitrogen resists both fixation and defixation is one of the most striking examples of perversity in nature. On the other hand, one can now see the connection between the stability of ruthenium-N_2 complexes and the effectiveness of ruthenium catalysts in both the fixation and defixation of nitrogen. Some of these ideas are discussed in the papers herein. We hope that this book serves to inspire further insights.

There were a number of people who helped make this symposium a success. We wish to call special attention to Dr. J. C. Schlatter. His attention to detail both in physical arrangements for the meeting and in the publication of this volume is gratefully acknowledged.

In addition, the help of Mr. K. Antonius for the meeting arrangements and Mr. D. Havelock for the book was indispensible. We would also like to single out Dr. K. C. Taylor, Mr. R. M. Sinkevitch, Mrs. A. Louth, and Mrs. M. Baldwin for their assistance.

Richard L. Klimisch
John G. Larson

CONTENTS

SESSION I

FUNDAMENTAL STUDIES
OF NITROGEN OXIDE/SURFACE INTERACTIONS

Session Chairman
G. A. SOMORJAI

University of California
Berkeley, California

EPR AND IR STUDIES

OF SURFACE NITROSYL COMPLEXES

J. H. LUNSFORD

Texas A&M University, College Station, Texas

ABSTRACT

Epr and ir studies provide convincing evidence that nitric oxide forms surface complexes with metal ions. The nature of the bonding and the stability of the complexes vary greatly from one metal ion to another. In complexes with Ni^{2+}, Cr^{2+}, and Fe^{2+} the NO moiety becomes cationic in character, and the epr spectra suggest a formal reduction of the metal ion. Such complexes are moderately stable at 25°C with respect to the loss of the NO ligand. Weak covalent bonding is observed when NO complexes with Cu^+ or Ag^+; however, a complex attributed to $[Ag(I)_2 NO]^{2+}$ is more stable. Nitrosyl complexes formed with Co^{2+} ions in Y-type zeolites are particularly interesting since it is possible to alter the anionic character of the NO group by the addition of electron – donating ligands such as NH_3. Three types of complexes may be formed, which have been described as $[Co(II) (NO)_2]^{2+}$, $[Co(II) NH_3 (NO)_2]^{2+}$, and $[Co(III) (NH_3)_5 NO]^{2+}$, based upon their infrared spectra. In the latter complex the anionic nitrosyl group reacts both with the ammonia ligands and with gas phase nitric oxide. The reaction with NH_3 produces N_2 and H_2O, whereas the reaction with excess NO yields N_2O and the nitro complex, $[Co(NH_3)_5 NO_2]^{2+}$.

INTRODUCTION

Metal nitrosyl complexes have been widely studied in both homogenous and heterogeneous systems with a view to understanding the relationship between bonding and the stability of the nitrosyl group. As pointed out by Ibers and co-workers (1), the mode of coordination of the nitrosyl group in a complex is dependent on the relative energies of the $\pi*$ orbital of the NO and the d orbitals of the metal, the two limiting cases being NO^+ and NO^-. If there exists an empty,

References p. 16.

low-energy, d-orbital on the metal, the pair of electrons forming the coordinate bond will be localized on the metal, and the ligand may be described as NO^+; if not, the electrons will fill the $\pi*$ orbital of the nitrosyl, and the species may be written as NO^-. The ligand NO^+ has a formal triple bond with sp hybridization; whereas, the ligand NO^- has a formal double bond with sp hybridization (2). The N-O bond strength is less in the latter case and the ligand is more likely to dissociate.

These differences in bonding are reflected in the epr, ir and photoelectron spectra of the complexes. Furthermore, where x-ray diffraction studies have been carried out, the M-N-O bond angle may be used to characterize the type of metal-ligand bond. Typically, a cationic nitrosyl ligand results in a nearly linear M-N-O bond, greater ls binding energy on the nitrogen, a larger N-O stretching frequency and clear epr evidence for the donation of an electron to the metal ion. An anionic nitrosyl ligand is usually characterized by an M-N-O bond angle near 120-125°, a smaller binding energy, etc. For example, $[Co(NO)(diars)_2](ClO_4)_2$ (diars = $C_6H_4[As(CH_3)_2]_2$) has an N (ls) binding energy of 402.3 eV and an N-O stretching frequency of 1852 cm^{-1}. An M-N-O bond angle of 174° has been observed in the analogous ruthenium complex in which two diphosphine groups have replaced the diarsine groups. The respective values are 400.7 eV, 1620 cm^{-1} and 119° for $[Co(NO)(NH_3)_5]Cl_2^2$.

Nitric oxide reacts with a variety of supported or exchanged transition metal ions, forming nitrosyl complexes which may be characterized by their ir spectra and in some cases their epr spectra. For certain reactions there appears to be a definite relationship between the type of bonding and the reactivity of the nitrosyl ligand. In this review spectroscopic evidence for surface nitrosyl complexes of Cr^{2+}, Cr^{3+}, Ni^{2+}, Fe^{2+}, Cu^+, Cu^{2+}, Ag^+, and Co^{2+} will be described for the purpose of demonstrating the influence of the metal ion on the nitric oxide molecule. In the course of such spectroscopic studies in our laboratory we have found that certain cobalt (II) complexes in zeolites have very reactive nitrosyl ligands.

NITROSYL COMPLEXES OF Cr^{2+} AND Cr^{3+}

Peri (3) has recently completed an infrared study of nitric oxide complexed with chromium ions on the surface of alumina. Fig. 1 is an example of the observed spectra, which indicate that several different types of nitrosyl complexes exist on a partially reduced surface. By varying such conditions as the reduction temperature, reduction with H_2 or CO, and concentration of supported chromium, Peri was able to resolve the seven bands listed in Table I where his assignments are also summarized. Although Peri does not make the observation, it is almost certain that the bands at 1820 cm^{-1} and higher wave numbers reflect a cationic form of NO. On the basis of the infrared spectra and other data Peri makes the distinction between chromium ions which are dispersed in the alumina surface and those which are present in a separate chromia phase. One of the surprising results is that Cr^{2+} apparently existed on a surface which had been subjected to oxygen at elevated temperatures.

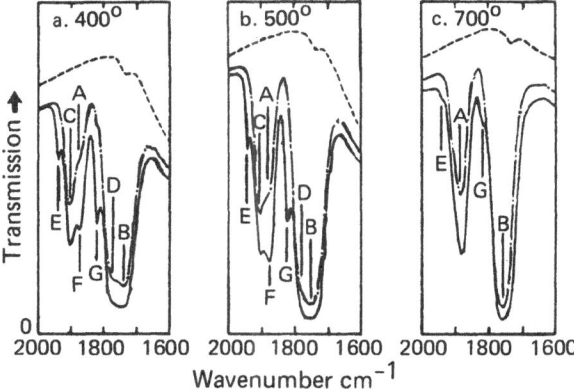

Fig. 1. Infrared spectra of adsorbed NO on 10% CrO_3/Al_2O_3, prereduced in H_2 at temperatures indicated (°C): (- - - -) after reduction; (———) after addition of $\sim$ 6 Torr of NO; (–·–) after evacuation (Ref. 3).

TABLE 1

Assignments for Adsorbed NO on Chromia/Alumina

Band, cm-1	Adsorption Sites	Open Coordination Positions/Cr	Adsorbed Species
A (1880)	Cr^{2+}	1	NO
C (1905)	Cr^{3+}	1	NO
B (1755)	$Cr^{2+}Cr^{2+}$ or $Cr^{2+}Cr^{3+}$ pairs	1	$(NO)_2$ dimer
D (1775)	$Cr^{3+}Cr^{3+}$ pairs	1	$(NO)_2$ dimer
E (1940)	Cr^{3+} in Al_2O_3 surface	2 }	ON NO
G (1820)	Cr^{3+} in Al_2O_3 surface	2 }	Cr^{3+}
F (1875)	Cr^{3+} in Al_2O_3 surface	1	NO
2260	$Al^{3+}O^{2-}$	1	NO+

In a related study Shelef (4) has observed an epr spectrum which is attributed to $[Cr(I)NO^+]^{2+}$ complexes on the surface of alumina. The signal with $g_I = 1.982$ and $g_{II} = 1.92$ (Fig. 2) was formed when NO was adsorbed on a pre-reduced sample of chromia on alumina. The g values are consistent with a low-spin d^5 configuration which is apparently formed by the donation of an electron from the nitrosyl ligand to Cr^{2+}. The axial symmetry is indicative of a linear M-N-O group as predicted by Gray *et al*(5). for a d^5 NO+ structure. The surface nitrosyl reacted very rapidly with molecular oxygen.

References p. 16.

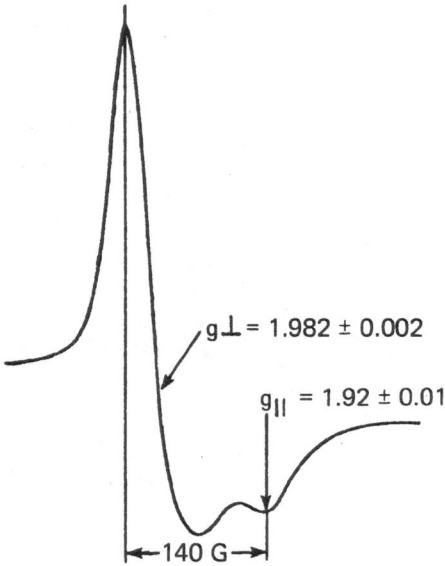

Fig. 2. Epr spectrum of NO on prereduced chromia/alumina (Ref. 4).

A combined epr and ir study of chromium nitrosyl complexes in zeolites has been carried out by Naccache and Ben Taarit(6). Adsorption of NO into a reduced zeolite resulted in an epr spectrum which was essentially identical to the one observed in the previous work of Shelef (4). The nitrosyl complexes were stable in the presence of oxygen at room temperature, presumably because they are hidden in the small cavities of the zeolite. The ir spectra of NO adsorbed in the chromium Y zeolite and mordenite, as shown in Fig. 3, reveal bands at 1900 and 1760-1780 cm^{-1}. Part of the band at 1760-1780 cm^{-1} is assigned to the $[Cr(I)NO^+]^{2+}$ complex, although the frequency is a bit lower than would normally be expected for a cationic nitrosyl ligand. The authors suggest that the band at 1900 cm^{-1} and part of the 1760-1780 cm^{-1} band are the spectra of terminal and bridging NO groups, respectively, in a complex depicted as

$$
\begin{array}{c}
N \\
O
\end{array}
- \; Cr
\begin{array}{c}
N \\
O \\
N \\
O
\end{array}
Cr \; -
\begin{array}{c}
O \\
N
\end{array}.
$$

It is tempting to speculate that the bands labeled B, C and D in Fig. 1 for chromia on alumina may have the same origin as the similar bands which have been observed in the zeolites.

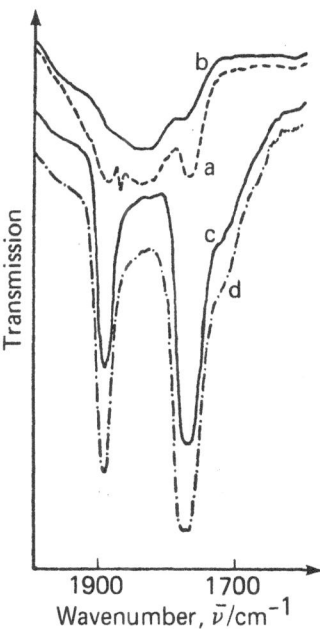

Fig. 3. Infrared spectra of NO adsorbed on a CrY zeolite: (a) 70 Torr NO on oxidized sample; (b) subsequently evacuated at 25°C for 5 min; (c) 70 Torr NO on reduced sample, subsequently evacuated at 25°C for 5 min; (d) further evacuation at 25°C for 4 hr (Ref. 6).

A NITROSYL COMPLEX OF Ni^{2+}

Perhaps one of the most convincing pieces of evidence for the donation of an electron from an NO ligand to a metal ion is found in the epr spectra of $[Ni(I)NO^+]^{2+}$. Kasai (7) first noted this electron transfer in Ni(II)Y zeolites. The epr spectra of Fig. 4 clearly indicate that d^9 Ni$^+$ ions are formed when NO is adsorbed in a Ni(II)Y zeolite. The two different values of g_{II} suggest that the nickel ions are located in different positions. Kasai, however, did not recognize that a nitrosyl complex was formed. Naccache and Ben Taarit (6) later assigned a nearly identical spectrum to the $[Ni(I)NO^+]^{2+}$ complex. Adsorption of NO on Ni(II)Y gave rise to a sharp strong band at 1892 cm^{-1} which was attributed to the nitrosyl complex.

NITROSYL COMPLEXES OF Fe^{2+}

The author and coworkers (8,9) have studied the formation of nitrosyl complexes in iron-exchanged zeolites. Nitric oxide reacted with ferrous ions yielding high-spin (S = 3/2) and low-spin (S = 1/2) complexes which were identified as $[Fe(I)NO^+]^{2+}$. The high-spin complex, which was formed upon adsorption of nitric oxide at low pressures, is characterized by an epr spectrum with $g_I = 4.07$ and $g_{II} = 2.003$ (Fig. 5)

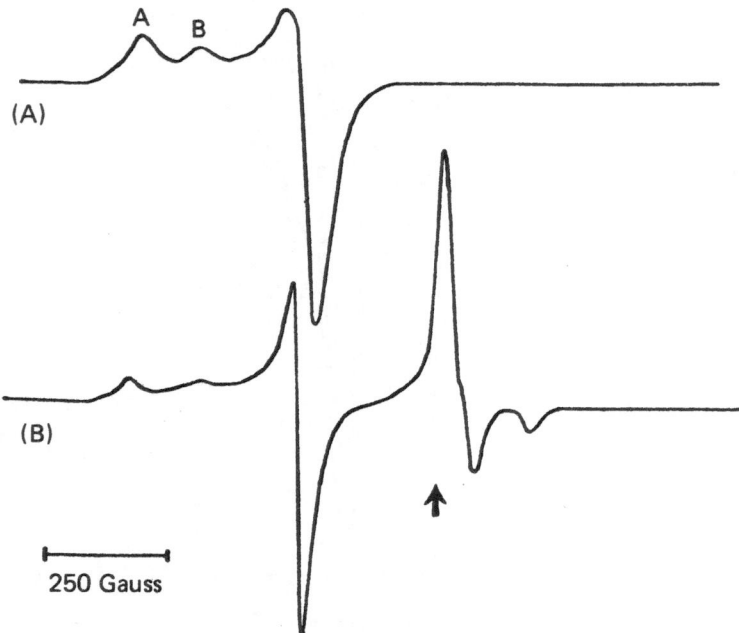

Fig. 4. Epr spectra of NO adsorbed on a NiY zeolite: (a) Ni(65%)-Y; (b) Ni(5%)-Y. The spectrometer gain for the trace (b) was 5 times that used to obtain trace (a). The arrow indicates the field corresponding to g = 2.0023 (Ref. 7).

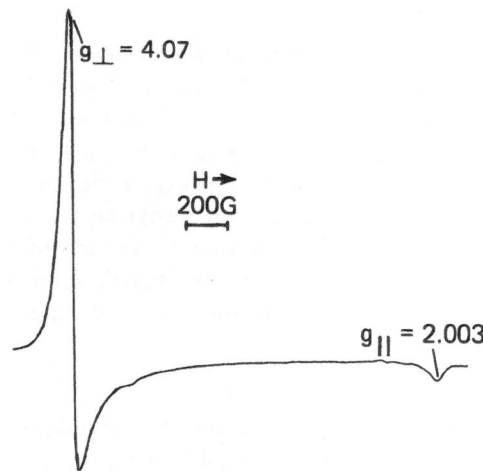

Fig. 5. Epr spectrum of S = 3/2 $[Fe(I)NO]^{2+}$ in a Y-type zeolite. Spectrum recorded with the sample at -196° (Ref. 8).

and an infrared band at 1890 cm^{-1} (Fig. 6). The low-spin complex was formed in a Y-type zeolite upon adsorption of nitric oxide followed by evacuation of the sample at 25°C. It is characterized by an epr spectrum with $g_{xx} = 2.015$, $g_{yy} = 2.055$ and $g_{zz} = 2.089$ (Fig. 7) and an infrared band at 1778 cm^{-1}. Both complexes were thermally stable at 25°C; however, at elevated temperatures the low-spin complex was more stable. The epr spectrum of a weakly held form of nitric oxide was attributed to a NO-Na$^+$ complex involving residual sodium ions. A slightly more stable NO species with an infrared band at 1822 cm^{-1}, but no comparable epr spectrum, was also observed.

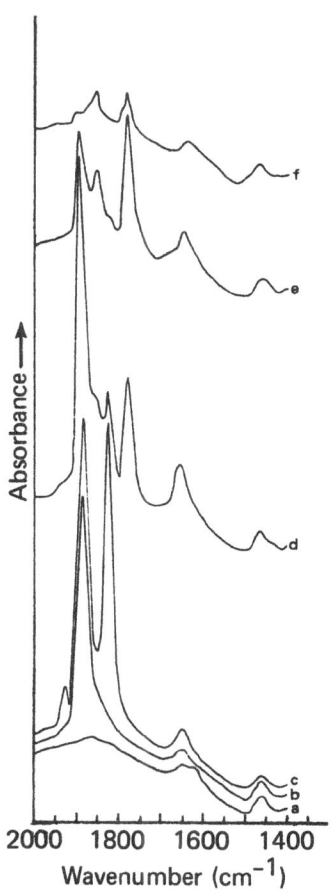

Fig. 6. Infrared spectra of an Fe(II)Y zeolite: (a) before addition of NO; (b) 1 Torr of NO; (c) 10 Torr of NO, evacuated 15 sec; (d) sample under vacuum 4 hr at 25°; (e) sample under vacuum 1 hr at 50°; (f) sample under vacuum 1 hr at 100° (Ref. 8).

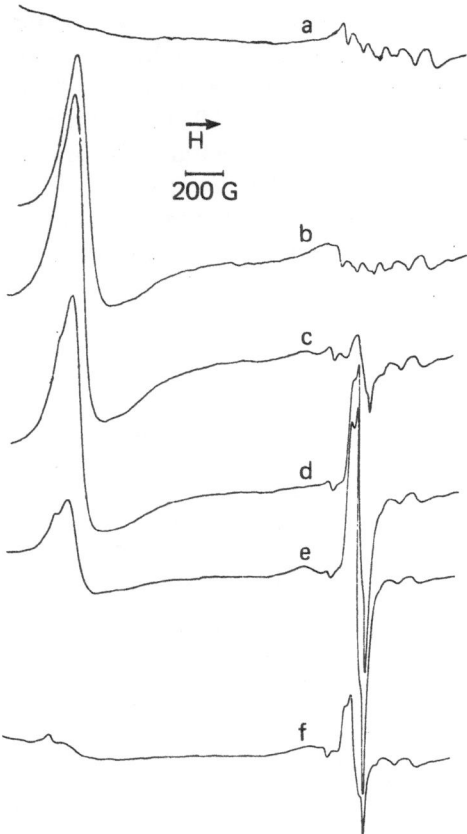

Fig. 7. Epr spectra of high-spin and low-spin $[Fe(I)NO]^{2+}$ complexes in a Y-type zeolite recorded with the sample at 25°: (a) before addition of NO, amplification of 2; (b) 1 Torr of NO, amplification of 2; (c) 10 Torr of NO, evacuated 15 sec, amplification of 2; (d) sample under vacuum 1 hr at 25°, amplification of 1,2; (e) sample under vacuum 1 hr at 50°, amplification of 1.0; (f) sample under vacuum 1 hr at 100°, amplification of 1.2 (Ref. 8).

Apparently the $[Fe(I)NO^+]^{2+}$ complex in the zeolites is near the crossover point, and the simple adsorption and desorption of nitric oxide at room temperature results in the formation of the low-spin isomer. The amplitudes of the spectra indicate that the high-spin species is not converted to the low-spin form, but rather new low-spin complexes are formed from Fe(II) which was previously coordinated in a different manner. Perhaps the location of the iron at two different sites in the zeolite is responsible for the two different spin isomers, although the influence of molecular oxygen on the spectrum suggests that both complexes are exposed to the supercage. A model is proposed in which the high-spin complex is in distorted tetrahedral (C_{3v}) symmetry, and the low-spin complex is in distorted square planar (C_{2v}) symmetry.

NITROSYL COMPLEXES OF Cu⁺ AND Cu²⁺

Nitric oxide forms a weakly bonded complex with Cu^+ in a Cu(I)Y zeolite which may be described as $[Cu(I)NO]^+$. The epr spectrum of the complex, reported independently by Chao and Lunsford and by Naccache, Che and Ben Taarit, is characterized by a hyperfine structure (Fig. 8) which reveals that the unpaired electron spends 20% of its time on the Cu^+ ion and is distributed about evenly between the $3d_{z^2}$ and 4s orbitals on the copper. The epr spectra indicate that the complex was unstable at temperatures above $-125°C$.

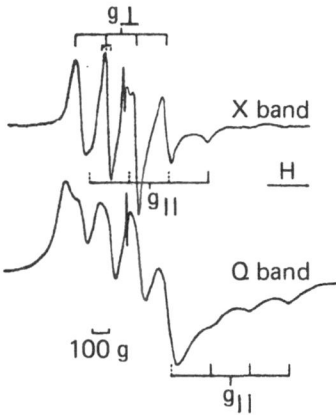

Fig. 8. Epr spectra of the $[Cu(I)NO]^+$ complex. The small sharp peak is due to DPPH (Ref. 9).

Nitric oxide reacts with Cu^{2+} ions in a Cu(II)Y zeolite forming $[Cu(I)NO^+]^{2+}$, which is characterized by an ir band at 1918 cm⁻¹. As expected, the epr spectrum of the Cu^{2+} ions in the zeolite disappeared with the formation of the nitrosyl complex since the d^{10} Cu^+ ion would be diamagnetic. Removal of the nitrosyl ligand by evacuation restored the Cu^{2+} spectrum.

NITROSYL COMPLEXES OF Ag⁺

Since both Cu^+ and Ag^+ are in a d (10) electronic configuration, it is expected that they would form similar nitrosyl complexes. In a Y-type zeolite a complex attributed to $[Ag(I)NO]^+$ is indeed formed upon the adsorption of NO at pressures less than 50 Torr (11). As shown in Fig. 9, this species is characterized by an epr spectrum with g_I = 2,000, g_{II} = 1.934, $^{Ag}A_I$ = 65G and $^{Ag}A_{II}$ = 72G. An infrared band at 1884 cm⁻¹ (Fig. 10) appears to be associated with this spectrum. As in the case of the $[Cu(I)NO]^+$ complex, the NO moiety could be removed by degassing the sample at room temperature. The unpaired electron is approximately 24% on the Ag^+ ion, with nearly equal distribution between the $4d_{z^2}$ and 5s atomic orbitals.

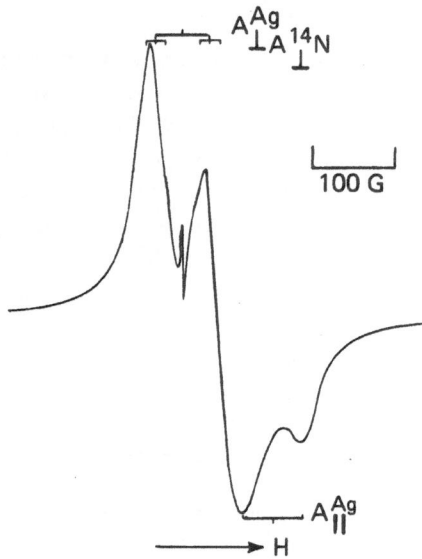

Fig. 9. Epr spectrum of the $[Ag(I)NO]^+$ complex. The sharp peak is due to a phosphorus-doped silicon marker (Ref. 11).

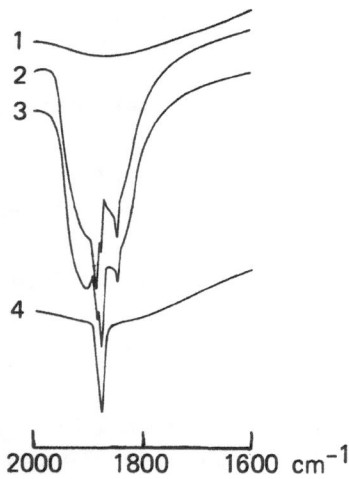

Fig. 10. Infrared spectra of an AgY zeolite: (1) after oxygen treatment and degassing at 500°C, (2) 5 min after introducing 320 Torr of NO in the cell, (3) 12 hr after introducing 320 Torr of NO into the cell, (4) after evacuating the cell for 1 hr at 25°C (Ref. 11).

Upon exposing the sample to approximately 300 Torr of NO a second more stable complex becomes the dominant species. The apparent stability of the complex, however, may be due more to the trapping of NO in the small cavities than to a stronger Ag-N bond. The rather complex epr spectrum shown in Fig. 11 has been interpreted in terms of an $[Ag(I)_2NO]^{2+}$ complex, in which the nitrosyl ligand is a bridging group between two silver ions (11). An infrared band at 1876 cm^{-1} (Fig. 10) is associated with this species. It is interesting to note that the position of this infrared band is at the same wavenumber as the fundamental band for free NO.

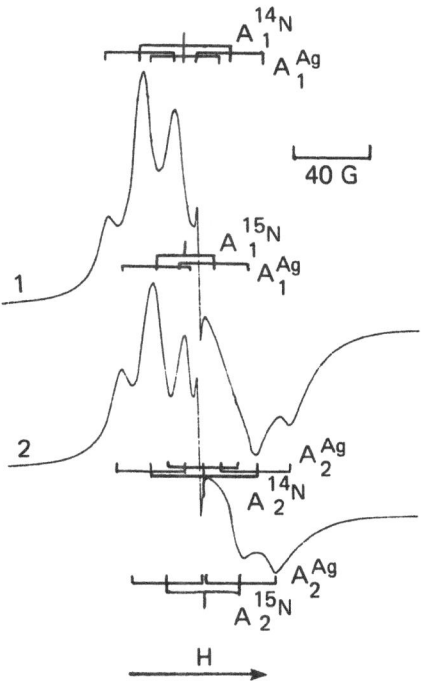

Fig. 11. Epr spectrum of $[Ag(I)_2NO]^{2+}$: (1) formed from ^{14}NO, (2) formed from ^{15}NO. The sharp peak is due to a phosphorus-doped silicon marker (Ref. 11).

NITROSYL COMPLEXES OF Co^{2+}

In most of the work described up to this point very little attention has been given to the reactivity of the nitrosyl ligand, except with molecular oxygen. Windhorst and the author (12) have recently completed a study in which an attempt was made to produce a reactive form of NO. It was felt that this could be achieved by synthesizing a nitrosyl ligand having anionic character, as reported for the $[Co(NH_3)_5NO]^{2+}$ complex (2). Nitric oxide was allowed to react with Co^{2+} ions, as well as with cobalt ammine complexes, in cobalt exchanged Y-type zeolites. The formation of $[Co(II)(NO)_2]^{2+}$, $[Co(II)NH_3(NO)_2]^{2+}$ and $[Co(III)(NH_3)_5NO]^{2+}$ was suggested from the infrared and adsorption data. As depicted in Figs. 12 and 13 infrared absorption bands for N-O stretching vibrations were observed at 1830 and 1910 cm^{-1} for

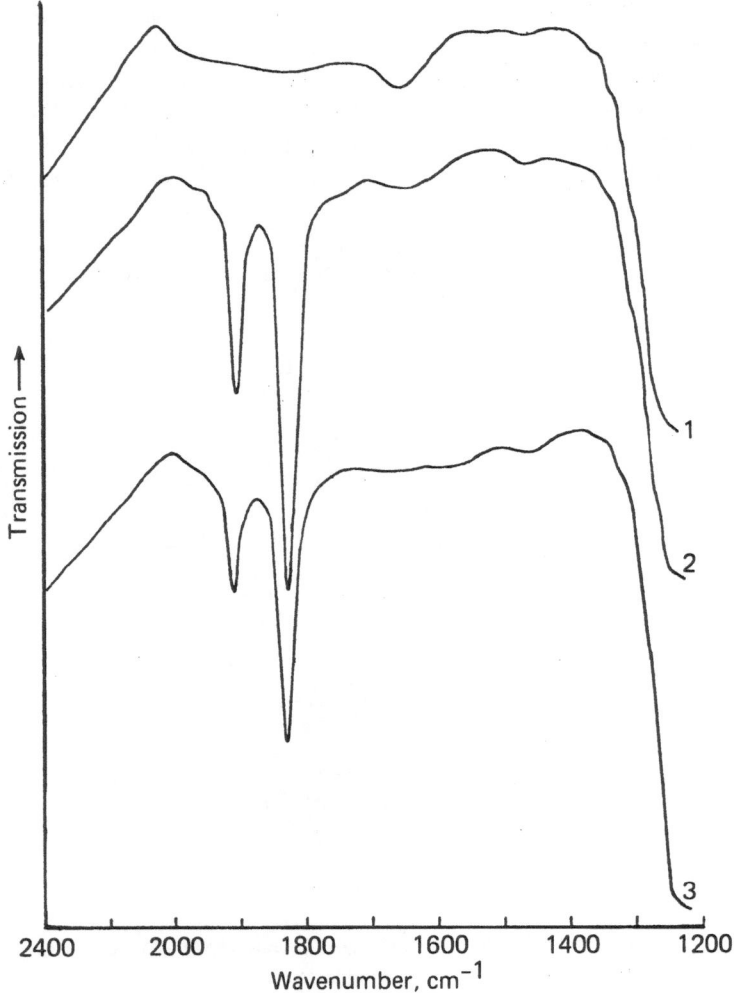

Fig. 12. Infrared spectra of a CoY zeolite: (1) after degassing at 500°C; (2) 1 hr after introducing 50 Torr of NO into the cell; (3) after evacuating the cell for 1 hr at 150°C.

$[Co(II) (NO)_2]^{2+}$, at 1800 and 1880 cm^{-1} for $[Co(II) (NH_3) (NO)_2]^{2+}$ and at 1710 cm^{-1} for the $[Co(III) (NH_3)_5 NO]^{2+}$ complex. A bond angle of 123° was calculated for the ON-Co-NO moiety in the dinitrosyl complexes from ir band intensities. The effect of the ammonia ligands on the electron density in the nitrosyl ligand is clearly seen in the spectra of these complexes.

When $[Co(NO)_2]^{2+}$ was thermally decomposed at 200°C, N_2, N_2O and NO were observed as products, although at 23°C only NO was present in the gas phase. The

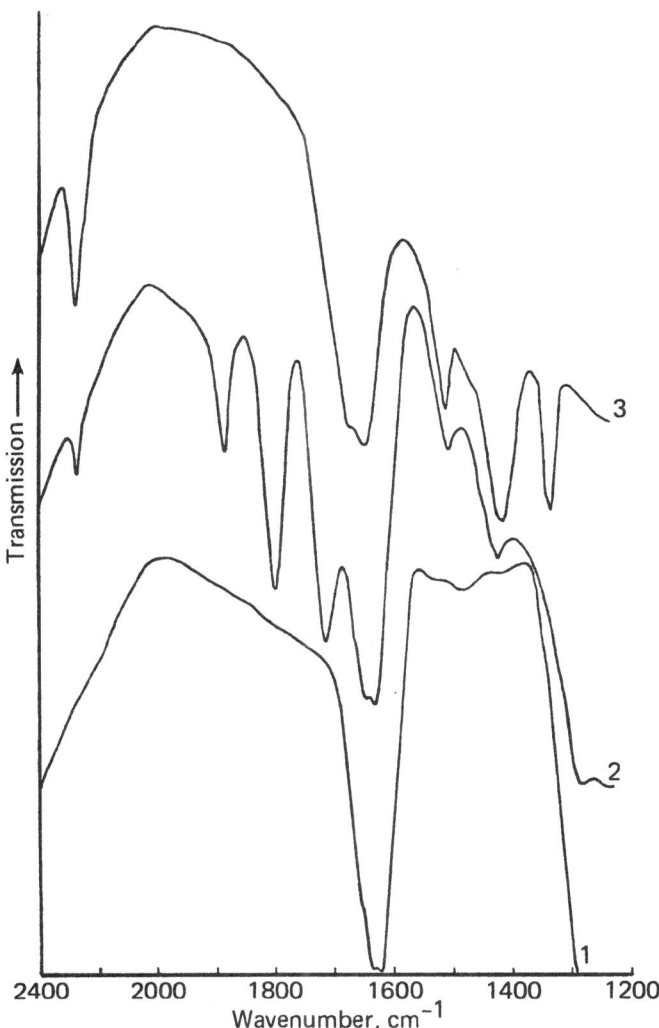

Fig. 13. Infrared spectra of a CoY zeolite: (1) after degassing at 500°C, addition of 200 Torr of NH$_3$ and the removal of excess NH$_3$ by brief evacuation; (2) 15 min after the introduction of 50 Torr of NO; (3) 5 hr after the NO addition.

intramolecular reaction of [Co(III) (NH$_3$)$_5$NO$^-$]$^{2+}$, as well as its intermolecular reaction with gas phase NO, was followed at 23°C, and the two reactions were found to yield N$_2$, N$_2$O and [Co(III) (NH$_3$)$_5$NO$_2$]$^{2+}$. Results from an isotope labelling experiment, prove that the N$_2$ was produced by the reaction of the NO$^-$ ligand with NH$_3$, whereas N$_2$O and NO$_2$ were produced by the disproportionation reaction of the NO$^-$ ligand with two nitric oxide molecules. It was observed that the

References p. 16.

disproportionation reaction was favored by a greater pressure of NO over the zeolite. The production of NO_2 and its coordination with cobalt prevented further formation of the nitrosyl complex, and hence destroyed the catalytic activity. Some very recent experiments suggest that the $[Co(II) (NH_3) (NO)_2]^{2+}$ complex may be active in the dissociation of nitric oxide at 25°C (13).

SUMMARY

The ir and epr spectra of surface nitrosyl complexes confirm that the nature of the metal ion and the type of other ligands significantly affect the bonding and the reactivity of nitric oxide. The relatively limited data on the effects of electron-donating ligands suggest that maximum reactivity is achieved by making the nitrosyl ligand anionic. If the dissociation of NO is important, one also should use a metal ion which will allow oxygen to desorb from the complex. Cobalt(II) ammine complexes in zeolites appear to meet these requirements.

ACKNOWLEDGEMENT

The author acknowledges the support of this work by The Robert A. Welch Foundation under Grant No. A-257.

REFERENCES

1. C. S. Pratt, B. A. Coyle and J. A. Ibers, J. Chem. Soc., 2146 (1971).
2. P. Finn and W. L. Jolly, Inorg, Chem., 11, 893 (1972).
3. J. B. Peri, J. Phys. Chem., 78, 588 (1974).
4. M. Shelef, J. Catal. 15, 289 (1969).
5. H. B. Gray, P. T. Monoharan, J. Pearlman and R. F. Riley, Chem. Commun., 62 (1965).
6. C. Naccache and Y. Ben Taarit, JCS Faraday Trans. I, 69, 1475 (1973).
7. P. H. Kasai and R. J. Bishop, J. Amer. Chem. Soc., 94, 560 (1972).
8. J. W. Jermyn, T. J. Johnson, E. F. Vansant and J. H. Lunsford, J. Phys. Chem., 77, 2964 (1973).
9. C. C. Chao and J. H. Lunsford, ibid., 76, 1546 (1972).
10. C. Naccache, M. Che and Y. Ben Taarit, Chem. Phys. Letters, 13, 109 (1972).
11. C. C. Chao and J. H. Lunsford, J. Phys. Chem., 78, 1174 (1974).
12. K. A. Windhorst and J. H. Lunsford, submitted for publication.
13. K. A. Windhorst and J. H. Lunsford, to be published.

DISCUSSION

K. H. Ludlum *(Texaco Research Center)*

Have you ever observed the disproportionation reaction of NO on your metal-ion supported by aluminum or on your metal-ion substituted in your Y-zeolites. I notice none of your infrared curves seem to correspond to the type of spectrum that you show on the last slide. Do you care to comment on whether you've seen any disproportionation on these materials?

Lunsford

Well, of course, we did some earlier work* in which we studied the disproportionation reaction of a sodium-Y and calcium-Y zeolite. On the latter zeolite a small amount of disproportionation reaction will go at room temperature with high pressures of NO. On sodium-Y it was necessary to cool the sample to -78°C. Now it appears that on the silver-Y and the copper-Y there is not a significant amount of the disproportionation reaction as long as the NO pressure is kept fairly low. Whereas it seems that the cobalt-Y catalyzes the disproportionation reaction, particularly when the cobalt has ammonia ligands. Incidentally there's some homogeneous work in which it was shown that disproportionation occurs at room temperature with related cobalt complexes. The complex is much more effective than the metal ions alone in the zeolite. I don't know of an example where the disproportionation reaction occurs on supported metal ions on alumina or other surfaces but such catalysts could well exist; I may just not be aware of them.

R. L. Burwell, Jr. *(Northwestern University)*

Jack, what does it mean when you say that you "almost completely transfer an electron" from or to nitric oxide, which would seem to me to make it diamagnetic. What happens in the spectrum? I assume NO^+ or NO^- wouldn't have any EPR spectrum. Is that right?

Lunsford

That's right. If you completely transfer the electron, the NO^+ or NO^- would be diamagnetic. The evidence for such an electron transfer, however, is a bit conflicting. In the case of the nickel nitrosyl complex, the nickel spectrum looks as if you took electrons and reduced the nickel from +2 to +1. So here it seems to be very conclusive that one has made nickel +1 and electron transfer has occurred. However, if you compare the infrared spectra of some extreme cases of NO^+ with that of the NO-nickel complex, you do not get the same bands. In other words, the NO^+ band in NOBFCl occurs at 2300 cm^{-1} and the band from NO coordinated to the nickel (11) occurs at about 1900 cm^{-1}. This indicates that you really do not have pure NO^+ in the nickel complex. Most people go through some sort of handwaving argument about back donation and this sort of thing to get out of that dilemma. But again, there seems to be (at least from the viewpoint of the EPR spectra) considerable donation of an electron into the d-orbitals of the metal ion.

T. P. Kobylinski *(Gulf Research)*

I noticed in your last slide you mentioned complexes of cobalt, where you have two NO molecules attached to one cobalt atom. We did some work on the selectivity of the conversion of nitric oxide to nitrogen instead of the ammonia with this type of complex. Theoretically, it would be very interesting because you can see the

*C. C. Chao and J. H. Lunsford, J. Am. Chem. Soc. 93 71 (1971).

possibility of easy formation of nitrogen-nitrogen bonds in relation to nitrogen-bonds to hydrogen. What also puzzles me here is the similarity of the spectrum between the nickel, chromium and iron complexes with nitric oxide. Now, what we found out from the reaction point of view is that these chromium ions behave very differently. In other words, chromium does not produce any ammonia even if you have enough hydrogen where iron does. Do you think that there is any possibility that over chromia you have the same situation as on the cobalt where you have two NO molecules which makes it much easier to form nitrogen-nitrogen bonds than on the iron?

Lunsford

Well, we have no evidence on the iron that there may be a dinitrosyl complex. One does not see these two bands that always rise and fall together. However, it is possible to reinterpret Naccache's data on chromium (II). This could be reinterpreted in terms of these dinitrosyl complexes. So it may be that chromium II, and cobalt II are alike while iron is different.

CHEMISORPTION AND REACTIVITY
OF NO ON (10$\bar{1}$0) Ru

R. Ku and N. A. GJOSTEIN

Ford Motor Company, Dearborn, Michigan

H. P. BONZEL

Exxon Research and Engineering Company, Linden, New Jersey

ABSTRACT

A combination of modern surface measurement techniques such as LEED, AES and Thermal Desorption Spectroscopy were used to study the chemisorptive behavior of NO on the (10$\bar{1}$0)Ru surface. The experimental evidence strongly favors an adsorption model in which NO adsorbs and rapidly dissociates into separate nitrogen and oxygen adsorbed phases, each exhibiting ordered structures: the C(2x4) and (2x1) structures at one-half and full saturation coverage, respectively. At temperatures as low as 200°C, the nitrogen phase begins to desorb, and continuous exposure to NO in this temperature range results in an increasing oxygen coverage until the surface is saturated with oxygen and no further NO dissociation can take place. There is evidence that once the surface is saturated with the dissociated NO phase further NO adsorption occurs in a molecular state.

The nitrogen desorption spectrum depends strongly on coverage and exhibits several peaks which are related to structure of the adsorbed phase. The implications of the results with respect to the catalytic reduction of NO by H_2 and CO and the N_2-selectivity of Ru catalysts are discussed.

INTRODUCTION

In order to understand the mechanism of NO reduction on the (10$\bar{1}$0)Ru surface, a combination of modern surface techniques such as low energy electron diffraction

(LEED), Auger electron spectroscopy (AES) and thermal desorption spectroscopy have been used to study the chemisorptive behavior of NO, O_2 and CO on $(10\bar{1}0)$Ru, as well as reactions of CO and H_2 with chemisorbed NO and oxygen. Since the results of this investigation will be published elsewhere (1) in detail, the purpose of this paper is to summarize the major experimental findings and to demonstrate that they lead to the conclusion that NO is dissociatively chemisorbed on $(10\bar{1}0)$Ru. Our findings on dissociative chemisorption of NO are in good agreement with those recently found by Klein and Shih (2). Furthermore, it is proposed that this feature of the NO chemisorption process influences the reaction path in the catalytic reduction of NO by CO and H_2, and in particular that it may be responsible for the high N_2-selectivity exhibited by Ru catalysts (3-5).

LEED AND AUGER ELECTRON SPECTROSCOPY OBSERVATIONS

When a $(10\bar{1}0)$Ru surface held at $\sim$ 90°C is exposed to *either* NO or O_2 gas in the 10^{-8} torr range and LEED patterns are observed, the same sequence of ordered surface structures is observed as function of exposure time, namely:

$$(1 \times 1) \rightarrow C(2 \times 4) \rightarrow (2 \times 1)$$

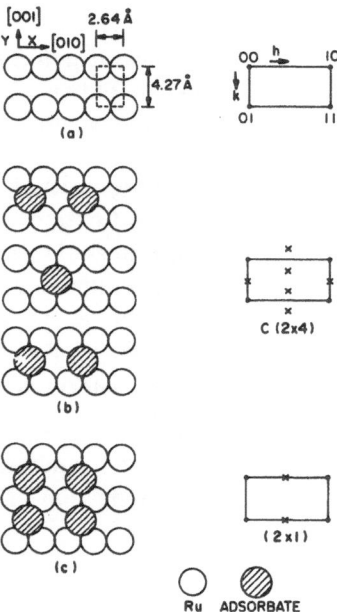

Fig. 1. Sequence of LEED patterns (right) observed as a function of increasing NO exposure (cf. Fig. 2). The corresponding proposed adsorbate structures are shown on the left.

Schematics of the LEED patterns and the corresponding postulated ordered arrangement of dissociated chemisorbed species (either oxygen or nitrogen atoms) for each LEED pattern are shown in Figs. 1a-c. The transition C(2x4) → (2x1) exhibits the interesting feature that first the 1/2 order diffraction spots in the k-direction fade, followed by a merging of the 1/4 order reflections to form the 1/2 order reflections of the (2x1) structure.

The Auger signal from the chemisorbed nitrogen and oxygen species can be followed as a function of exposure time, and typically, Fig. 2, the oxygen or nitrogen signal rises rapidly at first and then saturates at a coverage which we will designate as θ_s. During NO exposure the C(2x4) pattern is first seen at $\sim 0.2 \theta_s$ and reaches a maximum intensity at $\sim 0.5 \theta_s$. In the range $0.5 \theta_s \langle \theta \langle 0.6 \theta_s$, the 1/2 order reflections fade and finally for $\theta \rangle 0.6 \theta_s$ merging of the 1/4 order reflections occurs until a clear (2x1) is observed at θ_s (Fig. 2). It is important to note, however, that while the sequence is the same for both adsorbing gases, the absolute oxygen Auger signal for NO adsorption at saturation is only *one-half* that measured for O_2 adsorption at saturation (Table 1). The saturation Auger signals for nitrogen and oxygen are about equal for NO adsorption (Table 1).

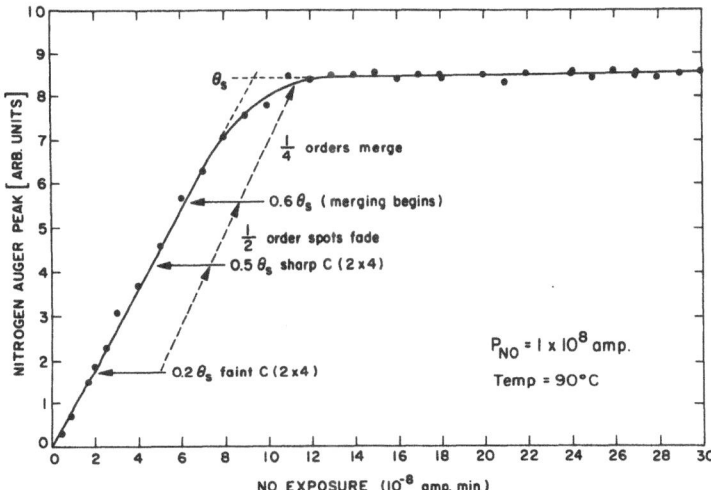

Fig. 2. Nitrogen Auger peak height (coverage) as a function of NO exposure. Saturation coverage is designated as θ_s. 1 x 10^{-8} amp-min is equivalent to 3.9 x 10^{-7} torr-sec.

LEED observations at 60°C indicate that with increasing NO exposure (beyond the knee in Fig. 2) the 1/2-order reflections associated with the (2x1) fade and are gradually replaced by very diffuse diffraction fractures near the (h, k $\pm$ 1/2) positions. During this transformation the (1x1) substrate reflections remain sharp. The diffuseness of the extra diffraction features makes it difficult to index the LEED pattern obtained in this case. We will demonstrate later from the nitrogen desorption

TABLE 1

Correlation of Auger Peak Heights with LEED Patterns

LEED Pattern Of Adsorbate (at $\sim$ 100°C)	Auger Peak Heights (Arbitrary Units)	
	N	O
O$_2$-Adsorption		
C(2x4)-O	–	55-64
(2x1)-O	–	93-120
NO-Adsorption		
C(2x4)-N+O	30	32
(2x1)-N+O	60	65
*(2x1)-N	57	6

spectra that during long NO exposures at T ⟨ 100°C, NO adsorbs in a molecular state, once the surface is saturated with dissociated species.

In this paper we are principally concerned with developing a model of the dissociative chemisorption process which precedes the formation of molecular NO. Our findings can be interpreted in terms of a model in which dissociative chemisorption of NO leads to the formation of separate islands of oxygen and nitrogen at θ_s each having a (2x1) structure. This model will be substantiated by other experiments described below, but at this point it is advantageous to develop the features of this model in more detail.

STRUCTURE OF NITRIC OXIDE AND OXYGEN CHEMISORBED LAYERS

The (10$\bar{1}$0)Ru surface, Fig. 1, is composed of closed-packed rows of atoms along the [010] direction which form potential energy troughs or channels that favor chemisorption. In the case of O$_2$-adsorption, for example, we assume that the molecular oxygen dissociates and each oxygen atom resides in the potential well formed by four underlying nearest neighbor Ru atoms. Both the C(2x4) and the (2x1) patterns imply that alternate sites along the channels are occupied, and in addition the C(2x4) pattern suggests that for θ ⟨ 0.5 θ_s alternate channels are vacant. These structures can be understood in terms of oscillatory, indirect adsorbate-adsorbate (A-A) interactions postulated by Grimley (7). Einstein and Schrieffer (8) have shown that the periodicity of these indirect interactions (via the substrate atoms) is simply related to the substrate periodicity. Adams (9) has recently sought to interpret some simple LEED structures due to adsorption of H$_2$, N$_2$ and CO on W surfaces in terms of oscillatory A-A interactions.

*After H$_2$ reduction.

Based on these theoretical findings, the origin of the C(2x4) structure can be understood in the following way: occupation of a given adsorption site along a channel will produce strong repulsive A-A interactions at the nearest neighbor channel sites and attractive interactions at the next nearest neighbor sites. These forces lead to an adsorbate structure with double the substrate periodicity along the channels, which is a characteristic of both the C(2x4) and (2x1) structures. In a similar manner, oscillatory A-A interactions in a direction perpendicular to the channels, [001], are also expected to exist. Repulsive A-A interactions at nearest neighbor sites in adjacent channels and attractive A-A interactions at next nearest neighbor sites would account for vacant alternate rows of the C(2x4) structure. Since the filled channels of the C(2x4) are alternately shifted by one substrate spacing along the channel, Fig. 1b, the attractive interactions probably have nodes along the [012] direction rather than the [001]. Furthermore, we presume that the repulsive interactions at nearest neighbor sites in adjacent channels are weaker than those along the channel since at higher θ the vacant channel sites can be occupied.

This model suggests that at low coverage alternate rows fill first in a staggered fashion until they become saturated at 0.5 θ_s, where the maximum intensity C(2x4) pattern is observed. Further adsorption then occurs on the empty (energetically less favorable) channels, and this results in a fading of the 1/4 order reflections in the k-direction. Finally, as the coverage nears θ_s, the channels which are nearly filled, but still staggered with respect to one another, begin to align by diffusion of atoms along the channels. This causes the 1/4 order reflections to merge into 1/2 order reflections characteristic of the final (2x1) structure. The fact that NO adsorption exhibits the identical sequence of ordered structures as O_2 adsorption is strong evidence that the model developed above is applicable to both the dissociated nitrogen and oxygen species.

The hypothesis that nitrogen and oxygen are not co-adsorbed in a sequence N-O-N-O along a channel, but rather in the form of separate islands, each with a doubly-spaced (2x1) structure, can be demonstrated by the following experiments: after NO adsorption to attain the saturated coverage at 100°C, the oxygen phase is selectively removed by H_2* at the same temperature. In this experiment no loss of nitrogen occurs but the oxygen coverage is reduced to nearly zero. The LEED pattern, however, remains (2x1). Then the surface is re-exposed to NO until saturation again occurs, followed by H_2-reduction. This cycle is repeated until no further uptake of NO is observed. At this point the nitrogen coverage is approximately double its initial value and the LEED pattern has remained (2x1) throughout the sequence (Table 1). This experiment clearly demonstrates that the nitrogen phase is ordered in a (2x1) structure and that it is not strongly coupled to the oxygen phase. Once the oxygen phase is selectively removed by hydrogen reduction the nitrogen atoms in the (2x1) ordered patches are now free to spread over

References p. 29.   *No evidence was found for H_2 chemisorption at 100°C.*

the surface and take up sites previously occupied by oxygen atoms. This does not occur rapidly at 100°C, but on slow heating one finds:

$$
(2x1)\text{-}N \xrightarrow[\text{heating}]{206°C} \quad \begin{array}{c} \text{Disordered} \\ \text{N-phase} \\ + \\ (1x1)\ \text{Ru} \end{array} \xrightarrow[\text{cooling}]{\langle 100°C} C(2x4)\text{-}N
$$

Since twice as many adsorption sites are now available to the nitrogen phase, a C(2x4) structure is expected and found on cooling below 100°C.

Analogously, selective removal of the nitrogen phase can be carried out successfully by thermal desorption. In this case, the sequence of LEED patterns is

$$
(2x1)\text{-}N+O \xrightarrow[\substack{\text{flash} \\ \text{heating} \\ N_2\text{-desorption}}]{\sim 450°C} \quad \begin{array}{c} \text{Disordered} \\ \text{O-phase} \\ + \\ (1x1)\ \text{Ru} \end{array} \xrightarrow[\text{cooling}]{\langle 100°C} C(2x4)\text{-}O
$$

Since selective removal of the N-phase necessitated heating to temperatures where the oxygen atoms are mobile, the metastable (2x1)-O structure could not be observed. Selective removal of nitrogen followed by readsorption of NO was carried out for several cycles until no further NO uptake occurred. Similar to the previously described cyclic experiments, the oxygen Auger signal at saturation was approximately double its initial value.

All of these experiements strongly support the view that NO chemisorbs on $(10\bar{1}0)$Ru, rapidly dissociates and forms separate nitrogen and oxygen islands that each exhibit an ordered C(2x4) structure at 0.5 θ_s and a (2x1) structure at θ_s. Further experimental support for this model is given in the following section.

KINETICS OF NITROGEN DESORPTION

Fig. 3 shows the thermal desorption spectra observed as a function of increasing NO exposure in the 10^{-8} torr range. Note that only NO, N_2 and O_2 desorption products were observed, and that the oxygen phase can be desorbed only partially at temperatures near 1000°C.

Three nitrogen desorption peaks, β_1, β_2 and β_3, were observed (Fig. 3), indicating three different desorption energies for the nitrogen phase. The β_3-peak occurs at $\sim$

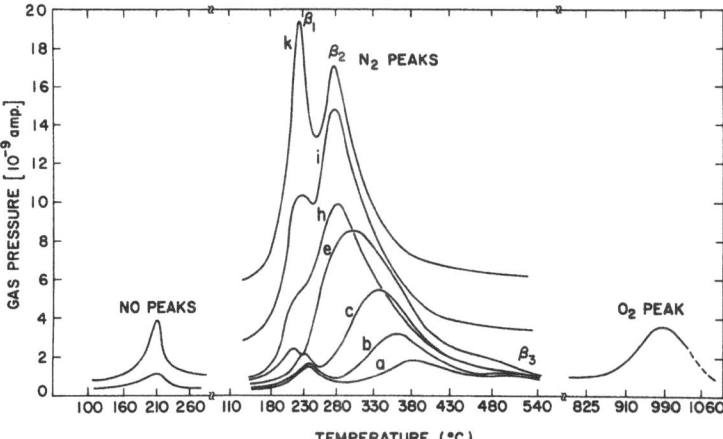

Fig. 3. Thermal desorption spectra for NO, N_2 and O_2 desorption products. N_2-peaks are shown as functions of increasing NO exposure (curves a through k). Curves i and h bracket in the exposure corresponding to the knee in Fog. 2 (10^{-7} amp-min.).

550°C and saturates at very low coverages; its origin is not understood. As the coverage increases, the β_2-peak emerges at lower temperatures and exhibits the characteristics of a second order desorption process, as shown by the strong peak temperature (T_p) shift with increasing θ. Fig. 4 gives a plot of $\ell\eta(n\,T_p^2)$ vs. $1/T_p$ where n is proportional to the number of desorbed nitrogen atoms associated with the β_2-peak (8). This plot shows that second order kinetics are obeyed at very low θ, but that significant deviations from second order kinetics occur at $\theta \geqslant 0.13\,\theta_s$, i.e., at a coverage near the value where the C(2x4) structure is first observed ($\theta = 0.2\,\theta_s$).

According to our structural model, the tendency to form ordered chains of N-atoms along channels will be less probable at very low θ. Thus, on heating to the desorption temperature, rapid mobility of N-atoms along the channel will cause a nearly random occupation of channel sites. In this case the probability of desorption is proportional to θ^2 and the activation energy for desorption should be independent of θ. However, as θ increases N-N interactions will cause ordered arrays to form and the probability of desorption will no longer be proportional to θ^2 and the activation energy will be strongly dependent on θ. Therefore, the deviations from simple 2nd order kinetics revealed by Fig. 4 are expected. Adams (9) has treated 2nd order desorption from ordered adsorbate arrays and has demonstrated that the deviations of the type shown in Fig. 4 can be explained by repulsive A-A interactions for nearest neighbor sites. However, the application of his analytical method to the present results is beyond the scope of this paper. It should be noted from Fig. 4 that the desorption area, the T_p-shift with θ, and the activation energy for desorption $Q_2 = 24$ kcal/mol are the same regardless of whether or not oxygen is present on the surface, again indicating the independence of the nitrogen and oxygen phases.

References p. 29.

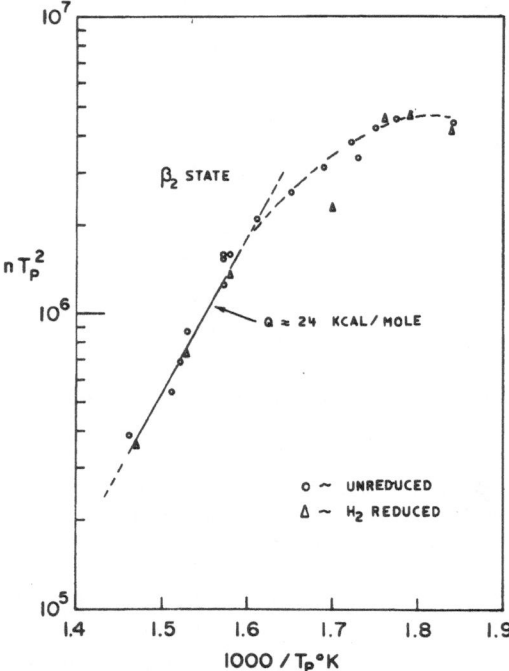

Fig. 4. A test of 2nd order nitrogen desorption kinetics, where n is proportional to the β_2 desorption area and T_p is the β_2-peak temperature. The linear portion yields an activation energy for desorption of 24 kcal/mol. At $\theta \rangle 0.13\ \theta_s$ deviations from simple 2nd order kinetics can be seen. seen.

When $\theta \cong 0.65\ \theta_s$, the β_1-peak emerges from the low temperature shoulder of the β_2-peak and grows with increasing θ. In this coverage range neither β_1 or β_2 exhibit large T_p-shifts. The β_1-peak develops during the transition C(2x4) → (2x1) and according to our model the relatively lower desorption energy (T_p) associated with the β_1-peak results from the repulsive N-N interactions associated with the occupation of nearest neighbor channel sites. When $\theta \langle 0.5\theta_s$ nitrogen atoms occupy only next nearest neighbor sites where attractive N-N interactions result in the relatively higher desorption energy (T_p) associated with the β_2-peak.

The amount of NO desorbed ($T_p \sim 210°C$) was negligible for $\theta \langle \theta_s$, indicating that the surface must be saturated with dissociated species before molecular adsorption of NO occurs. For long NO exposures beyond the knee in Fig. 2 at T $\langle 100°C$, the amount of NO desorbed increased to a maximum value of $\sim 8\%$ of the nitrogen desorbed. In the same exposure range, the nitrogen desorption area approximately doubled and the β_1-peak increased substantially until it was somewhat larger than the β_2-peak (Fig. 3). From these results it seems likely that NO adsorbed in the molecular state dissociates during the thermal desorption process and the resultant nitrogen contributes to the β_1-peak. The observation that molecular adsorption of NO follows dissociative adsorption has been confirmed by UPS measurements (6).

According to our model of the dissociated NO phase, the saturation coverage of both nitrogen and oxygen should be θ_s = 0.25. In principle the nitrogen saturation coverage can be determined from the p x t area associated with the nitrogen desorption spectrum. However, the possible presence of molecular NO, which dissociates during the thermal desorption process, can yield nitrogen desorption areas that are too large. Nitrogen desorption spectra were taken after NO exposures near the knee value in Fig. 2 at temperatures of $\sim$ 150°C. These conditions minimized the molecular NO contribution, and yielded θ_s values of $\sim$ 0.29, which are in approximate agreement with the model.

Nitrogen adsorption-desorption kinetics can also be monitored by AES as shown in Fig. 5. Note that at elevated temperatures (i.e., > 90°C) the nitrogen coverage rises to a maximum and then slowly decays to zero. Correspondingly, the oxygen (not shown) increases above the θ_s value observed at T < 90°C and reaches a value of approximately $2\theta_s$ (T = 90°C) when the nitrogen coverage reaches zero. This interesting behavior can be understood in terms of the following reactions:

$$NO + \sigma' \underset{k_{-1}}{\overset{k_1}{\rightleftharpoons}} \underline{NO} \tag{1}$$

$$\underline{NO} + 2\sigma \overset{k_2}{\rightarrow} \underline{N} + \underline{O} \tag{2}$$

$$\begin{array}{c}\text{N} \quad \text{N} \quad \text{N} \qquad \overset{\text{N---N}}{\diagup} \\ | \quad | \quad | \qquad \diagup \quad \vdots \quad k_3 \\ \text{Ru-Ru-Ru-Ru-Ru} \rightarrow \text{Ru-Ru-Ru} \rightarrow N_2 + 2\sigma \end{array} \tag{3}$$

where σ and σ' designate adsorption sites.

At T > 90°C, reaction 3 is extremely slow and the nitrogen and oxygen coverages reach a stable saturation value. At higher temperatures however, where nitrogen is readily desorbed, more free sites become available for NO dissociation. Thus, the oxygen coverage increases at the expense of the nitrogen coverage until the surface is saturated with oxygen (twice the low temperature θ_s) and no further NO dissociation can take place ($\sigma = 0$).

References p. 29.

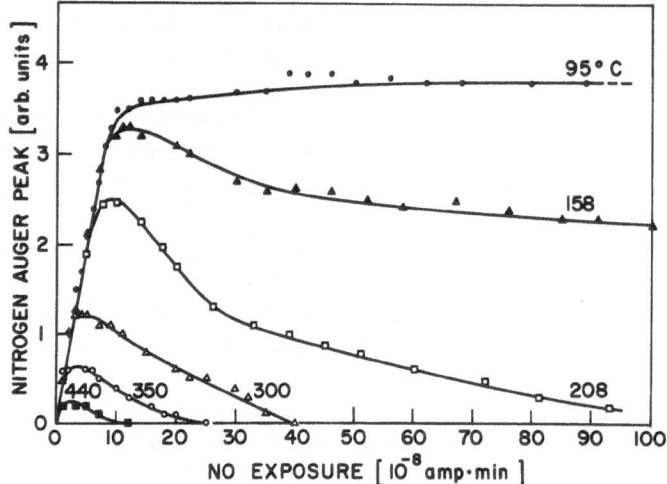

Fig. 5. Nitrogen Auger peak height (coverage) as a function of NO exposure at elevated temperatures.

CATALYTIC REDUCTION OF NO ON (10$\bar{1}$0)Ru

The dissociative adsorption model developed in the preceding sections can be used to understand the role of reducing agents such as CO and H_2 in catalytic reduction of NO on Ru. From reactions 1-3 we see that Ru surfaces are effective in breaking the N-O bond, and that at moderate temperatures nitrogen is readily desorbed, leading to a buildup of oxygen and finally a cessation of NO dissociation. If, however, H_2 or CO is present in the gas phase we have observed that adsorbed oxygen is removed, thus allowing the continuous dissociation of NO and evolution of N_2:

$$\underline{O} + \underline{CO} \rightarrow CO_2 \tag{4}$$

$$2\underline{H} + \underline{O} \rightarrow H_2O \tag{5}$$

Furthermore, one can conclude that since the residence time of N-atoms is very small at elevated temperatures, it seems likely that the probability of forming -NH or -NCO intermediates, from which NH_3 can be produced, will be small. In this way one can account for the high N_2-selectivly of Ru catalysts, particularly at temperatures ⟩ 300°C where higher pressure studies (5) show a decrease in the relative NH_3 formation with increasing temperature. Further support of this reaction model comes from Unland's (11) infrared studies, which show that Ru exhibits a much weaker -NCO band than most other noble metals.

In contrast to Ru, our studies indicate that only molecular adsorption of NO occurs on (110)Pt. At the same time, it is known that Pt favors NH_3 and N_2O formation at low temperatures (3-5,12). Since NO does not dissociate on Pt, the formation of intermediates such -NH and -NCO may be more probable.

ACKNOWLEDGMENTS

The authors are grateful to Dr. P. Wynblatt for reviewing the manuscript.

REFERENCES

1. R. Ku, N. A. Gjostein and H. P. Bonzel, to be published.
2. R. Klein and A. Shih, National Bureau of Standards, to be published.
3. M. Shelef and H. S. Gandhi, Ind. Eng. Chem. Prod. Res. Dev. 11, 393 (1972).
4. K. Otto and M. Shelef, Z. Physik Chem. N. F. 85, 308 (1973).
5. K. C. Taylor and R. L. Klimisch, J. Catal. 30, 478 (1973).
6. H. P. Bonzel, Exxon Research and Engineering Corp., private communication.
7. T. B. Grimley, Adv. Catalysis 12, 1 (1960).
8. T. L. Einstein and T. R. Schrieffer, Phys. Rev. B7, 3629 (1973).
9. D. L. Adams, Surf. Sci. 42, 12 (1974).
10. P. A. Redhead, Trans. Faraday Soc. 57, 641 (1961).
11. M. Unland, Science 179, 567 (1973).
12. T. P. Kobylinski and B. W. Taylor, J. Catal. 33, 376 (1974).

DISCUSSION

H. Wise *(Stanford Research Institute)*

I was very interested in the desorption spectra, that is the temperature programmed desorption spectra. You mentioned that there are two states of nitrogen, possibly three, one behaving like a second order state which can be explained by a model in the literature and one having first order characteristics at high coverage. How does one account for that second stage? I was wondering if possibly another reaction might occur at high coverage — one between N adsorbed plus NO which could possibly explain the first order characteristics.

Gjostein

In the text, we interpret the development of the β_1-peak in terms of interactions between adsorbed N atoms — repulsive interactions for 1st neighbor sites and attractive for next nearest neighbor sites. At the present time, it is not possible to determine whether this model would predict the "1st order characteristics" of the β_1-peak in the high coverage region. Calculations such as those carried out by Adams (9) would be needed to clarify this point.

It should also be pointed out that the "1st order characteristics" of the β_1-peak are not readily evident until it clearly stands well above the shoulder of the β_2-peak (as in curve k, Fig. 3). This requires long NO exposures, which as we discuss in the text, lead to the formation of a molecular NO adsorbed state, following saturation of the surface with dissociated NO. As you suggest, this could result in a reaction between adsorbed N (from the dissociated NO phase) and adsorbed molecular NO. The nitrogen desorption product would contribute to the β_1-peak. We have not seen any evidence of desorbed N_2O from such a reaction.

V. Haensel (UOP)

I was wondering about a competitive adsorption. If you start with a moderate temperature, with a mixture of NO and oxygen, how do you see that?

Gjostein

Below the desorption of nitrogen? I doubt if we've run that experiment. My guess is that it might displace the NO, but I'm not sure. I know NO displaces CO quite readily on the surface. Oxygen is very strongly bound to the surface, so it might perform a displacement reaction.

W. K. Hall (University of Wisconsin)

I was wondering how you explain these segregated patches. You must have sites of two different energies or to enable nitrogen and oxygen to separate into separate patches. Is there any surface reorganization?

Gjostein

Well, I think most of the results, or many of the results we've seen with low energy electron diffraction and auger spectroscopy indicate the formation of patches on the surface, that is, islands of material. In the case of oxygen, for example, we see diffraction features long before reaching saturated coverage, so that there is a tendency for clustering. This says that the energy does favor the formation of patches of ordered arrays. In this case, that kind of structure of two separate patches must have lower energy than the co-adsorbed species because there's no evidence for co-adsorption.

Hall

Yes, it's got to be considerable, because you've got to overcome the loss of entropy that's involved here.

Gjostein

Yes, correct.

Hall

So, you're working with a single crystal face but you're finding two different sites that allow this sort of segregation to occur. That's rather remarkable.

Gjostein

I don't think there have to be two different types of site. Each dissociated species has a preference for like neighbors; for example, nitrogen would prefer to have adjacent nitrogens rather than oxygen.

R. J. H. Voorhoeve *(Bell)*

First I would like to point out that recent work by Linnet has shown that on platinum you also get a dissociated adsorption with NO; the whole sequence of events is actually quite simple. There also the oxygen is only removed with great difficulty. I have a question. Have you ever found in your desorption spectra, or in dosing with NO, the formation of N_2O?

Gjostein

No.

M. Boudart *(Stanford University)*

Now that we have been told that NO also dissociates on platinum, I would like to be reassured that the dissociation on ruthenium or on platinum is not induced by the electron beam during the Auger work. How can you rule this out?

Gjostein

Well, the electron beam is on only momentarily to take an analysis. This is the case with the LEED also.

Frank Williams *(General Motors Research Laboratories)*

I'd like to comment on that same point. In fact we have observed electron beam dissociation of NO on many of these same surfaces and it's a matter that needs to be checked experimentally that the length of time you have the beam on and the current that you have don't affect this dissociation.

Gjostein

We've seen no evidence that the length of time that the beam is on influences the results. The UPS findings that molecular NO adsorbs once the surface is saturated with dissociated NO strongly support our contention that electron beam effects are not causing dissociation.

R. Klein *(National Bureau of Standards)*

We've done the same type of experiments with the NO on (10$\bar{1}$0) ruthenium and agree essentially with your findings. But one thing I want to mention is that in the

NO desorption, in the trailing edge of the NO desorption, there's a very sharp cutoff, and this appears just about where the first nitrogen peak starts to come off. So your model is probably correct, in that as the nitrogen comes off it frees up sites, and this NO which comes off only to the extent of about 3% then is adsorbed on these more active sites and dissociates. This accounts for this sharp trail off. Another point that I wanted to bring up is that we also agree, so far as the oxygen is concerned, that if you adsorb oxygen, you get half the oxygen desorbing from NO as you do from oxygen. But on the other hand, if you look at these desorption experiments, considerably less oxygen comes off than nitrogen. We found the same effect which you have on your slide. How do you account for this phenomenon?

Gjostein

We haven't explored the high temperature oxygen results at all. It could be coming off as another species for one thing. We don't see it come off until up around 800 to 1000°C, but we have not explored the details.

C. C. Chang *(General Motors Research Laboratories)*

Do you always introduce CO into the system after you observe nitrogen coming off the surface or do you put them in together? From infrared work, people propose that NO and CO react through an isocyanate species. Which of your experiments rule out the possibility of isocyanate species?

Gjostein

That was only a postulate based on our dissociative adsorption model and wasn't based on any direct experimental evidence.

FUNDAMENTAL APPROACHES TO THE DECOMPOSITION OF NITRIC OXIDE: THE TRANSITION METAL DINITROSYL MODEL

W. R. MOSER

Exxon Research and Engineering Company, Linden, New Jersey

ABSTRACT

A concept for the catalytic decomposition of nitric oxide to N_2 and O_2 has been developed and subjected to experimental verification. The model requires a *cis*-coplanar, *gem*-dinitrosyl transition metal microstructure, containing 1 bent (activated) and 1 linear nitrosyl, to overcome the M.O. symmetry restriction to the concerted decomposition of two NO molecules to N_2 and O_2.

gem-Dinitrosyl transition metal complexes were synthesized, and the mode of decomposition of the dinitrosyl moieties to N_2, NO, or N_2O was examined as a function of the metals' d-electron density, their non-bonding d-electron energy levels, their inner sphere coordination numbers, the ligand environment, and the dinitrosyl stereochemistry.

The results of these studies to date indicate that the dinitrosyl can be forced to selectively and rapidly decompose to N_2 although the evolved O_2 reacted in all cases with the ligand or metal. The thermal decomposition data utilizing dinitrosyls with well-defined X-ray structures supports the proposed *gem*-dinitrosyl structural requirements for a NO decomposition scheme.

INTRODUCTION

The efficient removal of nitric oxide from both stationary and mobile exhaust stream effluents remains today a paramount environmental problem.

The reduction of nitric oxide over catalysts utilizing a variety of reducing molecules, e.g. CO, H_2, CH_4 and NH_3 has been exhaustively studied (1). The

References p. 43.

catalytic decomposition of nitric oxide to molecular nitrogen and oxygen has been studied (2); although no efficient decomposition catalysts have been discovered to date. Since a successful nitric oxide disproportionation (decomposition) catalyst would afford the conversion of the toxic and polluting nitric oxide molecule to innocuous gaseous molecules, O_2 and N_2, without the requirement of a co-reducing gas, we have examined some fundamental chemical approaches to NO disproportionation.

Of the several approaches to nitric oxide disproportionation that have been studied in this laboratory, our concept utilizing *gem*-dinitrosyl transition metal moieties will be discussed here.

THE *gem*-DINITROSYL TRANSITION METAL MODEL

Nitric oxide is thermodynamically unstable by 21 to 17 kcal/mol between 300 K-1500 K favoring N_2 and O_2 formation. Despite this thermodynamic instability toward decomposition, shock tube studies on the homogeneous gas phase decomposition of NO have yielded activation energies ranging from 63 to 86 kcal/mol (3). This unusual kinetic stability of NO is due to the fact that the homogeneous, gas phase disproportionation of two NO molecules to N_2 and O_2 is molecular orbital symmetry forbidden. The *gem*-dinitrosyl model overcomes this symmetry limitation while simultaneously affording a complete catalytic cycle. The literature on transition metal nitrosyls shows that the main type of reaction observed when a single gaseous NO molecule reacts with a metal complex is the transfer of an electron from the π^* (antibonding) M.O. of NO onto the central metal atom of the complex. Thus, the NO moiety gains a positive charge resulting in a linear nitrosyl (<MNO~ 180°) and the metal is reduced. The reaction of this reduced metal with a second molecule of NO results in an electron transfer away from the metal into the (π^*(NO)) orbital of the NO molecule. An electronic reorganization of the resultant NO$^-$ causes a reduction in the NO bond strength leading to a bent (activated) nitrosyl where the M-N-O bond angle is ca. 120°. The stereochemistry of bonding of both NO$^+$ and NO$^-$ to a transition metal is precisely predicted by an examination of their respective M.O. diagrams. Furthermore, if the two nitrosyl groups can be forced into a *cis*-coplanar configuration, the oxygen of the negatively charged nitrosyl would be suitably located spacially to interact with the oxygen of the positively charged nitrosyl. Indeed, Eisenberg's group (4) showed by single crystal X-ray studies that the two oxygen atoms of the nitrosyls in $[RuCl(NO)_2(PPh_3)_2]^+$ were of adequate orientation and proximity to afford a direct, but weak, interaction. In this complex the two Ru-N-O bond angles were 136.0° and 179.5°.

Thus, the known coordination chemistry of metal nitrosyls supports the feasibility of the following schematic model for the catalytic decomposition of nitric oxide.

$$M^{d^n} \xrightarrow{\text{NO}} M^{d^{n+1}}N\equiv O \xrightarrow{\text{NO}} M^{d^n}N\equiv O \longrightarrow M^{d^n} + N_2 + O_2$$

$$\underline{1} \qquad\qquad \underline{2} \qquad\qquad\quad \overset{\displaystyle N=O}{} \qquad \underline{1} \quad \underline{4} \quad \underline{5}$$

$$\underline{3}$$

M = Transitional metal; d^n = no. of d-electrons.

This scheme provides a catalytic cycle since the metal's final and initial oxidation states are identical.

In order to experimentally realize a successful disproportionation system according to this model, several chemical parameters must be precisely adjusted. Fortunately, the known information on reactivity theory and coordination chemistry provides some clues to force molecules to adopt the electronic and stereochemical configurations demanded by the model. Some of the parameters which may be qualitatively dealt with are:

1. The central metal atom must be able to undergo at least a single electron redox reaction. It must have non-bonding d-electrons at suitable energy levels with respect to the vacuum level, contained in orbitals of suitable symmetry and stereochemistry so that the activation of the second nitrosyl group may be achieved. The number and energy levels of the d-electrons can be adjusted by proper selection of the metal, its oxidation state and its ligand environment.

2. The stereochemistry of the metal can be forced into a *cis*-coplanar arrangement by use of di- or tri-dentate chelating ligands.

3. The metal and its environment must be resistant to oxidation. This can be controlled by selection of difficultly oxidized metals and ligands, or by use of metals in their higher oxidation states.

4. The formation of the dinitrosyl from a starting metal complex must utilize an initial complex with either two vacant coordination sites or easily displaced ligands. Several ligands, including carbonyl, hydride and halogens are known to be easily replaced on metal centers by gaseous NO (5).

RESULTS AND DISCUSSION

To experimentally evaluate this model for NO disproportionation several different classes of metal compounds or complexes were synthesized. In some cases their catalytic activity for disproportionating a dilute NO gas stream was evaluated. In other studies the mode of decomposition of the pure dinitrosyls was examined to gain fundamental information relating to what must be done chemically to cause these molecules to decompose specifically to N_2 and O_2. Some of the molecules studied were:

1. Small metal particles.

2. Pure inorganic metal dinitrosyls

$$Co(NO)_2 X, \; Fe(NO)_2 X, \; Rh(NO)_2 X, \; Pd(NO)_2 X_2$$

$$\underline{6} \qquad\qquad \underline{7} \qquad\qquad \underline{8} \qquad\qquad \underline{9}$$

3. Pure anionic metal dinitrosyls derived from:

$$R_4 N\big[M(CO)_5 X\big], \; R_4 P\big[M(CO)_5 X\big], \; R_2 O\!\cdot\!Li\big[M(CO)_5 X\big]$$

$$\underline{10} \qquad\qquad\qquad \underline{11} \qquad\qquad\qquad \underline{12}$$

4. Pure cationic metal dinitrosyls

$$\big[Co(NO)_2 \; TMED\big]^+ PF_6^-$$

$$\underline{13}$$

5. Dinitrosyls in varied environments, $LM(NO)_2$ where the ligand L was:

   A. Poor π-acceptor, good σ-donor, L = amines, alkylphosphines.

   B. Good π-acceptor
   L = phosphites, arylphosphines

   C. di- or tri-dentate chelating ligand
   L = chelating amines phosphines, phosphine oxides

   D. Oxygenated ligand
   L = phosphine oxides, chelating ethers, chelating ketones

   E. Stereochemically well defined dinitrosyls

$$\big[RuCl(NO)_2 \; (PPh_3)_2\big] PF_6 \cdot C_6 H_6$$

$$\underline{14}$$

EXPERIMENTAL SYSTEM

The observation of how each adjustable parameter affected the decomposition of the $M(NO)_2$ moiety was accomplished by heating either the neat materials in an inert gas or suspending the materials in an inert solvent and measuring the compositions of the gases evolved.

$$M(NO)_2 \xrightarrow{\;25^\circ \text{-} 150^\circ C\;} N_2 + NO + N_2O$$

$$M = Co, \; Rh, \; Ru, \; Pd, \; Fe, \; W, \; Cr, \; Mo$$

INFLUENCE OF METAL ATOM ON MODE OF DECOMPOSITION

To determine the role of the central metal atom in the desired decomposition to N_2 and O_2, studies on pure dinitrosyls as well as those in the presence of σ-donor, chelating ligands were carried out as seen in Tables 1 and 2.

TABLE 1

Influence of Metal on Mode of Decomposition

Complex + PMDT(a) $\xrightarrow{\dfrac{150}{C_6H_4Cl_2}}$	% Yield		
	N_2	NO	N_2O
$[Co(NO)_2Cl]_2$ 15 + PMDT	11	29	60
$Pd(NO)_2Cl_2$ 16 + PMDT	1	99	0
$Rh(NO)_2Cl$ 17 + PMDT	100	0	0

(a) PMDT = Pentamethyldiethylenetriamine.

Table 1 demonstrates the dramatic divergence in results obtained utilizing different metals. Other results demonstrate the general phenomenon that the first row transition metal dinitrosyls mainly decompose to N_2O rather than to N_2.

TABLE 2

Decomposition of Inorganic Dinitrosyls

Complex	Conditions (Temp., solvent)	Products, % Yield		
		N_2	NO	N_2O
$Rh(NO)_2Cl$ 17	150°C, neat	30	20	50
$Rh(NO)_2Cl$ 17	150°, o-dichlorobenzene	13	54	33
$Rh(NO)_2Cl$ 17	150°, mesitylene	2	83	15
$[Co(NO)_2Br]_2$ 18	150°, neat	2	98	0
$[Co(NO)_2Cl]_2$ 15	150°, neat	8	83	9
$Pd(NO)_2Cl_2$ 16	150°, neat	0	100	0
$Pd(NO)_2Cl_2$ 16	150°, o-dichlorobenzene	1	99	0

The decomposition of the neat dinitrosyls, Table 2, showed very little difference between Co and Pd, although an ionic ligand variation of Br to Cl in the cobalt complexes 18 and 15 showed a modest change. The rhodium dinitrosyl 17 showed the greatest promise for decomposition to products other than NO. Within the rhodium series the interesting media effect, producing more N_2 and N_2O in the series

References p. 43.

mesitylene < dichlorobenzene < neat material, is likely the result of an increasingly more ionic environment. The formally Pd^{2+} complex *16* was principally examined because this central metal atom would be less subject to reaction with O_2 evolved from a $M(NO)_2$ decomposition. Naturally, this idea could not be tested since NO was the sole product from the decomposition of *16*.

The chelating effect of inner sphere coordinating ligands on the mode of decomposition of $M(NO)_2$ was examined utilizing the rhodium dinitrosyl *17*. The purpose of utilizing chelating ligands was to interact with the metal complex forcing the nitrosyl groups into a *cis*-coplanar configuration. Table 3 illustrates the dramatic effect observed when the ligand environment is a di- or tri-dentate, chelating polyphosphine (*19,20*) versus a di-, tri-, or tetra-dentate, chelating polyamine (*22,23* and *24*). It is suggested that this difference in reactivity results from a combination of the chelating and σ-donor effect of the amines compared to the strong π-acceptor strength of the phosphine derivatives. The chelating oxygenated compounds were examined since their coordinating group would be difficultly oxidized; in addition they are strong σ-donors. In all cases the polyether ligand resulted in very low rates of $M(NO)_2$ decomposition. Although the chelating polyamines (*22, 23* and *24*) all rapidly and quantitatively caused the $M(NO)_2$ structure to decompose to N_2, the

TABLE 3

Environmental Effect on Decomposition-Neutral Ligands

$Rh(NO)_2Cl$ + Ligand	$\dfrac{150°C}{\text{Dichlorobenzene}}$	N_2	+	NO	+	N_2O
17						
$Ph_2PCH_2CH_2PPh_2$ *19*		25		0		75
$PhP(CH_2CH_2PPh_2)_2$ *20*		9		5		86
$Bu_3P=O$ *21*		15		1		84
TMED[a] (100 C) *22*		100		0		0
PMDT *23*		100		0		0
HMTT[b] *24*		100		0		0
Bu_3N *25*		43		0		57
Diglyme *26*		16		37		47
Acetylacetone *27*		14		55		31
No Ligand *28*		13		54		33

(a) TMED = N,N,N',N' - tetramethylethylenediamine.

(b) HMTT = N,N,N',N",N" ', N" ' - hexamethyltriethylenetetramine.

relative rates of decomposition were in the series HMTT, *24*, (tetramine) >PMDT, *23*, (triamine) > TMED, *22*, (diamine). That a chelating effect was actually observed was evidenced by the fact that the non-chelating amine tri-*n*-butyl amine, *25*, resulted in decomposition mainly to N_2O at a much lower rate. The effect of atoms of different trans directing influence on the $M(NO)_2$ decomposition was examined in Table 4 where the trans activators (6) are in the series $I > Br > Cl$.

TABLE 4

Environmental Effect on Decomposition-Charged Ligand

$$\text{HMTT} + \left[\text{Co(NO)}_2\text{X}\right]_n \xrightarrow[\text{Mesitylene}]{150°\text{C}} \qquad N_2 \quad + \quad NO \quad + \quad N_2O$$

24	6			
	X			
	−Cl	6	27	67
	−Br	6	6	88
	−I	2	4	94

An examination of the molecular orbital correlation diagrams for transition metal nitrosyls led Eisenberg (7) to conclude that a metal nitrosyl possessing more than 20 electrons in its inner coordination sphere should result in an NO orbital rehybridization in which the (π^*NO) electrons are transferred to an (SP_N^2) orbital. The result of this reorganization is the formation of a bent nitrosyl ($<MNO \sim 120°$) with a -NO bond order of 2 for complexes with 22 electrons or more. Table 5 shows the results of some attempts to substantially activate the NO bond *further* by building complexes with increasing numbers of electrons in their inner coordination sphere. An inspection of the correlation diagram for a 26 electron, six coordinate system of the type, $\left[L_4Co(NO)_2\right]^+$, shows that the orbital of dz^2 character in an octahedral ligand field moves closer to the vacuum level as compared to the same orbital in a tetragonal complex. The net result is to make the $(\pi^*(NO))$ orbital energetically accessible. A 26 electron system of this type would result in population of precisely this $(\pi^*(NO))$ orbital leading to a reduction of the already weak NO bond order to less than 2. This electron transfer also accommodates the complex compatibility with the EAN-18 electron rule. Thus, Table 5 investigates the possibility of inner sphere electron population on the activation of the NO bond in which the number of electrons for the expected intermediates $\left[Co(NO)_2Cl\right]_2$, $\left[TMED\ Co(NO)_2\right]^+$, $\left[PMDT\ Co(NO)_2\right]^+$, and $\left[HMTT\ Co(NO)_2\right]^+$ are 21, 22, 24 and 26 respectively.

In other studies utilizing $Ph_2PCH_2CH_2PPh_2$, $Ph_2PCH_2CH_2AsPh_2$, $CH_3C(CH_2PPh_2)_3$, and $PhP(CH_2CH_2CH_2PPh_2)_2$ ligands reacting with $Rh(NO)_2Cl_2$, 95 to 100% selectivity was observed for the formation of N_2 and N_2O over NO

References p. 43.

expulsion. Dinitrosyl intermediates formed from these ligands tested the effect of molecules of high electron counts in that they would have 24, 24, 26 and 26 interacting electrons respectively. We believe that these data indicate a relationship between the activation of coordinated NO and the population of its π^* antibonding orbital.

TABLE 5

Effect of Inner Sphere Electron Population on NO Activation

Complex	Interacting Electrons	% Yield[b]		
		N_2	NO	N_2O
$(Co(NO)_2Cl)_n$	21	8	83	9
$[TMEDCo(NO)_2]^+PF_6$	22	7	93	0
$[TMEDCo(NO)_2]I^{(a)}$	22	10	52	38
$[PMDTCo(NO)_2]^+Cl^{(a)}$	24	7	16	77
$[PMDTCo(NO)_2]^+I^{(a)}$	24	7	12	81
$[HMTTCo(NO)_2]^+Cl$	26	7	27	66

(a) Complex generated *in situ.*

(b) Relative yields based on metal complex in mesitylene at 150°C.

STUDIES ON OXIDATION RESISTANT METALS

During the course of this work, it was recognized that oxygen evolved during the disproportion step reacted in some cases with the metal atom. This would likely lead to catalyst deactivation when applied to a real catalytic system. Thus, we examined dinitrosyls of the difficultly oxidizable, divalent metal Pd^{2+}.

The decomposition of $Pd(NO)_2Cl_2$ promoted by hard σ-donor ligands like ethers and tertiary amines as well as poor π-acceptors like alkyl phosphines caused little activation of coordinated nitrosyl. Table 6 shows that these ligand environments around the nitrosyl led mainly to expulsion of NO.

However, the good π-acceptors like phosphites and arylphosphines as well as the chelating arylphosphines activate the NO molecule leading in most cases to N_2O with varying amounts of N_2.

In nearly all of the cases involved in our studies, the π-accepting ligands both activated the NO moiety more effectively and led kinetically to higher rates of $M(NO)_2$ decomposition when compared to poor π-acceptors.

TABLE 6

Decomposition with a Difficulty Oxidized Metal

Complex	Ligand	Product Composition, %		
		N_2	NO	N_2O
$Pd(NO)_2Cl_2$	PBu_3	3	72	25
$Pd(NO)_2Cl_2$	PPh_3	13	18	69
$Pd(NO)_2Cl_2$	$P(OPr)_3$	1	9	90
$Pd(NO)_2Cl_2$	$OPPh_3$	25	75	0
$Pd(NO)_2Cl_2$	$(Ph_2PCH_2CH_2)_2PPh$	71	1	88
$Pd(NO)_2Cl_2$	$(Ph_2PCH_2)_3CCH_3$	23	9	68
$Pd(NO)_2Cl_2$	PMDT,25°C	27	67	6
$Pd(NO)_2Cl_2$	$(CH_3OCH_2CH_2)_2O$, 50°C	13	87	0

STUDIES ON COMPLEXES WITH cis-COPLANAR DINITROSYLS

To further test our hypothesis that a cis-coplanar metal dinitrosyl transition state is necessary for disproportionation to N_2 and O_2 a dinitrosyl 14 with a well defined single crystal X-ray structure was examined. $RuCl(NO)_2(PPh_3)_2$ PF_6 (14) is best described as a distorted tetragonal pyramid containing a nitrosyl in the basal plane coordinated linearly (<Ru-N-O = 178°) and a second bent nitrosyl in the apical position with a Ru-N-O bond angle of 138°. Their infrared absorptions appear at 1845 cm^{-1} and 1687 cm^{-1} respectively. The atoms in the $Ru(NO)_2$ part of the structure are coplanar and the two oxygen atoms of the nitrosyl moieties are separated by 3.915 A. Although compared to an O-O bond length of 1.45 A in benzoyl peroxide (8) and 1.62A in $[IrO_2(diphos)_2]^+$ (9), the oxygens of the nitrosyls are not within bonding distance in the ground state. However, they certainly are close enough to interact with one another.

The same type of orientation of a bent nitrosyl has been observed in other cases where the oxygen of the nitrosyl can react with a good π-acceptor. In $[IrCl(NO)(CO)(PPh_3)_2]^+$ (10) the bent NO group is oriented toward the CO group and in $IrCl_2(NO)(PPh_3)_2$ (11) it is oriented toward one of the basal phosphine groups. This forced orientation of the NO toward the π-acceptors is best explained (4) by a donor-acceptor interaction between an sp^2 lone pair on oxygen with either the $(\pi^*(CO))$ orbital or an empty 3d phosphorus orbital. Likewise the cis-coplanar microstructure for the dinitrosyls is likely due to a similar interaction with the $(\pi^*(NO))$ orbitals.

In all cases the decomposition of $RuCl(NO)_2(PPh_3)_2$ $^+$ resulted in rapid, quantitative yields of molecular nitrogen. Although no molecular oxygen was observed in these studies, shown in Table 7, we believe that these results support our concept of the importance of a *cis*-dinitrosyl microstructure for a successful disproportionation reaction to N_2 and O_2 products.

TABLE 7

Decomposition of $\left[RuCl(NO)_2(PPh_3)_2\right]PF_6$

$\left[RuCl(NO)_2L_2\right]^+$	$\xrightarrow[\text{Solvent}]{150°}$	N_2	+	NO	+	N_2O
dichlorobenzene		100		0		0
mesitylene		100		0		0

14

DECOMPOSITION OF $M(NO)_2$ TO N_2 VERSUS N_2O

This study has uncovered several interesting phenomena in the decomposition of transition metal dinitrosyls. In the cases where the NO moiety was activated, differing amounts of either N_2 and N_2O were formed.

An intriguing question is: what are the fundamental factors which cause the $M(NO)_2$ microstructure to fall apart selectively on the one hand to N_2 and in other cases exclusively to N_2O?

From the data contained in this paper the following conclusions are drawn, and it is believed that this information could be applied to catalyst synthesis when selectivity for either $N_2 + O_2$ or $N_2O + O_2$ is desired.

For disproportionation either to N_2 or N_2O metals with partially populated d π-orbitals are necessary. If the dinitrosyl can be forced into a *cis*-coplanar structure, it will lead to $N_2 + O_2$. However, *trans*-dinitrosyls or dinitrosyls of the first row transition metal elements, where the two NO groups are spatially well separated due to their usual tetrahedral stereochemistry, afford N_2O. The second row transitional elements with accessible trigonal bipyramidal (tbp) structures favor N_2O formation with good π-accepting ligands. This is due to the high *trans*-effect of these ligands leading to stabilization of the tbp transition state resulting from the attack of the oxygen of the coordinated nitrosyl. This interaction affords a metal nitrosyl nitride which decomposes rapidly to N_2O. Unless the metal atom has an unusually high d-electron population, good σ-donors are necessary for observing the N_2 product. In the case of metals with an excess of this electron density, π-acids may lead to more facile and selective decomposition to N_2.

REFERENCES

1. F. R. Taylor, *Air Pollution Foundation, Rept. No. 28 (1959); J. F. Roth and R. C. Doerr, Ind. Eng. Chem. 53, 293 (1961); R. A. Baker and R. C. Doerr, Ind. Eng. Chem. Process Design Develop. 4, 188 (1965); R. J. Ayen and M. S. Peters, ibid., 1, 205 (1962); K. H. Schmidt and V. Schluze, German Pat., 1259298 (Jan. 25, 1968); M. S. Peters, AEC TID-18423 (1963).*

2. a.   C. S. Howard, F. Daniels, *J. Phys. Chem., 62, 360 (1958).*
   b.   M. Shelef, K. Otto, H. Gandhi, *Atmospheric Environment, 3, 107 1969).*
   c.   J. M. Fraser and F. Daniels, *J. Phys. Chem. 62, 215 (1958).*
   d.   S. Sourirajan and J. L. Blumenthal, *Proc. Intern. Cong. on Catalysis, Paris 2, 2521 (1960).*
   e.   C. H. Riesz, E. L. Moritz, K. D. Franson, *Air Pollution Foundation Report, No. 20, (1957).*

3. F. Kaufman, J. R. Kelso, *J. Chem. Phys., 23, 1702 (1955).*

4. C. G. Pierpont and R. Eisenberg, *Inorg. Chem. 11, 1088 (1972).*

5. W. Hieber and J. J. Anderson, *Z. Anorg. Allgem. Chem., 208, 238 (1931); T. I. Eliades, R. O. Harris, and M. C. Zia, Chem, Commun., 1709 (1970); W. Hieber, W. Beck and H. Tengler, Z. Naturforsch 16B, 68 (1961).*

6. A. Cotton and G. Wilkinson, *Advanced Inorganic Chemistry, Interscience, New York, 1972,* p. 668.

7. C. G. Pierpont and R. Eisenberg, *J. Am. Chem. Soc., 93, 4905 (1971).*

8. V. Kasatochkin, S. Perilina and K. Ablesova, *Daklady Akad. Nauk, SSSR 47, 37 (1945); CA, 40, 4044 (1946).*

9. J. A. McGinnety and J. A. Ibers, *Chem. Commun. 235 (1968).*

10. D. J. Hodgson and J. A. Ibers, *Inorg. Chem., 7, 2345 (1968).*

11. D. M. P. Mingos and J. A. Ibers, *Inorg. Chem., 10, 1035 (1971).*

DISCUSSION

M. Shelef *(Ford Motor Co.)*

I would like to ask you about the complexes leading to nitrogen which were also the ones that contained the nitrogen ligands. Is there any interaction that would link the nitrogen between the nitrogen containing ligands and NO?

Moser

Well, there is some data in the literature that says a coordinated nitrosyl can react with an amine molecule. But we showed that this was not the case in our particular work. I might indicate that we also studied the effect of nitric oxide, i.e., gaseous nitric oxide, on these reactions and showed that the rates of decomposition were in fact inhibited by nitric oxide. Thus we know that we are not performing reactions between a gaseous nitric oxide and a coordinated nitric oxide which has also been mentioned in the literature.

J. W. Hightower *(Rice University)*

Can you comment on what happened to the oxygen?

Moser

I looked at that several ways. In the case of phosphines, I observed phosphine oxides emitted. And also in some other cases I looked at the metal by photo-electron spectroscopy and I can observe that the metal was in fact oxygenated too. I did some work utilizing palladium compounds which I showed you there. The intent of that was to utilize palladium-II which is in a high oxidation state, which should eliminate that particular problem with the disproportionation.

F. Williams *(General Motors Research Laboratories)*

You have shown us transient selectivity during the decomposition of these compounds and I wonder if you could comment on the rates for the various classes of decomposition that you found? Would you also comment on the ease of regeneration for the different classes?

Moser

To answer your last question, I haven't really examined the ease of regeneration. I will say that the rates of the bidentating versus the tri- versus the tetra dentating chelating amines increase significantly in that order. It turns out that the rates are increased for very good pi-acceptors like trialkyl or a triarylphosphites, over the rates for poor to moderate pi-acceptors like a triarylphosphites. These rates are all significantly greater than the rates for oxygenated ligands like diglyme.

USE OF NO CHEMISORPTION FOR THE STUDY
OF CATALYST-SUPPORT INTERACTIONS

H. C. YAO and M. SHELEF

Ford Motor Company, Dearborn, Michigan

ABSTRACT

Previously it was shown that nitric oxide chemisorbs on a series of supported transition metal surface ions at room temperature, with one NO molecule being adsorbed per surface ion. The study of NO chemisorption on supported cobalt and nickel oxides has been pursued further to characterize the catalyst-support interactions. On cobalt aluminate the chemisorption was found to be very small. This was explained by the shielding of the tetrahedrally coordinated cobaltous ions from the surface by the oxygen ions in this spinel compound. The explanation is supported by the ion-scattering spectra of the $CoAl_2O_4$ surface. Measurement of the NO uptake of cobalt and nickel oxide catalysts variously supported, after heat-treatments at successively increasing temperatures, has revealed the different character of the support-catalyst interaction. On zirconia, the area of the NO uptake reaches the value of the BET area. On alumina, there is diffusion of the cobalt or nickel ions into the bulk of the support accompanied by spinel formation. Diffusion into the bulk occurs also, to some degree, on preformed cobalt-aluminate support. The supported cobalt and nickel catalysts represent complex systems and at low metal ion loadings on a zirconia support, multiple NO adsorption on a surface metal ion was noted.

SYNOPSIS OF PREVIOUS WORK

The chemisorption of nitric oxide on transition metal oxides has been the subject of study at the Scientific Research Staff of the Ford Motor Co. in the last several years (1−6). In these studies the chemisorption was measured both volumetrically and gravimetrically using supported and non-supported catalysts as adsorbents. In all cases the adsorption isotherms were of the Freundlich-type and the adsorption

References p. 58.

kinetics were well described by the Elovich equation. Whenever the chemisorption of NO and CO on the same adsorbent was compared (3,6) it was found that the NO is adsorbed more strongly than CO. This is in agreement with IR chemisorption studies (7) and with the behavior of transition metal complexes in solution (8). For instance, on Co_3O_4 at room temperature NO co-adsorbs on a surface covered with pre-adsorbed CO with very little impediment; reversal of the adsorption sequence blocks the adsorption of CO almost completely (6).

In some instances the adsorption is much faster and probably stronger on a reduced surface, as for instance on Cr^{3+} ions vs. Cr^{6+} (1), Fe^{2+} vs. Fe^{3+} (2), or Mn^{2+} vs. Mn^{4+} (5). Conversely, in the case of copper, Cu^+ ions do not chemisorb NO, while the Cu^{2+} ions do (4). This is associated with the (d^{10}) configuration of the Cu^+ ions which does not permit the accommodation of the antibonding electron in the NO molecule upon chemisorption. Such behavior parallels also the behavior of copper ions in solution with respect to the formation of nitrosyl complexes (9). Neither do Zu^{2+} ions (d^{10}) chemisorb NO (6), with the exception of the very low coverages associated with the EPR signals (10).

Thus, the NO molecule is suitable as an adsorbent for the measurement of the dispersion of transition metal ions on insulator supports, since the uptake on the supports is very small and easily corrected for. This has been demonstrated for Cr^{3+}, Ni^{2+}, Cu^{2+}, Fe^{2+}, Fe^{3+}, and Co^{3+}. There are, however, exceptions to the applicability of the NO chemisorption method and care should be taken in its application. The method is inapplicable to adsorbents which are easily oxidized such as completely dispersed metals or ions such as Mn^{2+} (5) or Co^{2+}. Here, irreversible surface oxidation takes place accompanied by the reduction of the adsorbate. Further chemisorption on the oxidized surface then follows. These complications make interpretation of the results difficult.

The suitability of the NO molecule as a selective probe for surface transition metal ions can be demonstrated by considering two extreme situations: very diluted supported catalysts and pure transition metal oxides.

In the first case when there is less than 1% by weight of the transition metal oxide present in a catalyst (which was prepared by impregnation of a support with a specific surface area of $\sim$ 100 m^2/g), it is assumed that every transition metal ion is present on the surface. The expected ratio of chemisorbed NO molecules to the total amount of transition metal ions is unity. Some experimental values are shown in Table 1 which bear out the expectations. However, there are special situations where multiple NO adsorption on single ion sites is observed; these will be mentioned subsequently in this paper.

If the ratio of one chemisorbed NO molecule to one surface transition metal ion still holds for pure oxide powders, the uptake of NO should be the same as the average surface density of metal ions. The usual procedure to estimate the latter value

TABLE 1

NO Chemisorption on Dilute, Alumina-Supported
Transition Metal Catalysts

Active Oxide	Support	BET Area m^2/g	Wt % Metal	Surface Density of ions, $10^{18}/m^2$ (BET)	NO Molecules Adsorbed, $10^{18}/m^2$ (BET)	NO Molecules Per Ion	Ref.
CuO	Kaiser KA201 γ-Alumina	171	0.02	.45	.42	0.93	4
NiO	Kaiser KA201 γ-Alumina	180	.69	.39	.40	1.02	3
Fe_2O_3	95% Alumina 5% Silica	199	.85	.46	.45	0.99	2
Fe_2O_3	95% Alumina 5% Silica	199	.15	.081	.068	0.84	2
Co_3O_4	Linde B γ-Alumina Pre-Calcined at 1000°C	69	.15	.22	.19	0.87	This Study

is to average the theoretical density on the three low index planes since there is no direct method for measurement. Table 2 gives the uptake of NO on pure oxides. Comparing the NO uptake with the surface density of ions we observe that it is somewhat higher than the ion density on the stable, half-populated surfaces. Since this evaluation is only a crude estimate, we tend to retain the notion that the 1:1 ratio holds for pure oxides, at least within a first approximation. Table 3 summarizes the variations in the heat of chemisorption of NO in the range of middle coverages, more specifically at $\theta = 0.37$. Since the adsorption follows the Freundlich isotherm the heat of adsorption decreases logarithmically with coverage. Heats of chemisorption are a comparative measure of the chemisorption bond strengths at the surface. It is seen from Table 3 that the heat of adsorption is highest on the transition metal oxides with the spinel structure.

Chemisorption of NO on spinels is a useful approach to characterize the surface of such adsorbents, especially when chemisorption is done in conjunction with other methods of surface investigation. The chemisorption of NO was studied on Co_3O_4 and other cobalt containing spinels, such as $CoAl_2O_4$ and $ZnCo_2O_4$ (6). The uptake of NO by $ZnCo_2O_4$ is the same as by Co_3O_4. On the other hand $CoAl_2O_4$ is almost totally inert for this adsorption. The divalent cobalt ions were found by ion-scattering

References p. 58.

TABLE 2

NO Chemisorption on Non-Supported Oxides

Oxide	*Surface density of metal atoms, $10^{18}/m^2$	NO uptake at monolayer, $10^{18}/m^2$	**Ratio of NO to Surface ions	Area occupied by NO molecule, A^2	Ref.
CuO	11.0	5.03	.91	19.9	4
NiO	11.0	5.78	1.05	17.3	3
Fe_2O_3	9.8	6.26	1.27	16.0	2
Cr_2O_3	9.8	6.5	1.32	15.4	1
Co_3O_4	–	7.83	–	12.8	6

*Estimated by averaging the three low index planes;
**The more stable half-populated surface is assumed (Ref. 22). The half-populated surface should
be construed as representing the lower bound for the surface density. In some cases, such as the
(110) and (100) planes in cubic NiO, where both oxygen and metal ions are in plane, the
fully-populated surface is stable. Generally, the surface density of metal ions will fall somewhere
between the values of column 2 and one-half of these.

TABLE 3

Heat of NO Chemisorption on Oxides of
First Transition Series Elements at $\theta = 0.37$

Oxide	Crystal Lattice	H_m kcal/mole	Ref.
Fe_3O_4	Spinel	16.5	2, 6
Co_3O_4	Spinel	13.4	6
NiO	NaCl	8.6	3
CuO	Tenorite	5.1	4
Fe_2O_3	Corundum	4.9	2
Cr_2O_3	Corundum	4.9	1

spectroscopy to be largely shielded from the surface (11). It was concluded that the
tetrahedral co-ordination of the cobaltous ions in $CoAl_2O_4$ is responsible for the
inactivity in chemisorption, rather than their oxidation state.

We have also followed the NO chemisorption on some other spinels (11) and
compared NO uptake to that on pure oxides (Fig. 1). Among the aluminate spinels,
$CuAl_2O_4$ adsorbs as much (or even slightly more) as CuO, while $NiAl_2O_4$ adsorbs only
1/4 of the amount of NiO, and $CoAl_2O_4$ takes up a relatively negligible amount as
mentioned above. Copper chromite adsorbs approximately two-thirds as much as CuO
within the employed pressure range, but it must be noted that the slope of the
isotherm is not as flat as the slope of the isotherms for the other adsorbents.

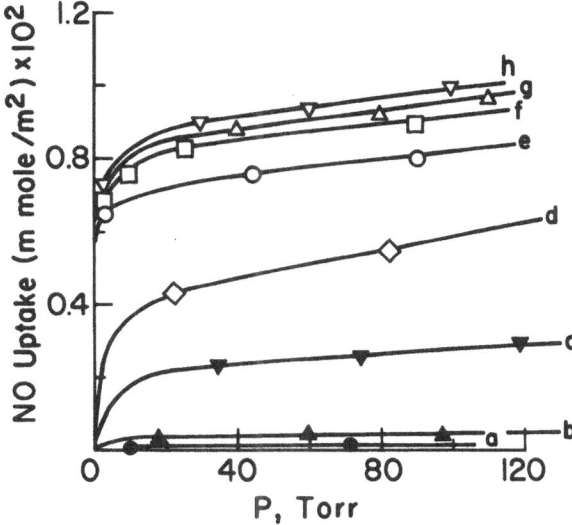

Fig. 1. NO adsorption isotherms at 25°C on (a) γ-Al_2O_3, (b) $CoAl_2O_4$, (c) $NiAl_2O_4$, (d) $CuCr_2O_4$, (e) CuO, (f) $CuAl_2O_4$, (g) NiO and (h) Co_3O_4.

The chemisorption behavior of the spinels can be explained by the distribution of the transition metal ions. In a normal spinel, such as $CoAl_2O_4$, the large majority of the cobalt ions are in tetrahedral co-ordination sites which are shielded from the surface by oxygen anions. On the other hand, Ni ions in $NiAl_2O_4$ have a large octahedral site preference, although this tendency is diminished on the surface (12). Copper-containing spinels have distorted structures due to the preference of the Cu^{2+} ions for square-planar co-ordination. Ion scattering spectra show the dense surface population of copper ions in copper spinels (11).

The deactivation of supported transition metal oxide catalysts by solid state interaction with the support is frequently associated with the formation of spinels or migration into the bulk to form solid solutions with the support (13,14). Since the presence of transition metal ions on the surface of the fresh supported catalyst on one hand and on spinels, on the other, can be characterized by NO chemisorption, we have adopted this method to study the interaction between the active components and the supports.

EXPERIMENTAL

Adsorbents — The interaction of Co_3O_4 was examined for the following supports: zirconia, α-Al_2O_3, γ-Al_2O_3, and $CoAl_2O_4$. The interaction of NiO with zirconia and γ-Al_2O_3 was also studied.

The ZrO_2 was made by decomposition of $Zr(NO_3)_2$ (Johnson Matthey, specpure grade) at $\sim$ 300°C followed by calcining at 800° for 6 hrs. The resulting white powder

References p. 58.

had a BET surface area of 23 m^2/g. The aluminas, α - Al$_2$O$_3$ (Linde A, particle size 0.34μ) and γ - Al$_2$O$_3$ (Linde B, particle size 0.05μ) were preheated to 1000° in air before use. Their BET areas were 11.4 and 66.0 m^2/g, respectively. X-ray diffraction analysis showed only minor amounts ($\langle$10%) of α - Al$_2$O$_3$ in the Linde B sample both before and after calcining. CoAl$_2$O$_4$ was prepared by heating the γ - Al$_2$O$_3$, impregnated with a stoichiometric amount of Co(NO$_3$)$_2$, dried and ground together beforehand, at 1000°C for 7 hrs. The blue solid was examined by x-ray diffraction and showed only spinel structure lines. The BET area was 40 m^2/g.

The unsupported Co$_3$O$_4$ was prepared by precipitation of an aqueous nitrate solution with a solution of NH$_4$OH. The thoroughly washed solids were dried and calcined at 400°C for 8 hrs giving a final BET area of 17 m^2/g. The unsupported NiO was prepared by slow decomposition of Ni(NO$_3$)$_2$ · 6H$_2$0 at temperatures up to 550° and degassed further under vacuum for 3 hrs at 550°. The BET area of this NiO was 5.4 m^2/g.

The supported adsorbents were prepared by impregnation from the corresponding nitrates and calcined at 400°C for Co$_3$O$_4$ and at 550°C for NiO.

Adsorbate — The nitric oxide was purified by several freezing evaporation cycles (2) retaining 2/3 of the middle fraction for the chemisorption experiments. The argon used for the BET area determination was of ultra-pure grade ($\rangle$99.9% pure).

Heat Treatments — The interaction between the active transition metal oxide and the support was engendered by heat-treatments carried out at successively increasing temperature in vacuum for three hours at a time. After each heat-treatment a BET measurement was taken, followed by an NO adsorption isotherm measurement and a 3 hour degassing at room temperature. Then another NO adsorption isotherm was taken.

Adsorption Measurements — NO uptake was measured at 25°C in a conventional constant volume adsorption apparatus equipped with a Texas Instruments Model 145 precision pressure gage linked to a fused-quartz Bourdon spiral. The BET areas were measured in the same apparatus.

We have used as a measure of the NO uptake the amount irreversibly adsorbed at 60 torr pressure, i.e., the difference between the two measurements described under heat treatments. The use of this value instead of the total amount adsorbed eliminates the NO adsorbed on the support which can differ from one support to another. Secondly, it also eliminates the pressure effect as it is the reversible part that is dependent on the pressure. To assess the amount of area on the supported cobalt and nickel oxide adsorbents, the irreversible uptake on the pure oxides, under the same adsorption conditions, was used which was 0.17 cc/m^2 (0.46 x 10^{19} molecules/m^2) and 0.15 cc/m^2 (0.40 x 10^{19} molecules/m^2) for Co$_3$O$_4$ and NiO, respectively.

RESULTS

Pure Co$_3$O$_4$ and NiO — Table 4 shows the irreversible uptake of NO per m^2 of BET area which was measured on two different samples of non-supported Co$_3$O$_4$. One of the samples was heated to successively higher temperatures which caused a $\sim$ 5-fold shrinkage of the area between 400 and 700°C. The specific uptake of irreversibly adsorbed NO remained constant and the average value was used for the measurement of the area covered by NO on the supported Co$_3$O$_4$, as pointed out in the Experimental Section. Similarly, for NiO the specific uptake of irreversibly adsorbed NO remained fairly constant with the heat-treatment temperature. NiO prepared at 550° is a much more refractory oxide than Co$_3$O$_4$ as seen from the constancy of the surface area in the 550-850°C temperature range.

TABLE 4

Irreversible Uptake of NO on Non-Supported Co$_3$O$_4$ and NiO

Sample	Heat-treatment Temperature °C	BET Area, m^2/g	NO Uptake at 25°, 60 torr mℓ(STP)/m^2
Co$_3$O$_4$ (1)	400	25.3	0.168
Co$_3$O$_4$ (1)	550	15.1	0.172
Co$_3$O$_4$ (1)	700	4.6	0.178
Co$_3$O$_4$ (2)	400	13.5	0.169
NiO	550	5.4	0.132
NiO	650	5.5	0.141
NiO	750	5.5	0.149
NiO	850	5.1	0.150

It is worth pointing out that the amount of irreversibly held NO on Co$_3$O$_4$, at 25°C and 60 torr pressure, is 56% of the monolayer referred to in Table 2. For NiO, the irreversibly held NO amounts to 54% of a monolayer. The monolayer is comprised of both reversibly and irreversibly adsorbed NO at the pressure corresponding to the point of intersection of a family of Freundlich isotherms and is usually obtained by extrapolation of experimental isotherms to this pressure (1).

Effects of Heat-Treatment on Zirconia-Supported Co$_3$O$_4$ — Two samples of Co$_3$O$_4$ on ZrO$_2$ were used. The amount of Co$_3$O$_4$ incorporated into the more concentrated sample (B), 2.32 wt. % Co, was calculated to provide an amount sufficient to cover all of the support area ($\sim$ 23 m^2/g initially) with a monolayer of Co$_3$O$_4$, with a slight excess of 25-30%. The other sample (A) was much more diluted, 0.1 wt % Co, so that only $\sim$6% of the area could be associated with Co ions.

References p. 58.

The results of NO uptake on both samples are given in Table 5. On the diluted sample one notices an increase of the uptake with heat-treatment temperature which reaches a maximum value at $\sim$ 600°C and remains fairly constant up to 800°C. Quite unexpectedly it was noted that in the diluted Co_3O_4 on ZrO_2 support, the ratio of NO taken up per one Co ion present is four, instead of one, as noticed previously in diluted samples of transition metal ions on alumina supports (Table 1).

TABLE 5

Relationship Between the Irreversible Uptake
of NO and the Total Amount of Cobalt Ions in
Heat-Treated Co_3O_4 on ZrO_2 Samples

Heat treatment, temp. °C	Sample A*		Sample B**	
	NO uptake molecules, $10^{18}/m^2$	NO molecules / Total Co ions	NO uptake molecules, $10^{18}/m^2$	NO molecules / Total Co ions
400	1.263	2.61	2.015	0.18
500	1.881	3.89	3.842	0.33
550	–	–	4.678	0.42
600	1.961	4.06	4.488	0.41
650	–	–	4.944	0.45
700	2.013	4.16	–	–
750	–	–	4.756	0.36
800	1.999	3.8	5.052	0.29

*0.10 wt % Co, BET Area 22.8 m^2/g
**2.32 wt % Co, BET Area 21.2 m^2/g

In the more concentrated sample the amount of surface that can be covered by the cobalt ions is larger by a factor of 15-20 than in the diluted sample, while the uptake of irreversibly adsorbed NO is larger only by factor of 2 to 2.5 per unit of surface area. The amount of NO molecules irreversibly adsorbed per one Co ion in the concentrated sample reaches a maximum of 0.45 after treatment at 650°C. This agrees with the ratio between the irreversibly adsorbed NO on non-supported Co_3O_4 and the monolayer coverage, which is 0.56, if one takes into account the 25% excess of Co ions which cannot be present at the surface (0.56/1.25 $\sim$ 0.45). With increasing temperature one observes an increase in the uptake of NO, which is then, at the higher end of the temperature range, followed by a decrease. This decrease parallels the decrease in the total BET area shown in Fig. 2. The ratio between the irreversible NO uptake (0.17 ml/m^2) and the total amount of area (the R-value) remains, however, constant as plotted in the same Figure; and for sample B it is very close to unity. For sample A the ratio also remains constant as shown in Fig. 2. The curve of the R-value given in Fig. 2 for sample A is shown only for comparison purposes, since it is difficult to visualize that a Co surface ion in the diluted sample will adsorb more NO reversibly, in addition to the 4 NO molecules adsorbed irreversibly. Obviously, the chemisorption yardstick derived from pure non-supported Co_3O_4 is not applicable to dilute cobalt ions supported on ZrO_2.

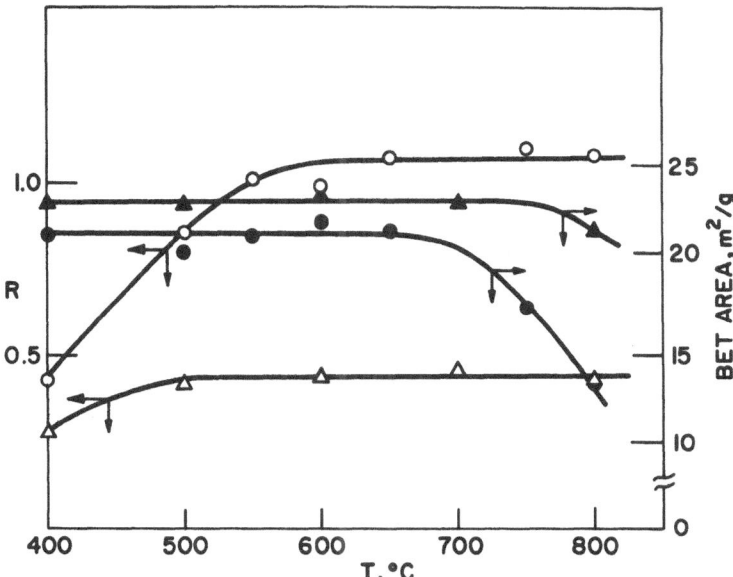

Fig. 2. Heat-treatment Effect on BET Area and the Ratio of NO Adsorption to BET Area (R) on 2.3 wt % (○,●) and 0.10 wt % (△,▲) Co on ZrO_2 samples.

Effect of Heat-Treatment on Alumina-Supported Co_3O_4 — Fig. 3 shows the result of increasing heat-treatment temperatures on two alumina samples of different structure and surface area. The amount of Co_3O_4 in both cases is in excess of that required to cover the whole surface. In the case of the α-Al_2O_3 support this excess is large. As could be expected the changes in the BET area caused by the heat-treatment in this range are not large. The ratio between the NO adsorption and the BET area is always smaller than 0.5 for the sample with the smaller BET area and smaller than 0.25 for the sample with the larger area. This ratio increases slowly as the heat-treatment temperature increases to 600° and then decays for higher heat-treatment temperatures.

Effect of Heat-Treatment on $CoAl_2O_4$ — Supported Co_3O_4 — Fig. 4 shows the changes in the BET area and the R-values for two samples of Co_3O_4 supported on presynthesized cobalt aluminate. In one sample the amount of Co_3O_4 was sufficient

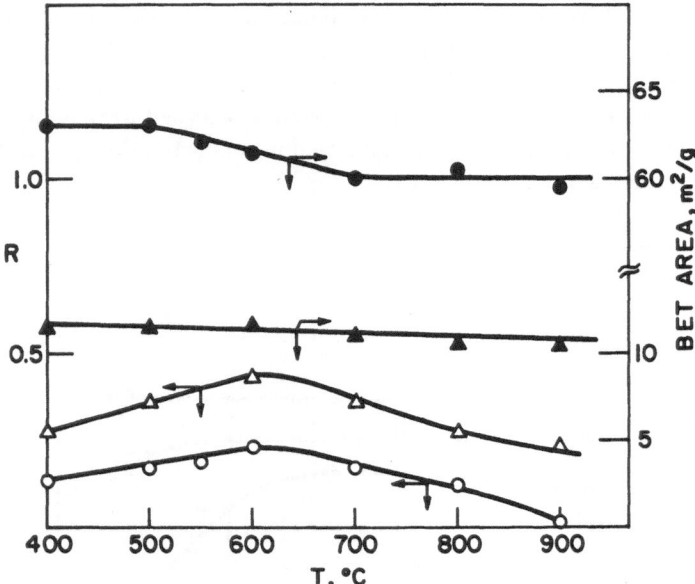

Fig. 3. Heat-treatment Effect on BET Area and the Ratio of NO Adsorption to BET area (R) on Co_3O_4 (5.46 wt % Co) supported on γ-Al_2O_3 (O, ●) and (4.63 wt % Co) on α-Al_2O_3 ($\triangle$,▲).

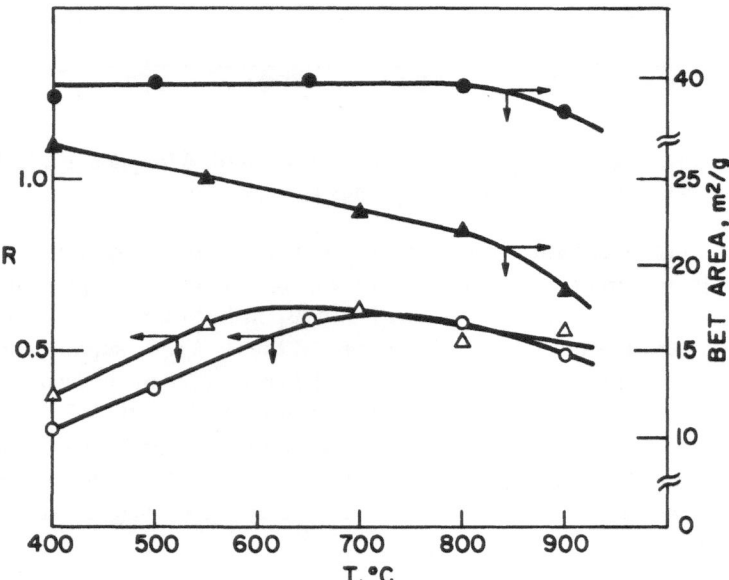

Fig. 4. Heat-treatment Effect on the BET Area and the Ratio of NO Adsorption to BET Area (R) on Co_3O_4 supported on Pre-synthesized $CoAl_2O_4$. 5.72% Co (in Excess over $CoAl_2O_4$) − O,●; 32% Co (in Excess over $CoAl_2O_4$) − $\triangle$, ▲.

to cover the support surface with a small excess. In the other sample a very large excess of Co_3O_4 was used. The decrease in the BET area of the first sample with heat-treatment is small, as expected, since the support itself was synthesized at 1000°C, as mentioned above. In the second sample the BET area decrease reflects the sintering of the excess Co_3O_4 (compare Table 4). The sintering is slower, obviously, than that of pure Co_3O_4. In this case the R-value passes through a maximum with increasing heat-treatment temperature as was observed in the alumina-supported samples. The decrease of the R-value at the higher temperatures is less steep than in the case of the alumina-supported samples.

TABLE 6

Relationship Between the Irreversible Uptake of NO
and the Total Amount of Ni Ions on
a Heat-Treated Diluted (0.15 % wt Ni), Zirconia-Supported Sample

Treatment Temp. °C	Ni concn., ions, $10^{18}/m^2$	NO uptake, molecules $10^{18}/m^2$	NO molecules Total Ni Ions
600	0.752	1.73	2.3
800	0.782	2.59	3.3

Effect of Heat-Treatment on Zirconia and Alumina-Supported NiO – It was of interest to extend the study of active oxide-support interaction, as monitored by NO chemisorption, to other oxides besides Co_3O_4. To this end two samples of NiO on zirconia were prepared. One dilute, with 0.15% Ni, and the other with enough NiO (4.45 wt. %) to cover the surface of the support. Similarly to the diluted sample of Co_3O_4 on ZrO_2, multiple adsorption of NO was observed in this case also as is shown in Table 6. The ratio observed here is lower than the maximum value of 4 NO molecules per Co ion. There is a definite increase in the ratio with the increase of the heat-treatment temperature from 600 to 800°C.

Fig. 5 shows the behavior of the more concentrated NiO-on-ZrO_2 sample. The R-value is always quite close to unity and changes very little with heat-treatment. On the other hand the R-value of a NiO (10.8 wt. % Ni) on alumina sample, also shown in Fig. 5 for comparison, decreases at the higher end of the temperature range. The temperature of heat-treatment required for the onset of this decrease is $\sim$ 100° higher for the NiO on γ-Al_2O_3 samples than for Co_3O_4 on γ-Al_2O_3.

References p. 58.

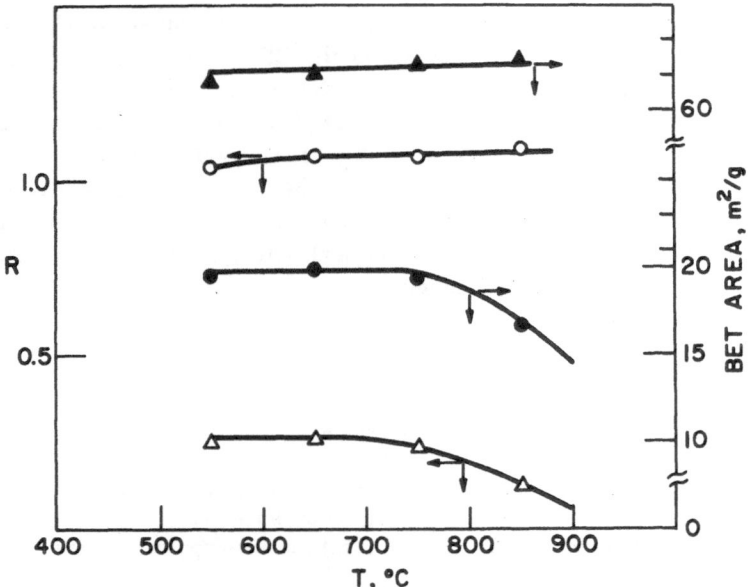

Fig. 5. Heat-treatment Effect on the BET Area and the Ratio of NO Adsorption to BET Area (R) on 4.5% Ni Supported on ZrO_2 (○, ●) and 10.8% Ni on γ - Al_2O_3 (△, ▲).

DISCUSSION

The interaction of the active oxides, Co_3O_4 and NiO, with supports was followed with two radically different types of support. One kind, the aluminas, interact very readily with the metal oxides while the other, zirconia, does not. As a result the chemisorption behavior of the resultant adsorbent (or catalyst) is vastly different, as characterization by NO chemisorption reveals.

On zirconia both Co_3O_4 and NiO, when present in amounts sufficient to cover the surface of the support, spread out with increasing heat-treatment temperature until the area covered by NO chemisorption corresponds to the BET area. Further heating lowers the BET area but the R-value remains close to unity for both Co_3O_4 and NiO supported on zirconia. It should be noted that one can postulate a completely different distribution of the transition metal ions on the zirconia surface which may also account for the correspondence between NO chemisorption and BET area. Thus, we can consider the transition metal oxide as consisting of two phases in equilibrium with each other. One phase consists of discrete ions (or ordered two-dimensional patches of ions) and the other phase consists of three-dimensional crystallites. Multiple adsorption on the dispersed phase, such as takes place on the diluted samples, and the regular adsorption on the crystallite phase add up to a complete coverage of the BET area by NO. Such a two-phase model was proposed by Tomlinson and co-workers (15) from magnetic measurements in silica- and alumina-supported cobalt catalysts and references to similar systems are quoted in the cited work. M. Bettman at the Ford Scientific Research Staff (16) has obtained experimental evidence, from x-ray and catalytic activity measurements, supporting the two-phase view.

Multiple NO adsorption in the diluted, zirconia-supported samples is the only such instance observed. It has been noted first by M. Bettman (16) and confirmed in the present study. It is in contrast to the adsorption behavior in dilute, alumina-supported adsorbents and further confirms the non-penetration of cobalt and nickel ions into the zirconia structure. The ions are situated on the surface in a fashion which leaves several free co-ordination sites for NO chemisorption. With the increase in the heat-treatment temperature of the diluted samples the adsorption ratio also increases. Some small three-dimensional clusters left after impregnation spread out into two-dimensional patches or isolated ions increasing the ratio of adsorbed NO molecules per transition metal ion in the sample to the limiting value. In cobalt-on-zirconia samples this limit appears to be four. Since NiO is more refractory than Co_3O_4, higher temperatures are required for the diffusion. It is possible that for nickel ions the limiting value has not been reached under our experimental conditions, or that it is lower than for cobalt ions, or both.

The solid-state interaction of Co_3O_4 with alumina is a very thoroughly investigated subject (17-20). It has been recently stated (21) that the onset of surface $CoAl_2O_4$ formation might occur as low as 200°C. Indeed, this high reactivity is the prime reason for the catalytic deactivation of alumina-supported cobalt (and other transition-metal oxide) catalysts. Since cobalt ions which enter into cobalt aluminate do not chemisorb NO the change in the amount of active ions on the surface of the catalyst is conveniently followed by this method (Fig. 3). At the lower end of the temperature range there is some spreading by surface diffusion which increases the R-value. Above 660°C, diffusion of the cobalt ions into the bulk becomes the dominant process and the resultant de-activation of the catalyst is demonstrated by lowered NO uptake. With NiO on alumina no surface diffusion is noted at the lower end of the temperature range, owing to the higher refractivity of NiO, and only diffusion into the bulk is observed at higher temperatures. Since $NiAl_2O_4$ does chemisorb some NO, the spinel formation may be somewhat higher than could be deduced from the decrease of the R-value.

Co_3O_4 supported on pre-formed cobalt aluminate represents an intermediate case. The pre-forming of the spinel slows down the penetration of Co_3O_4 into the support but does not prevent it entirely. Obviously, there is a possibility of the formation of a continuous series of solid solutions between the two spinels to the Co-rich side of $CoAl_2O_4$, and such solid solutions chemisorb less NO than Co_3O_4.

ACKNOWLEDGMENT

Many stimulating discussions with M. Bettman are gratefully acknowledged.

References p. 58.

REFERENCES

1. *K. Otto and M. Shelef, J. Catal., 14, 226 (1969).*
2. *K. Otto and M. Shelef, J. Catal., 18, 184 (1970).*
3. *H. S. Gandhi and M. Shelef, J. Catal., 24, 241 (1972).*
4. *H. S. Gandhi and M. Shelef, J. Catal., 28, 1, (1973).*
5. *H. C. Yao and M. Shelef, J. Catal., 31, 377 (1973).*
6. *H. C. Yao and M. Shelef, J. Phys. Chem., in print.*
7. *A. V. Alekseev and A. N. Terenin, Prob. Kinetiki i Kataliza, (AN USSR, Inst. Fiz Khim), Moscow, No. 12, 220 (1968).*
8. *F. A. Cotton and G. Wilkinson, Advanced Inorganic Chemistry, 2nd Ed., Wiley, New York, 1966, p. 748.*
9. *C. C. Addison and J. Lewis, Quart. Rev., 9, 115 (see p. 144) (1955).*
10. *J. H. Lunsford, J. Chem. Phys., 72, 2141 (1968).*
11. *M. Shelef, M. A. Z. Wheeler and H. C. Yao, to be published.*
12. *M. LoJacono, M. Schiavello, and A. Cimino, J. Phys. Chem., 75, 1044 (1971).*
13. *Y. F. – Y. Yao, J. Catal., 33, 108 (1974).*
14. *M. Schachner, Cobalt, (a) No. 2, 37 (1959); (b) No. 9, 12 (1960).*
15. *J. R. Tomlinson, R. O. Keeling, Jr., G. T. Rymer and Joanne M. Bridges, Actes du Deuxieme Congres International de Catalyse. Ed. Technip. Paris 1960, p. 1831.*
16. *M. Bettman, to be published.*
17. *F. S. Stone and R. J. D. Tilley, Fifth Int. Symp. on the Reactivity of Solids, Elsevier, Amsterdam, 1959 p. 583.*
18. *A. Navrotsky and O. J. Kleppa, J. Inorg. Nucl. Chem., 29, 2701 (1967); 30, 479 (1968).*
19. *J. Birch Holt, Sintering and Related Phenomena, Proceedings of an International Conf., Univ. of Notre Dame, June 1965, p. 169.*
20. *J. S. Armijo, Oxidation of metals, 1, 171 (1969).*
21. *P. H. Tewari and N. S. McIntyre, Int. Symp. on Characterization of Adsorbed Species in Catalytic Relations, Univ. of Ottawa, Canada, June 17-20, 1974.*
22. *J. T. Kummer and Y. F. – Y. Yao, Can. J. Chem. 45, 421 (1967).*

DISCUSSION

J. H. Lunsford *(Texas A & M University)*

In view of Peri's results* for NO on chromia-alumina, I wonder if it wouldn't be more correct to say that on the average the NO adsorbs one molecule per metal ion, recognizing the possibility that you might have some multiple adsorption sites.

Yao

Yes.

Lunsford

Do I understand you correctly? Do you say that there is 1 NO per metal ion on the surface of alumina?

Yao

In the concentrated form or the diluted form?

J. B. Peri, J. Phys. Chem. 78, 588 (1974).

Lunsford

Well, I had more in mind the dispersed form.

Yao

Yes, in diluted form we have found this on the alumina support. This was shown in one of our tables.

R. J. H. Voorhoeve *(Bell Laboratories)*

Do you have any idea how that cobalt sits on the zirconium oxide, in the case where you have multiple adsorption?

Yao

For cobalt oxide on alumina, it was reported** that there are two phases of dispersion. One in which the cobalt oxide is in cluster form, the other is a mono-layer dispersion; therefore, it's very diluted. Such dispersion, according to this data I would imagine involves isolated spots with the cobalt ion and the oxygen anion. Is that what you mean?

Voorhoeve

No. My question was not whether it's a mono-layer or not, but what is the atomic arrangement of the cobalt? It seems to me that this whole question whether you have multi-adsorption or single adsorption is very much connected with how the cobalt sits there. In Prof. Lunsford's results it seems to me the cobalt with ammonia has lifted the cobalt out of the surface to form a supported cobalt ammonia complex with zirconia since the oxygen lattice is expected to be very rigid. Cobalt hopefully still has a limited number of ligands.

Yao

Yes. Actually, our data cannot demonstrate how the cobalt sits on the surface. Of course, we can take a guess, but I think this may be because the adsorption can lift the cobalt ion to the upper surface and form a complex like in cobalt-ammonia in zeolites.

J. R. Katzer *(University of Delaware)*

Is this ion scattering with single crystals or with powders? And if it's with the single crystals, have you varied the angle to see if there's any dependence at zero time with angle?

Yao

We have done these in the pressed powder form.

**Tomlinson et al.*, Actes du Deuxieme Congres Internationale De Catalyse, Paris, 1960 p. 1831.*

SESSION II

RELATED CATALYTIC CHEMISTRY

Session Chairman
W. K. HALL

University of Wisconsin
Milwaukee, Wisconsin

CURRENT STATUS OF THE CATALYTIC DECOMPOSITION OF NO

J. W. HIGHTOWER* and D. A. VAN LEIRSBURG**

Rice University, Houston, Texas

ABSTRACT

Except at very high temperatures, NO is thermodynamically unstable relative to N_2 and O_2. Simple cleavage of the N-O bond is all that should be necessary for its decomposition. However, to date no solid catalyst has been developed that will effect this direct decomposition rapidly in an oxidizing atmosphere under relatively mild conditions. This paper will review the types of materials that have been tested, the factors that limit their activity, and the kinetics and mechanism of the reaction. Noble metals (mainly platinum, rhodium, and ruthenium) either in pure or alloyed states, as well as pure and mixed oxides (e.g. cobalt oxide, nickel oxide, zirconia, certain perovskites, etc.) will be discussed. It appears that most of these materials are active in a reduced state, but the oxygen released from the decomposed NO remains strongly attached to the surface and poisons the activity. High temperatures and/or gaseous reductants are required to remove the surface oxygen and regenerate the catalytic activity. Furthermore, oxygen from the gas phase competes with NO for the adsorption sites, and the kinetics over many of these catalysts can be adequately described by simple Langmuir-Hinshelwood kinetics involving competition between NO and O_2.

INTRODUCTION

Even though according to thermodynamics NO is highly unstable relative to its molecular elements, from a kinetic viewpoint the compound is extremely difficult to decompose either thermally or in the presence of catalysts. Several studies (1) about the turn of the century were conducted to explore the homogeneous formation and decomposition of NO in the temperature range 1500-3000°C, but it was Jellinek (2)

References pp. 89-90.   **Presented paper and to whom correspondence should be addressed.*
**Present address: Oregon Graduate Center, Beaverton, Oregon.

who first reported that the noble metals platinum and iridium would catalyze the decomposition at temperatures as low as 670°C. In the 68 years since that paper no solid material has been found that will effectively catalyze the reaction at temperatures dramatically lower than this.

The extreme resistance of NO to decomposition is highlighted by an experiment begun in 1918 and continuing to the present time. As a senior at Worcester Polytechnic Institute, C. S. Howard (3) prepared samples containing 28 different potentially catalytic materials and contacted them with NO. The glass tubes were sealed, heated to 300° for eight hours, cooled to room temperature, and saved for observation. The appearance of brown colored NO_2 formed in the tubes by reaction of product O_2 with unreacted NO would be a very sensitive visual test for occurrence of even the smallest amount of decomposition. By 1957 (4) none of the samples gave any indication of a brown color, and the remaining tubes have been offered to the Smithsonian Institution for preservation and observation in the future. The catalysts in this negative test included transition metals, noble metals, rare earth oxides, and organic polycyclic molecules, and it is among this group of materials that research has continued until today to find a successful catalyst.

With NO (or its atmospheric product NO_2) being linked to smog formation and to adverse health effects (5), increasingly stringent standards are being imposed that will decrease excessive release of NO_x into the atmosphere from automobiles, power generators, and chemical plants. While there are no ideal means for removing NO_x from these sources once it is formed, the most effective measure used to date involves *reduction* of NO into N_2 and/or N_2O (plus CO_2, H_2O, etc.) with some reducing agent such as H_2, CO, NH_3, or hydrocarbons. But introduction of these agents into exhaust streams is expensive and represents an uneconomical use of fuel at a time when concern is mounting over energy supplies. Hence, there is considerable incentive to develop a catalyst that will carry out the decomposition of NO directly without involving these other potential fuels. Despite such an incentive, success in this endeavor has not been overwhelming. Furthermore, some of the severe chemical factors that play a role in the decomposition over conventional materials make it unlikely that ideal catalysts for this reaction will ever be developed unless there is a breakthrough in terms of a radically different approach.

There have been several excellent papers dealing with various aspects of the interaction of NO with solid catalysts, the most comprehensive being one by Shelef and Kummer (6). The present paper will attempt to summarize the salient features of these papers, and it will present some limited qualitative recent data from our own laboratories. Because of the preference of NO to react with reducing agents present in the gas phase instead of undergoing direct decomposition (even when the gas phase may be net oxidizing), most of the material discussed herein will deal with systems devoid of gaseous reductants. Reduction will then be covered in detail in papers later in this Symposium.

THERMODYNAMIC CONSIDERATIONS

In addition to the direct decomposition reaction leading to the molecular products N_2 and O_2

$$NO \rightarrow \tfrac{1}{2}N_2 + \tfrac{1}{2}O_2 \tag{1}$$

there are two disproportionation reactions that can also result in decomposition through subsequent reactions of the primary products

$$NO \rightarrow \tfrac{1}{2}N_2O + \tfrac{1}{4}O_2 \tag{2}$$

and

$$NO \rightarrow \tfrac{1}{2}NO_2 + \tfrac{1}{4}N_2 \tag{3}$$

Nitrous oxide formed in reaction (2) is relatively easily decomposed into N_2 and $\tfrac{1}{2}O_2$ over a wide range of catalysts, whereas the nitrogen dioxide in reaction (3) will usually revert back to NO with release of $\tfrac{1}{2}O_2$. All these reactions are exothermic; the enthalpy for reaction (1) is about -10 kcal/mol NO over the temperature range from 0 to 1205°C (7). As indicated by the equilibrium constants shown in Table 1 (7), these reactions are all favorable at least up to temperatures as high as 650°C. The equilibrium constant for direct NO decomposition does not become unity until temperatures are well above 2000°C. NO can be formed directly from N_2 and O_2 at such elevated temperatures, and once formed the gas must be rapidly quenched (in the range of 20,000°C/s) to avoid decomposition of the NO as the temperature is lowered (8). In the temperature range of catalytic interest (below 1000°C), none of these reactions is excluded by thermodynamics.

TABLE 1

Equilibrium Constants for NO
Decomposition and Disproportionation

Temp. (°C)	Equilibrium Constants for Products*		
	$\tfrac{1}{2}N_2+\tfrac{1}{2}O_2$	$\tfrac{1}{2}N_2O+\tfrac{1}{4}O_2$	$\tfrac{1}{2}NO_2+\tfrac{1}{4}N_2$
0	4.4×10^{16}	8.1×10^6	6.8×10^{12}
93	1.7×10^{12}	3.1×10^4	1.9×10^8
371	4.8×10^6	27.	1.6×10^4
649	3.0×10^4	1.7	78.
927	1.9×10^3	0.48	8.3
1205	3.5×10^2	0.15	2.1

*K_p with pressure measured in atmospheres.

Reduction reactions are even more thermodynamically favored than decomposition, as noted by the equilibrium constants in Table 2 (7) for a few selected reduction reactions listed below:

$$NO + H_2 \longrightarrow \tfrac{1}{2}N_2 + H_2O \tag{4}$$

$$NO + 5/2\, H_2 \longrightarrow NH_3 + H_2O \tag{5}$$

$$NO + CO \longrightarrow \tfrac{1}{2}N_2 + CO_2 \tag{6}$$

$$NO + CH_4 \longrightarrow \tfrac{1}{2}N_2 + CO + 2H_2 \tag{7}$$

$$NO + \tfrac{1}{2}CO \longrightarrow \tfrac{1}{2}N_2O + \tfrac{1}{2}CO_2 \tag{8}$$

Although these reactions are important in the presence of chemical reductants, this paper will concentrate on the direct decomposition (or disproportionation) reactions (1) − (3).

TABLE 2

Equilibrium Constants for NO Reduction
with Selected Reducing Agents

Temp. (°C)	Equilibrium Constants for Products*				
	$\tfrac{1}{2}N_2+H_2O$	NH_3+H_2O	$\tfrac{1}{2}N_2+CO_2$	$\tfrac{1}{2}N_2+CO+2H_2$	$\tfrac{1}{2}N_2O+\tfrac{1}{2}CO_2$
0	3.6×10^{60}	1.3×10^{64}	1.7×10^{66}	2.3×10^{32}	5.0×10^{31}
93	2.4×10^{44}	5.0×10^{45}	1.1×10^{48}	1.9×10^{26}	2.5×10^{22}
371	7.1×10^{23}	1.4×10^{22}	1.1×10^{25}	7.2×10^{18}	4.2×10^{10}
649	4.2×10^{15}	4.2×10^{12}	8.3×10^{15}	1.1×10^{16}	8.9×10^{5}
927	1.5×10^{11}	2.8×10^{7}	1.0×10^{11}	3.7×10^{14}	2.8×10^{3}
1205	2.5×10^{8}	4.8×10	9.5×10^{7}	4.5×10^{13}	8.1×10
Reductants:	H_2	$5/2\ H_2$	CO	CH_4	$\tfrac{1}{2}CO$

*K_p with pressure measured in atmospheres.

TYPES OF CATALYSTS

The types of catalysts that have been tested for NO decomposition can be conveniently divided into two groups: those containing noble metals and "others". Shelef and Kummer (6) have summarized data on non-noble metal catalysts through 1970, and these are shown in Table 3. In 1971 Winter (18) published data about NO decomposition on 30 different oxides (including the noble metal oxides Rh_2O_3 and IrO_2), as shown in Table 4. Most data on noble metals have been limited to platinum, although in some cases iridium and platinum-rhodium alloys have been used, as

TABLE 3

Summary of Data on the Decomposition on Various Catalysts

Catalyst	Pressure Range	NO Concentration Range	Temperature Range °C	Activation Energy kcal/mol	Order of Reaction	Ref.
Quartz			$\rangle$ 600	21.4	2	9
Ga_2O_3	atm	10%	$\langle$ 1000	19.2	0	
Al_2O_3	atm	10%	$\langle$ 1000	31.6	0	
CaO	atm	10%	$\langle$ 1000	23.0	0	
ZrO_2	atm	10%	$\langle$ 1000	27.8	0	10
Cr_2O_3	atm	10%	$\langle$ 1000	40-60	0	
Fe_2O_3	atm	10%	$\langle$ 1000	16	0	
ZrO_2	atm	10%	$\langle$ 1000	31.5	0	
Al_2O_3	atm	5-15%	700-1100	31.0	0	11
ZrO_2	atm	5-15%	700-1100	24.5	0	
Fe oxide*	atm	0.25%	325			12
Copper Chromite*	atm	0.62%	320			
Supported Pt*	1-15 atm	0.4%	430-540		2	13
$CuO:SiO_2$	atm	900ppm	510	39	1	14
Al_2O_3	atm	10-50%	664-807	20.0	2	15
$CuO:Al_2O_3$ (1:1)	atm	720-2200ppm	304-520	31.6	0	16
$CeO_2:Al_2O_3$ (1:1)	atm	720-2200ppm	30-520	42.2	0	
Co_3O_4	100-200 torr	100%	250-350	29.0	1.5-1.8	
NiO	100-300 torr	100%	450-550	20.0	2	
CuO-I**	Not Given	100%	350-700	$\sim$ 12.0	2	
CuO-II***	Not Given	100%	300-450	$\sim$ 12.0	2	17
Fe_2O_3	100-400 torr	100%	300-400	22.0	2	
Cr_2O_3	150-270 torr	100%	350-475	20.0	2	
ZrO_2	100-225 torr	100%	650-750	35.0	2	

*Commercial Catalysts.
**Calcined at 700° in air.
***Calcined at 450° in air.

Taken from reference (6).

TABLE 4
Summary of Winter's NO Decomposition Results over Oxide Catalysts

Oxide	T ($^\circ$C)[a]	E_0[b]	$Log_{10}(A_0)$[c]	Pressure Dependence	E_1[b]	E_N[b, d]
MgO	590-760	37	20.66	1.0	38	35
NiO	430-600	30	20.16	1.0	45	39.5
ZnO	680-770	41	21.29	1.0	36	42
CaO	580-720	28	18.74	—	30	34
SrO	620-750	15	17.38	—	15	23
CuO	370-490	9	14.81	1.0	22	24
Al_2O_3	650-750	38	20.33	—	45	43
Fe_2O_3	520-660	16	15.00	1.0	27	22
Cr_2O_3	680-780	23	16.25	1.0	38	40
Ga_2O_3	610-760	29	18.09	1.0	40.5	40.5
Rh_2O_3	400-560	14	16.45	—	10	34
CeO_2	640-800	18	16.56	—	26	26
HfO_2	730-830	18	15.71	—	30	30
ThO_2	580-840	14	15.02	1.0	22	30
SnO_2	650-790	19	16.00	—	27	32
TiO_2	670-870	19	15.99	—	35	39
IrO_2	330-450	16	17.91	1.0	24	30.5
Sc_2O_3	540-710	31	20.43	—	43	46
Y_2O_3	550-700	24	18.23	1.0	18.5	27.5
La_2O_3	630-830	16	16.40	—	11	20
Nd_2O_3	580-680	25	18.71	—	12	28
Sm_2O_3	550-760	14	15.85	1.0	19	23
Eu_2O_3	550-720	23	17.85	—	16	27.5
Gd_2O_3	470-650	19	16.59	—	12	26.5
Dy_2O_3	650-760	25	18.69	—	18	36
Ho_2O_3	590-760	32.5	20.97	1.0	37	40
Er_2O_3	570-700	28	18.49	—	36	39.5
Tm_2O_3	630-760	28	18.89	—	34	32
Yb_2O_3	600-720	29	18.62	—	30	32
Lu_2O_3	540-690	26	18.59	—	29	35

[a] Reaction temperature range.
[b] kcal mol[-1].
[c] Molecules cm[-2] s[-1] at 200 torr of NO.
[d] Activation energy for N_2O decomposition.

illustrated by Laidler's (19) summary which is included on Table 5 along with more recent results on these noble metals. More recently, catalysts containing another noble metal, ruthenium, have been extensively tested for NO reduction (25-29), and some of these materials may also show promise for NO decomposition. Certain perovskites with various combinations of ruthenium, lanthanum, manganese, potassium, and strontium have shown some decomposition activity, but to date no optimum composition has been found that is vastly superior to others.

TABLE 5

Summary of NO Decomposition on Noble Metal Catalysts

Surface	Temperature Range, °C	Pressure Range, torr	Rate Law	Activation Energy, kcal/mol	Ref.
Pt	882-1403	100-500	$\dfrac{k\,(NO)}{(O_2)}$	14	20
Pt	1210	201-479	$\dfrac{k\,(NO)^2}{(O_2)}$	–	21
Pt-10%Rh	1040-1390	193-477	$\dfrac{k\,(NO)^2}{(O_2)}$	27	21
Pt	860-1060	–	$\dfrac{k\,(NO)}{1 + b\,(O_2)}$	22-25	22
Pt/Al$_2$O$_3$	386-532	760	$k\,(NO)^{0.95}$	20	23
NM-UOP	393-761	760	$k\,(NO)^{0.3-0.45}$	13	23
Pt-Ni/Al$_2$O$_3$	427-536	1-15 atm	$\dfrac{k\,(NO)^2}{(1 + B\sqrt{O_2})^2}$	variable	13
Pt/Al$_2$O$_3$	600-700	10-150	$\dfrac{k\,(NO)}{1 + b\,(O_2)}$	18	24
Rh$_2$O$_3$	400-560	50-400	$\dfrac{k\,(NO)}{1 + b\,(O_2)}$	14	18
IrO$_2$	330-450	50-400	$\dfrac{k\,(NO)}{1 + b\,(O_2)}$	16	18

Another class of catalysts includes alkali or alkaline earth metal oxides that can operate in an oxidizing atmosphere (30-32). For example, sodium oxide will form sodium nitrate by reaction with gaseous NO_2 (formed by homogeneous oxidation of NO), and the nitrate may then be thermally decomposed to give N_2 and/or N_2O. However, these materials are usually sensitive to poisoning by water and sulfur, a property that renders them useless for most commercial applications.

References pp. 89-90.

KINETICS OF DECOMPOSITION

General Problems — In spite of the relative simplicity of the molecules involved, kinetic rate expressions for NO decomposition are not at all certain, and in some cases the experimental data on similar catalysts from different laboratories are conflicting. One factor contributing to the confusion is the fact that many "catalysts" require high temperatures not much below the point where homogeneous thermal decomposition becomes important. This is illustrated by batch experiments conducted by Wise and Frech (9) in a quartz reactor between 600 and 1000°C. At all temperatures the decomposition was second order, but they showed that it was actually made up of two components: a homogeneous reaction that dominates at the highest temperatures and a heterogeneous wall-catalyzed reaction that controls at the lower temperatures. In other words, the rate expression took the form

$$-\frac{d(NO)}{dt} = (k_{homo} + k_{hetero})(NO)^2 \tag{9}$$

with activation energies of 82 kcal/mol for k_{homo} and 21.4 kcal/mol for k_{hetero}. The Arrhenius curve in Fig. 1 shows a smooth transition from the homogeneous to the heterogeneous region as the temperature is lowered. In some cases contributions from the homogeneous reaction have been ignored in treating experimental data, and this is usually indicated by an abnormally large observed activation energy.

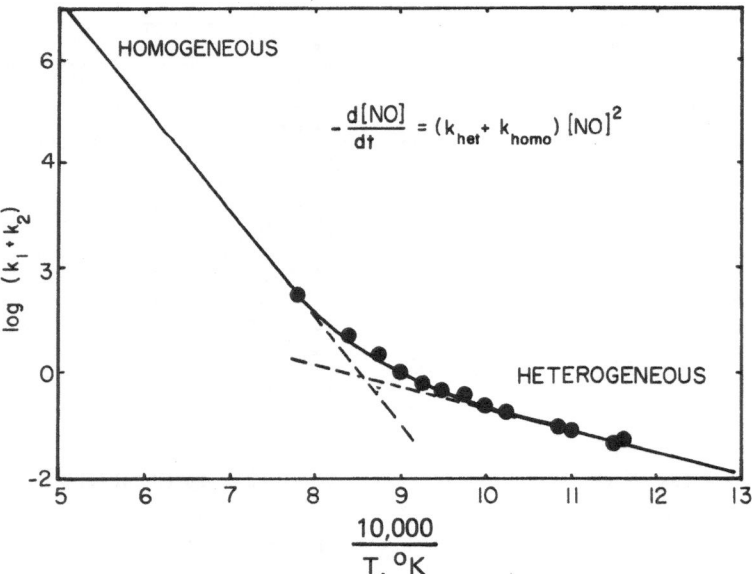

Fig. 1. Homogeneous and heterogeneous regions for NO decomposition in a quartz reactor (ref. 9).

A second complicating factor stems from the exothermic nature of the reaction. While the enthalpy is only 1/3 the heat released by oxidation of one mole of CO, it can still be a factor in causing temperature control problems at high concentrations in flow reactors.

The effects of reaction products as well as impurities in the feed stream may have a significant influence on the reaction rate. As early as 1925 Hinshelwood and co-workers (20, 33) found that oxygen inhibited the rate of decomposition of NO and N_2O over platinum. However, in both the homogeneous and heterogeneous regions Wise and Frech (34) found that oxygen *enhanced* the decomposition rate in quartz vessels, although Kaufman and Kelso (35) were unable to reproduce this autocatalytic effect in their reactors.

A fourth complication is introduced by the wide variety of compounds that can be formed under different conditions, and this is magnified by the analytical difficulties involved. For example, in the presence of oxygen, NO_2 can be formed, and its formation can be controlled by either thermodynamics or kinetics (36, 37). N_2O_4, N_2O, and N_2O_3 may also be present. Even though a reactor may be operated at temperatures sufficiently high to minimize formation of these compounds, they may be formed in the analyzer train, as this part of the apparatus is usually at lower temperatures than the reactor. Unless all these compounds are accounted for, erroneous conclusions can be drawn from correct but incomplete analytical data. Usually a combination of techniques is required for complete analysis.

Finally, the chemical state of the catalysts can greatly influence the reaction state; e.g., in some cases (see next section) the solids can behave as stoichiometric reactants instead of as catalysts (38). This possibility makes it essential to do thorough material balances on the product and reactant streams. In the remainder of this section rate equations for various classes of catalysts will be discussed. Many of the parameters are listed in Tables 3-5.

Noble Metals — Although many of the early studies of thermal NO decomposition were later shown to involve some contributions from heterogeneous wall-catalyzed reactions, the first documented catalytic reaction was over a Pt wire (2). For this reaction Jellinek suggested a second-order rate equation. Since 1926 (20) almost all investigators have agreed that product oxygen poisons the reaction over noble metals, but there remains disagreement about the reaction order with respect to the concentration of NO, as shown in Table 5. The most definitive experiments are those by Amirnazmi, Benson, and Boudart (24), who used the following Langmuir-Hinshelwood model to describe the reaction

$$r \text{ (molecules NO/cm}^2 \text{ s)} \quad = \frac{Nk\,(NO)}{1 + aK\,(O_2)} \tag{10}$$

where N is Avgadro's number, k is the reaction rate constant in cm/sec, K is an adsorption equilibrium constant for O_2 in atm^{-1}, a is a dimensional correction factor in atm cm^3/mol, and the concentrations of NO and O_2 are in mol/cm^3. By systematically varying the inlet concentrations of NO and O_2 and the temperature in a differential flow reactor, the authors were able to determine all the parameters for this reaction. With the rate and adsorption equilibrium constant given by the equations

$$k = A \exp (-E/RT) \tag{11}$$

and

$$K = \exp (\Delta S°/R) \exp (-\Delta H°/RT) \tag{12}$$

the values of A = 19 cm/s, E = 18.4 kcal/mol, $-\Delta H° = 14.0$ kcal/mol, and $-\Delta S° = 2.85$ cal/mol K were obtained. Two supported Pt catalysts with quite different dispersions, 0.037 and 0.39, gave almost identical turnover numbers (molecules NO converted per surface Pt atom per second) of 0.18 and 0.17 s^{-1}, respectively, at 600°C when the partial pressure of O_2 and NO were 0 and 40 torr, respectively. These numbers have been confirmed in our laboratories using a highly dispersed Pt/alumina catalyst.

Apparently Pt is the most effective of the noble metals for decomposing NO. Data for other supported noble metals are not found in the literature, although Winter (18) has studied Rh_2O_3 and IrO_2. The only other systems mentioned are iridium (2) and a commercial ammonia oxidation wire catalyst, 90%Pt-10%Rh by weight (21). The rate expression for the latter is given in Table 5 and is of the same form the authors obtained for a pure Pt wire. However, very likely neither of these rate equations is correct.

Mixed Noble Metal-Base Metal — Corcoran (13) and his co-workers studied the kinetics of NO decomposition on an alumina catalyst containing 0.1%Pt and 3.0% Ni (by weight) in a tubular reactor between 800 and 1000°F at pressures of 1-15 atmospheres. The rate equation that best fit their data was

$$r = \frac{A \, (NO)^2}{(1 + B\sqrt{O_2})^2} \tag{13}$$

where

$$A \, (\text{g mol NO/g cat. hr atm}^2) = 39.7 \exp (-3080/T)$$
$$- 17.3 \exp (-3740/T) \, (P_{N_2})^{\frac{1}{2}}$$
$$+ 3.18 \exp (-3010/T) \, (P_{N_2}) \tag{14}$$

and

$$B \, (\text{atm})^{-\frac{1}{2}} = 0.0135 \exp (11110/T) \tag{15}$$

Temperatures are given in °R and the pressure is in atm. Since the Stanford group (24) has shown that Pt is much more active than NiO, it can probably be safely assumed that the catalytic activity is due mainly to the Pt.

Metal Oxides — Müller and Barck (39) were the first to report NO decomposition over a whole host of transition metal and alkaline earth oxides. While no kinetic data were presented, these authors suggested that in some cases the decomposition appeared to proceed via an N_2O intermediate.

In 1958 Fraser and Daniels (10) used a flow reactor to investigate NO decomposition over Cr_2O_3, Fe_2O_3, Ga_2O_3, ZrO_2, TiO_2, and ZnO catalysts. In all cases their data could be explained by *zero order* kinetics. Wikstrom and Nobe (16) also stated that zero order kinetics were observed for NO_2 dissociation over CuO/Al_2O_3 and CeO_2/Al_2O_3. However, according to their plots shown for the CuO/Al_2O_3 (see Fig. 2), the fractional conversion was independent of inlet NO_2 concentration for a given temperature and W/F condition, a fact that clearly indicates *first order* kinetics. Yur'eva et al (17) found *second order* kinetics for NO decomposition, but their data are also subject to criticism because of the type of reactor they used (18). As if to strike an average position, Sourirajan and Blumenthal (14) found first order kinetics were obeyed over Co_2O_3 and CuO, and Shelef et al. (23) agree with this conclusion, as seen in Table 6. Also, extensive data on 30 oxides, many of them rare earths, collected by Winter (18) are interpreted by a first order equation in NO and inhibition by O_2 (similar to equation (10)); the data are in Table 4.

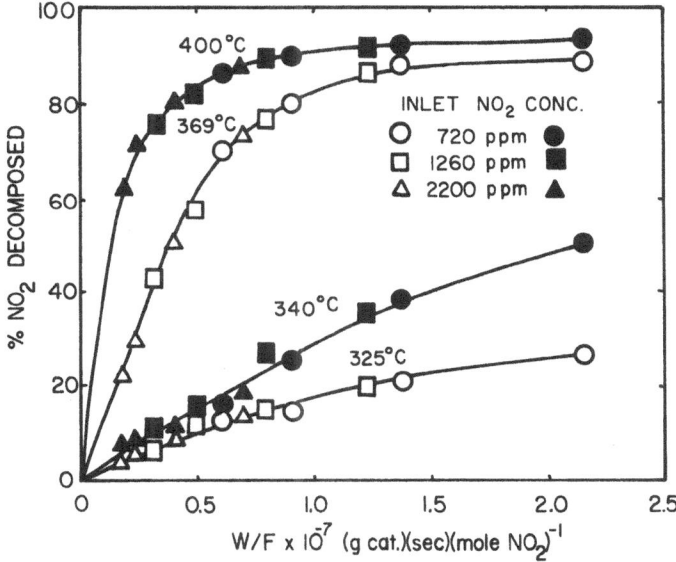

Fig. 2. Kinetic plots for NO_2 decomposition over a CuO/Al_2O_3 catalyst (ref. 16).

TABLE 6

Summary of NO Decomposition Data Obtained by Ford Researchers

Designation of Catalyst	Amount used (g)	Surface* area (m²/g)	Temp. range (°C)	Act. Energy (kcal mol)	Reaction Order		Decomp. rate at 500°† [mol/m² x min) x 10¹⁰]
					Measured (at °C)	Assmd.	
1 Al_2O_3-SiO_2	43	204	548-716	25.9	0.8 (750)	1	2.67
2 Cu-Si	41	21.5	661-938	33.2	1.2 (501)	1	2.98
7 Co	47	175	402-709	22.5	1.2 (654)	1	16.2
9 Co	60	359	399-498	23.2		1	37.1
11 Co+Cr	52	197	419-752	17.0		1	37.6
8 Co	48	206	434-585	14.2	0.75 (572)	1	45.4
13 Co+Li	52	136	443-761	19.4		1	47.2
5 Cu-Cr	46	183	424-737	14.4	1.09 (533)	1	48.7
12 Co+Li	54	187	440-740	18.1		1	55.0
4 Pt	35	231	386-532	20.1	0.95 (504-532)	1	74.8
6 Cu-Cr	44	175	496-699	10.6	0.88 (579)	1	116
3 UOP	24	153	393-761	13.2	0.3-0.45 (532-645)	1/2	147
10 Co_3O_4	26	8.0	279-368	28.3		1	954.2

*After use.

†At constant NO concentration as given at inlet (p = 1 atm, T_0 = 25°).

Taken from reference (23).

As was the case with Pt catalysts, the most thorough investigation of the kinetics of NO decomposition over Fe_2O_3, Co_3O_4, NiO, CuO, ZrO_2-Se_2O_3, and ZrO_2-CaO has been carried out by Boudart and his group (24). Over all these catalysts they found strong inhibition by O_2 and first order kinetics in NO concentration, i.e. the rate could be described by equation (10). Values for most of the kinetic parameters for these materials are included in their paper.

Alumina, Quartz — Most of the theoretical treatments of catalytic NO decomposition have attempted to explain kinetic results obtained over alumina or quartz. The *second order* results of Wise and Frech (9) over quartz have already been discussed, and they are in agreement with the low concentration Langmuir-Hinshelwood equation based on a bimolecular surface reaction of two adsorbed NO molecules, viz.

$$r = \frac{(NO)^2}{(1 + K(NO))^2} \qquad (16)$$

that explained the data reported by Peters (15) for an Al_2O_3 catalyst. In sharp contrast, Daniels (10) and co-workers suggest *zero order* kinetics over alumina catalysts and alundum spheres in an alundum reactor; they also said the rate was retarded by the presence of CO_2 and other similar components in the gas phase. On their support "A" (a Cyanamid 95% Al_2O_3-5% SiO_2 catalyst), Shelef et al. (23) found *first order* kinetics for the same reaction.

Solbakken (40) and Golodets (41) have performed theoretical calculations in an attempt to explain the very low transmission coefficient (about 10^{-8}) for adsorption and reaction, respectively. In his analysis of the pre-exponential factor for the heterogeneous reaction, Golodets chose zero order kinetics. While his result is interesting, considerable uncertainty about the actual order of the reaction makes his calculations somewhat tenuous. Until the reaction orders and rates are properly measured, we feel that these theoretical calculations are premature.

General Observations — There is obviously considerable uncertainty and disagreement about the rate expressions that describe heterogeneous NO decomposition. A few of the apparent contradictions can be attributed to different temperature and concentration regimes used by the various investigators, but much of the data simply reflect use of reactors that are not properly designed for such studies. In some cases the investigators simply misinterpreted the kinetic data. Furthermore, material balances have not always been carefully checked, and the product analyses have rarely been complete.

It is interesting to note that different groups tend to observe the same types of kinetic expressions regardless of the catalytic material used. This becomes apparent when one examines the data collected by individual groups, such as in Tables 3-4. Some of these generalizations have been summarized chronologically in Table 7 not according to catalyst but by investigators. In our opinion the most accurate rate expressions that can be applied to almost all catalysts are those based on Langmuir-Hinshelwood kinetics, such as used by Bachman and Taylor (21), Winter (18), and by Boudart et al. (24). In the most recent investigations there is general consensus that the reactions are first order in NO, but the inhibition by oxygen (whether due to dissociated O atoms or O_2 molecules) is still open to question.

Shelef and Kummer (6) have shown that NO adsorption isotherms are better described by Freundlich than with Langmuir equations. This might suggest that a $(P_{NO})^{1/a}$ term be used in the rate equation instead of the concentration dependency in equation (10). However, such an equation would not account for the poisoning by oxygen.

Almost all investigators agree that the reaction has a very low apparent activation energy and that the slow rate is due to an abnormally small pre-exponential factor. Investigators who have studied both NO and N_2O decomposition (18) find that

TABLE 7

Rate Expressions Most Frequently Used by Various Investigators

Investigator	Location	Rate Expressions	Ref.	Year
Jellinek	Germany	Second order in NO	2	1906
Hinshelwood, et al.	England	First order in NO, inhibited by O_2	20	1926
Bachman, et al.	du Pont	Second order in NO, inhibited by O_2	21	1929
Zawadzki, et al.	France	First order in NO, inhibited by O_2	22	1934
Wise, et al.	JPL	Second order in NO, inhanced by O_2, slightly retarded by N_2	34	1952
Daniels, et al.	Wisconsin	Zero order in NO, retarded by CO_2, N_2	10	1958
Corcoran, et al.	Cal Tech	Second order in NO, inhibited by dissoc. O_2	13	1961
Sourirajan, et al.	UCLA	First order in NO	14	1960
Peters	Colorado	Second order in NO, inhibited by NO	15	1963
Yur'eva, et al.	USSR	Second order in NO	17	1965
Nobe, et al.	UCLA	Zero order in NO_2, but data actually suggest first order	16	1965
Shelef, et al.	Ford	First order in NO	23	1969
Winter	England	First order in NO, inhibited by O_2	18	1971
Lawson	Canada	First order in NO	60	1972
Boudart, et al.	Stanford	First order in NO, inhibited by O_2	24	1971

generally the former has the lower activation energy *and* the slower rate. This may be seen in Table 4 where E_N refers to N_2O and E_o to NO decomposition.

Reactors — A few words about different kinds of reactors that have been used might be appropriate here. Since there is no change in the number of moles according to reaction (1), pressure measurements cannot be used as indicators of the extent of reaction unless some of the products undergo subsequent reactions. At high temperatures (above 600°C) no other reactions are thermodynamically favorable. However, at lower temperatures the product O_2 can react with the remaining NO according to the equilibrium reactions

$$NO + \tfrac{1}{2}O_2 \rightleftarrows NO_2 \rightleftarrows \tfrac{1}{2}N_2O_4 \qquad (17)$$

In addition, NO_2 can react with NO to form N_2O_3. Under conditions where formation of any of these products is not thermodynamically forbidden, their presence must be taken into account. Even if the reactor temperature is high enough to prevent their occurrence, these reactions can occur in the sampling lines between the reactor and the analytical devices. Whether or not true equilibrium is achieved will depend in part on the time in and temperature of lines between the reactor and the analytical devices. Thus it is obvious that reactor and analytical design are of fundamental importance for meaningful kinetic measurements.

Over many catalysts the temperature required approaches that where thermal decomposition begins to occur. To minimize contributions from the homogeneous gas phase reaction, one must keep the void volume in the heated part of the reactor small. This was done by Bachman and Taylor (21) by using a reactor shown in Fig. 3. Most of the gases are contained in the large mixing volume, and the circulation is achieved by convection through the vertical arm containing an electrically heated wire catalyst. The resistance of the wire is a measure of its temperature. Since most of the fluids are thermostated for relatively long times at a much lower temperature, it may be assumed that the reactions shown in equation (17) are in equilibrium. This will cause an overall decrease in the pressure, and from pressure measurements (after corrections for the equilibrium reactions) the rate of disappearance of NO can be calculated.

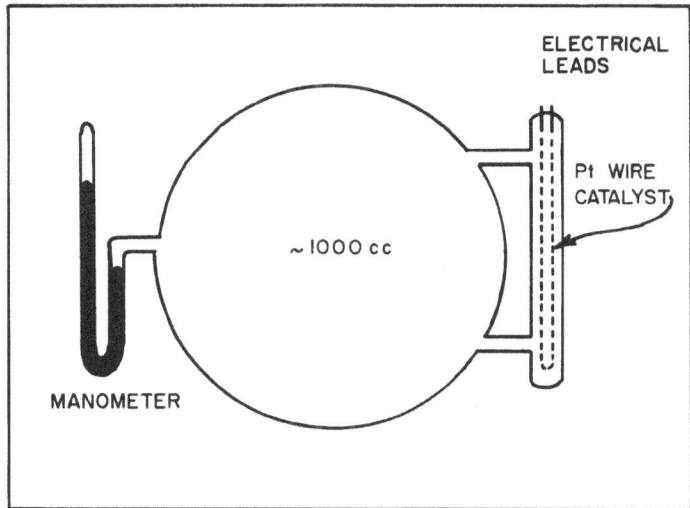

Fig. 3. Recirculation reactor of Bachman and Taylor (ref. 21).

Of course uncertainties about the *fluid* temperatures near the catalyst wire, as well as the extent of homogeneous decomposition in that zone, limit the accuracy of data from such a reactor.

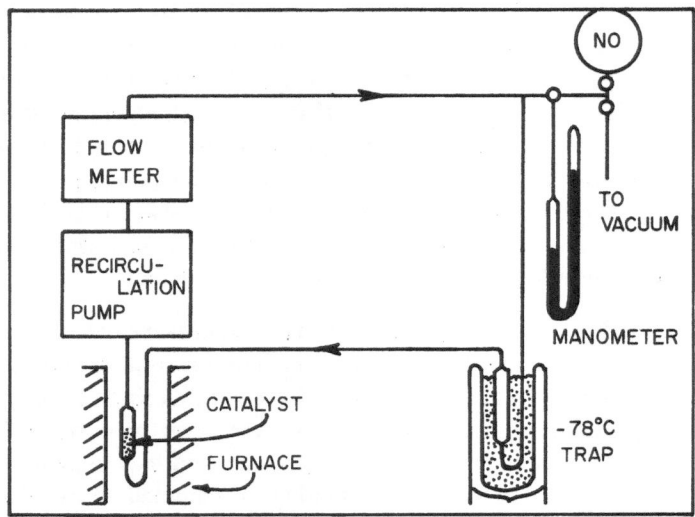

Fig. 4. Recirculation reactor of Yur'eva, Popovskii, and Boreskov (ref. 17).

Another rapid recirculation batch-type reactor that has been popular with the Russian researchers (17) is shown schematically in Fig. 4. The O_2 product is converted into NO_2 which can be collected in a trap thermostated at $-78°C$. However, formation of N_2O_3 through reaction with NO may give an apparent NO decomposition rate that is too large. As Winter (18) has suggested, this may account for the high reaction order (second order) they reported over several catalysts. Of course such a reactor scheme is unsuited for measurements of O_2 inhibition.

The most appropriate experiments for establishing kinetic parameters for NO decomposition involve steady state flow measurements in either backmix (rapid recirculation) gradient-less reactors or differential flow reactors run at very low conversions. The latter has the disadvantage of placing great demands on analytical sensitivity (if looking at the small amount of products formed) or accuracy (if looking at the small amount of reactant decomposed). To the inlet streams for either reactor type can be added continuously variable concentrations of reactants, products, inerts, and potential poisons (or promoters). It is absolutely essential that some standard set of reaction conditions be repeated every few test points in order to assure that there has been no change in catalytic activity with time. This is most important for temperature changes, as sintering and other deactivation processes are usually accelerated at higher temperatures.

Analytical equipment should ideally be designed to give instantaneous readings of all the components under conditions similar to those in the reactor, e.g. at high temperatures. This will avoid the necessity of uncertain assumptions about dimerization or oxidation reactions that may or may not have gone to completion at the lower temperatures. Even a GLC sampling loop down-stream from the reactor does not always avoid these problems, and some reactions (e.g. NO_2 decomposition into NO and O_2) can actually occur in a GLC column. A multiplicity of spectroscopic analyzers, each calibrated for a particular component and checked for interference by all other species, represents the ultimate analytical system.

No kinetic experiments are complete without tests for mass and heat transport effects. For one of their porous catalysts Shelef et al. (23) have published a simple plot, shown in Fig. 5, that allowed their rates to be classified into transport limited or

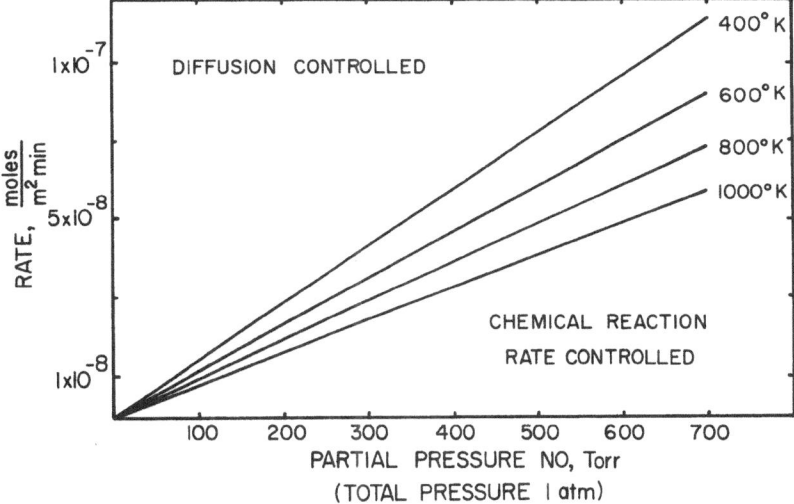

Fig. 5. Test for the influence of transport limitations on the rate of catalytic NO decomposition (ref. 23).

chemical reaction rate limited regions. Similar plots may be made for other catalysts. Weisz and Prater (42) have suggested another simple method for determining the contribution of diffusion in porous media, and this is shown by the inequality

$$\left(\frac{V_p}{S_x}\right)^2 \frac{1}{D_e C_{A_o}} \left(-\frac{1}{V_c}\frac{dn}{dt}\right) < 1 \tag{18}$$

If the left hand side is less than unity, diffusion can usually be ignored. An experimental test for possible diffusional influences in particulate catalysts involves varying the particle size and looking for changes in the reaction rate, kinetic order, and/or activation energy. Care must be taken in interpreting data from catalysts containing a non-uniformly distributed active phase on a porous support when the large particles were simply ground to vary the particle size. Since the activity of most catalysts tested to date is not very great, Shelef et al. (23) concluded in their work that diffusion did not present a significant limitation.

A precise knowledge about the reaction kinetics is fundamental to a complete understanding of the reaction mechanisms. However, it appears that much more work needs to be done in this area before kinetic data can be applied with confidence to confirm or reject a mechanistic proposal.

REACTION MECHANISMS

In contrast with ideas about the reaction rate equations, there is general agreement among investigators that it is removal of the strongly adsorbed product oxygen from the surface that limits the NO decomposition rate on most catalysts. Several reduced metals or partially reduced metal oxide catalysts are quite reactive towards dissociation of the NO molecules even at ambient temperatures. In such a condition N_2 (or N_2O) is rapidly desorbed, but the oxygen remains tenaciously attached to the surface. Once all the surface is saturated with such species, the low temperature catalytic activity ceases and the solid becomes inert for the decomposition except at high temperatures where presumably oxygen removal can be thermally stimulated (38).

It is the absence of a chemical medium to remove the oxygen and maintain the surface in a reduced state that renders most catalysts ineffective. The presence of a reducing species (e.g. H_2, CO, NH_3, or hydrocarbons) in the gas phase will serve to keep the surface oxygen species removed through formation of H_2O and/or CO_2. It is not absolutely necessary for the overall gaseous mixture to be *net* reducing under all conditions, as long as there is sufficient reductant to react stoichiometrically with the NO (43, 44). In fact, this is the basis of a patented method (45, 46) for removing NO from very fuel-lean stack gases resulting from nitric acid plants. If the temperature of a Pt catalyst is kept within well-defined limits, addition of a small amount of NH_3 (slightly in excess of that required for the stoichiometric reactions

$$6NO + 4NH_3 \rightarrow 5N_2 + 6H_2O \tag{19}$$
$$6NO_2 + 8NH_3 \rightarrow 7N_2 + 12H_2O)$$

will almost completely remove the NO. At these mild temperatures (about 200°C), the activity is too low for the ammonia oxidation reaction to occur, thus leaving the NH_3 to react selectively with the NO rather than with the O_2. These points will be discussed in greater detail in a later paper in this Symposium.

An example of the reaction of a reduced catalyst with NO is shown in Fig. 6. About 4.6 grams of a synthetic clay mineral catalyst containing 34.6% Ni (by weight) and 0.1% Pd (kindly provided by the Baroid Division of NL Industries) were placed into a plug flow reactor, the effluent from which flowed through a Beckman 315A infrared NO analyzer. The catalyst was reduced in flowing H_2 for 1 hour at 600°C, flushed with He, and then cooled in He to various constant temperatures. At time t=0 at each temperature, a stream containing about 99% He and 1% NO was introduced to the reactor at 60 cc/min and allowed to continue flowing until NO appeared in the exit stream. The curves at 100, 200, 300, and 400°C for the NO_{out}/NO_{in} are shown in Fig. 6. N_2 was the only reaction product observed, and the area between the

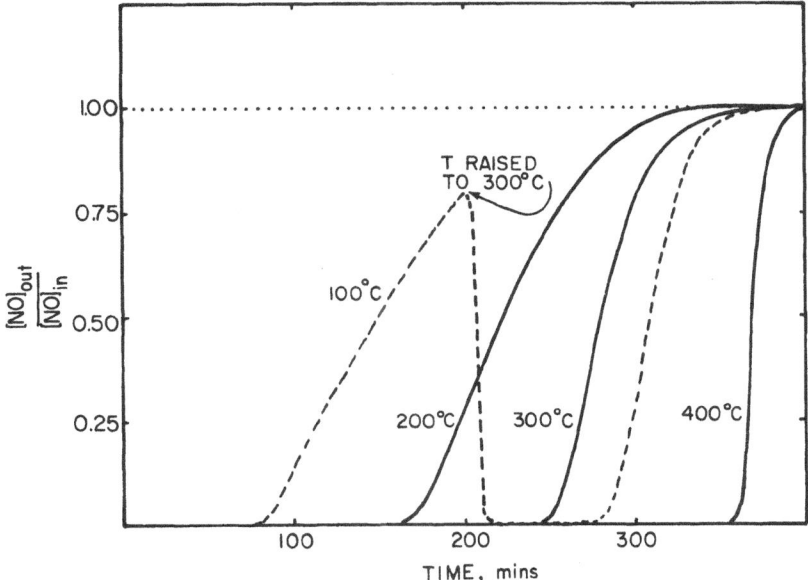

Fig. 6. Stoichiometric reaction of NO with a reduced Ni-Pd/synthetic clay catalyst.

dotted zero conversion curve and the NO detector curve is proportional to the total NO decomposed (plus a small amount adsorbed) at each temperature. At the highest temperature shown in the figure, 400°C, the ratio of the (total NO molecules decomposed)/(total Ni + Pd atoms) is 0.27; the break-through curve is quite steep. However, at lower temperatures the product curves are not nearly so steep; the break-through at 100°C corresponds to an NO/metal atom ratio of 0.07. After each experiment the same catalyst sample could be reduced in H_2 (or CO) and the same results duplicated over and over again.

We interpret these results as meaning that dissociative chemisorption of NO occurs on the reduced metallic sites. N_2 is released into the gas phase, but the oxygen is retained as NiO (or a small amount of PdO). Once all the available metal atoms have been oxidized, the dissociative chemisorption ceases and the decomposition stops. Only at temperatures in excess of 500°C does a small amount of true catalytic activity appear, but its low activation energy makes the reaction rate increase rather sluggish as the temperature is increased. At temperatures above 650°C the catalyst begins to undergo irreversible solid state changes. While this material is not a good candidate for a decomposition catalyst, it does very dramatically illustrate the difference in reactivity towards NO of reduced and oxidized substances. We believe this phenomenon is common to many potentially active materials. The reverse of this surface reaction has been reported by Roth and Doerr (38) on a copper chromite catalyst. Once the catalyst has been oxidized by NO, a feed stream containing CO was allowed to flow through the reactor and the effluent stream analyzed for CO. Fig. 7 shows the same kind of a "break-through" curve as seen with the NO, and they interpreted this as indicating removal of the surface oxygen atoms by the CO. Once thoroughly reduced, the catalyst no longer removed the CO from the stream by converting it to CO_2.

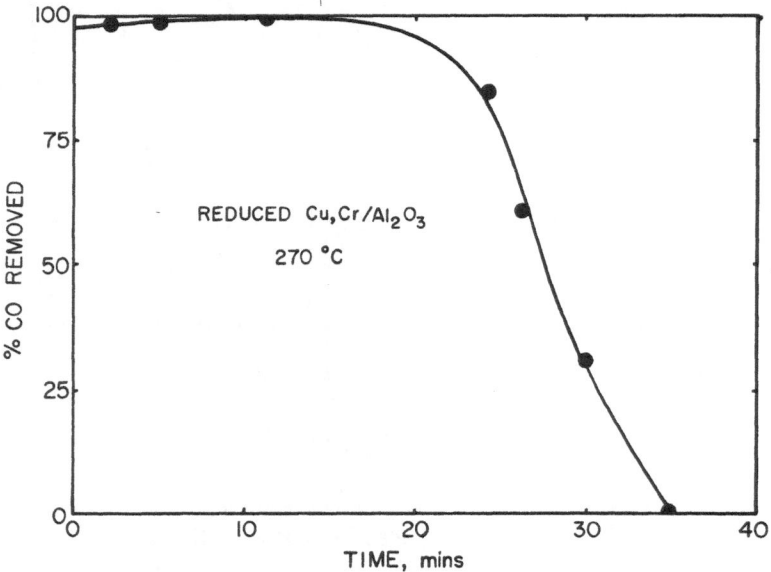

Fig. 7. Stoichiometric reaction of CO with an oxidized copper chromite/alumina catalyst (ref. 38).

While over the Ni-Pd/clay catalyst the NO was adsorbed dissociatively, its disappearance from the gas stream may be due to associative adsorption in some cases. The work of Nobe et al. (16) on NO_2 decomposition is a good example. Once steady state had been achieved over a CuO catalyst at some moderate temperature, the temperature was allowed to increase slowly, as shown in the Curve A in Fig. 8. The $(NO_2)_{out}/(NO_2)_{in}$ curve at first showed an increase due to desorption of some of the adsorbed NO_2, but then the enhanced activity due to the higher temperature caused the curve to stabilize at a lower level, i.e. there was more chemical conversion and less adsorption. A single measurement of the effluent NO$_2$ concentration taken before steady state had been re-established might have indicated *less* conversion at the higher temperature. Thus, these examples highlight the necessity for measuring the concentrations of *all* species in the reactor effluent as a function of time to make sure the activity is catalytic and not stoichiometric or transient.

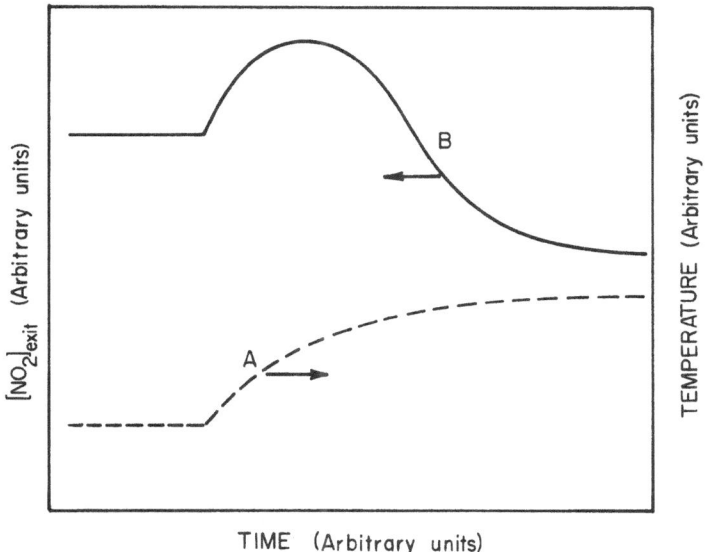

Fig. 8. Desorption of NO$_2$ from a CuO catalyst in a flow reactor (ref. 16).

Similar results have been obtained with cobalt oxides. In the steady state these materials show moderate activity, as indicated by the data of Shelef et al. (23) in Table 6. However, the catalyst can be made temporarily much more active by high temperature treatment in vacuum, a process that presumably creates a partially reduced surface by removal of some surface oxide ions. Exposure to NO soon restores this removed oxygen and decreases the activity to its original value. Also, such thermal cycling can cause bulk solid state changes in some systems. If supported on alumina, the cobalt can irreversibly form the inactive spinel CoAl$_2$O$_4$ at high temperatures (47).

With these considerations in mind, it might not be suprising to find a correlation between an oxide's catalytic activity for NO decomposition and its ability to promote homonuclear exchange of oxygen isotopes via the reaction

$$^{18}O_2 + {}^{16}O_2 \rightleftarrows 2{}^{18}O{}^{16}O \qquad (20)$$

since both reactions involve desorption of oxygen species from the surface. Such a correlation was found by Boreskov (48) and is shown in Fig. 9. Proponents of this theory suggest that the rate of this reaction is a measure of the ease of removal of adsorbed surface oxygen atoms. Boreskov (49) has also suggested that NO decomposition should be inversely related to the metal-oxygen bond strengths for lattice oxides, but this theory has not been proved and has not been very popular among many researchers (18).

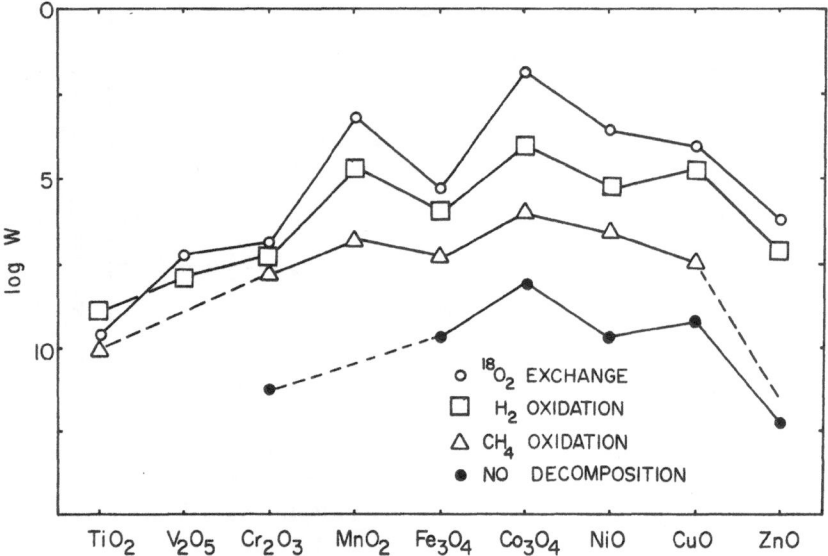

Fig. 9. Comparative activity measurements for selected reactions over transition metal oxide catalysts (ref. 48).

The most thorough treatment of the modes of adsorption of NO on catalysts has been published by Terenin and Roev (50). Table 8 shows the different types of species they claim may occur on different metals, oxides, and sulfates. Note that in most cases the NO is adsorbed with the N atoms nearest the surface, although sometimes the molecules may be present lying down in a bridged structure. Never is the oxygen atom nearest the surface when the adsorbed species is "standing up". In fact, on a Ni catalyst Blyholder and Allen (51) have identified an IR band in the 1750-1620 cm^{-1} region as being due to an Ni-N stretch when NO is chemisorbed.

Using a LEED-mass spectrometer apparatus to study NO interactions with Ni(100), Onchi and Farnsworth (52) found NO strongly chemisorbed on both clean and contaminated surfaces. The work function of the surface was found to decrease (in contrast to increasing when CO is adsorbed on the same surface) thus indicating a *positive* surface charge with electrons being transferred into the metal.

TABLE 8

Assignment of Observed Vibration Frequencies
of Adsorbed NO to Different Types of Bonding

Metals	Oxides	Sulfates
$Fe]^- .. \overset{N}{\underset{O^+}{\mathbin{\|\|\|}}}$ 2008	$Cr_2O_3]\left\{\begin{array}{l}{}^N\!\!-O^+\\[4pt]{}_N\!\!-O^+\end{array}\right.$ 2093, 2028	$Co^+ .. \overset{N}{\underset{O^+}{\mathbin{\|\|\|}}}$ 2035
$Cr]^- .. \overset{N}{\underset{O^+}{\mathbin{\|\|\|}}}$ 2010		$Mn^+ .. \overset{N}{\underset{O^+}{\mathbin{\|\|\|}}}$ 2000
$Cr]^-: N \equiv O^+$ 1905 $\left.\right\}$ $Cr]^- = \overset{+}{N} = O$ 1830	$Fe_2O_3]: N \equiv O^+$ 1927 $Cr_2O_3] = \overset{+}{N} = O$ 1842	$Cr^{2+} .. \overset{N}{\underset{O^+}{\mathbin{\|\|\|}}}$ 2010
$Ni]: NO$ 1830	$Fe_2O_3]: NO$ 1805 $NiO]: NO$ 1805	$Co^+: N \equiv O^+$ 1900 $\left.\right\}$ $Co^+ = \overset{+}{N} = O$ 1805
$Ni] - N = O$ $Fe] - N = O$ $Cr] - N = O$ $\left.\right\}$ ca. 1735	$Fe_2O_3] - NO$ $NiO] - NO$ $Cr_2O_3] - NO$ $\left.\right\}$ ca. 1735	$Cr^{2+}: N \equiv O^+$ 1920 $\left.\right\}$ $Cr^{2+} = \overset{+}{N} = O$ 1830
		$Ni^{2+}: NO$ 1850 $Fe^{2+}: NO$ 1850
		$Ni^{2+} - N = O$ $Fe^{2+} - N = O$ $Cr^{3+} - N = O$ $Co^{2+} - N = O$ $\left.\right\}$ ca. 1735

Taken from reference (50).

These experimental observations are in sharp contrast with the suggestions first made by Bachman and Taylor (21) and later used by Winter (18) that NO is adsorbed with the *oxygen* atom fitting into anion vacancies on the surface. Winter compared N_2O and NO decomposition by envisioning adsorption and bond cleavage illustrated for the oxide lattices shown in Fig. 10. Charge transfer of an electron into the antibonding orbital of the NO molecule reduces the order of the NO bond from 2½ to 2 and should thus make the molecule less stable than it was in the neutral state. Solids that are good electron donors should then make the most effective catalysts, according to this scheme. Unfortunately, no experimental evidence exists that suggests formation of an NO^- species; the limited data indicate that charge transfer, if it occurs at all, is in the other direction to form NO^+ species with a bond order of 3, i.e. it is even a more difficult entity to dissociate than was the neutral NO. In any case, the scheme in Fig. 10 might also account for N_2O being formed as an intermediate in NO decomposition by appropriate manipulation of the bonds, as was suggested by Sachtler (53) who reported N_2O being formed on fully oxidized NiO. As seen in the data in Table 9, Shelef and Kummer (6) have evidence that N_2O is an intermediate in NO decomposition over an initially reduced Pt catalyst.

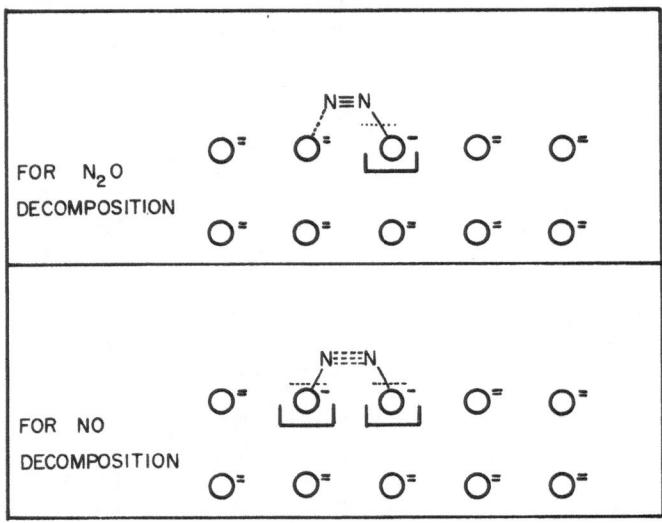

Fig. 10. Proposed active sites for adsorption and decomposition of NO and N_2O on oxide catalysts (ref. 18).

TABLE 9

Appearance of N_2O during Contact
of NO with a Reduced Platinum Catalyst

Temperature °C	Time of Contact Hrs.	% of Surface Reoxidized	N_2/N_2O Ratio in Gas Phase
25	70	10	1.1
95	14	Not determined	0.18
270	70	92	2.42

Taken from reference (6).

Similar mechanisms have been postulated by Voorhoeve and co-workers (25) for NO reduction over perovskite-like ruthenates and manganites. According to their scheme NO could either adsorb dissociatively

$$M\text{-}\square\text{-}M + NO \rightarrow M\text{-}O\text{-}M + N_{ads} \tag{21}$$

or associatively

$$NO \rightarrow NO_{ads} \quad (\text{on } M\text{-}\square\text{-}M) \tag{22}$$

To avoid the uncertainty associated with the direction of NO adsorption in the latter case, the authors have cleverly declined to suggest a "picture" of the adsorbed state!

The strength of NO adsorption on NiO has been examined by mutual displacement studies using IR to monitor the surface species. Alexeyev and Terenin (54) found that NO will rapidly displace adsorbed CO but that NO and CO_2 can co-exist for considerable time on the surface. Similar results were obtained by Onchi and Farnsworth (52). Since some bands due to adsorbed CO_2 are shifted upon adsorption of NO, the authors (56) concluded that the adsorption strengths of the adsorbates follow the order

$$NO \rangle CO_2 \rangle CO$$

on NiO, which means NO is a more effective "poison" for such catalysts than are the other two compounds.

Nitric oxide is also a poison for certain reactions over platinum catalysts. In some preliminary studies of NO reduction by CH_4, we (55) have recently shown that the presence of NO stops the normally rapid $CH_4 - CD_4$ isotopic exchange reaction over supported Pt. As soon as the NO is completely removed by reduction (if the unequilibrated mixture of methane isotopic species is in excess), the exchange reaction re-commences.

Another example of strong NO adsorption has been given by Poling and Eischens (56) during a study of corrosion inhibition on iron. They found that an intense band at 1820 cm^{-1}, which they attributed to Fe^{++}:$N=O^+$, was quite stable and could not be displaced by O_2 or by water. Butyl nitrite also formed this band, and the authors concluded "that the corrosion inhibition properties involve both oxidizing action of nitrogen oxides and the strong chemisorption of neutral or cationic inhibitor species on the oxide surface".

While the exact nature of the sites on which NO decomposition occurs is not well understood, it seems clear that NO is strongly adsorbed on these catalysts and that removal of the released oxygen from the surface plays an important role in determining the overall reaction activity.

APPLICATIONS

A catalyst that would be effective for direct NO decomposition *in an oxidizing atmosphere* would be tremendously useful. When fuels that have relatively high organic nitrogen contents (e.g. certain coals) are used, a significant fraction of this nitrogen appears as NO in the exhaust stacks of combustion furnaces, and the environment is highly oxidizing (57). An NO decomposition catalyst that would not be poisoned by sulfur or other compounds and that would function at moderate temperatures could correct this problem. Of course such a catalyst would eliminate the need for adding NH_3 as an expensive reductant to the tail gases from nitric acid plants, as is the current practice (43).

To help remove NO_x emissions from the exhausts of motor vehicles such a catalyst would be most welcome. The demand is not so great for exhausts from spark ignited engines where CO (which could serve as a reducing agent) is usually present in concentrations considerably in excess of the NO concentrations. However, for diesels, which operate quite lean, frequently the NO concentration exceeds both the CO and HC emissions, and the *only* way to remove NO in such systems (short of adding reductants to the exhaust) is by decomposition.

According to the results of Shelef (23), cobalt oxides are the most active catalysts for NO decomposition (see Table 6). But even with such activity, calculations have shown that at 500°C (assuming first order kinetics) 670 kg of the material would be required to achieve 90% conversion at 60 miles per hour with an engine emitting 2.6 g NO/mile! Thus, this catalyst is not sufficiently active to handle the task by several orders of magnitude.

A low temperature approach to NO decomposition that shows some promise under very limited conditions for mixtures of air and NO has been reported by Lawson (58). At temperatures below 100°C and in the presence of water, nitrous acid (HNO_2) becomes a stable gas-phase species that can be very selectively decomposed over a $Ag-Ag_2O/Al_2O_3$ catalyst. If the temperature exceeds 100°C, however, there is a tendency to form $AgNO_3$, a species that acts as a poison for the decomposition reaction.

Considering the difficulties involved and the absence of complete success to date, we are not optimistic that a thermally stable solid catalyst will ever be found to decompose NO in an oxidizing environment at mild temperatures. Such an achievement will likely require some radical new ideas based on a fundamental understanding of chemical interactions between NO and solids. It is indeed encouraging to see increasing research activity in this important field both in industrial laboratories and in universities. In his introductory remarks at the opening of the Fifth International Congress on Catalysis, Chairman Haensel (59) facetiously remarked that a single atom was separating academic and industrial researchers, and that was an N atom. While the former were busily converting N_2O, the latter needed to eliminate NO. It looks like that pesky N atom barrier is rapidly becoming smaller.

To conclude on a note of cautious optimism, there are rumors that researchers associated with Cummins Engine Company (60) (the largest producers of diesel engines in the US) may be on the verge of developing a catalyst that will work on their exhaust systems at temperatures below 600°C, but we have seen no data to support this claim.

ACKNOWLEDGMENTS

We gratefully acknowledge financial assistance from the National Science Foundation, which is supporting research on this subject in our laboratories as a part of the US-USSR Joint Technology Exchange Program. We also appreciate a grant from the Baroid Division of NL Industries, and we thank Gulf Research and Development Corporation for loan of the NO analyzer.

REFERENCES

1. W. Nernst, Z. Anorg, u. Allgem. Chem. 49, 213 (1906).
2. K. Jellinek, Z. Anorg, u. Allgem. Chem. 49, 229 (1906).
3. C. S. Howard, B.S. Thesis in Chemistry, Worcester Polytechnic Institute, Worcester, Mass., 1918.
4. C. S. Howard and F. Daniels, J. Phys. Chem. 62, 360 (1958).
5. National Academy of Sciences Report No. 93-24, Vol. I, prepared by the Coordinating Committee on Air Quality Standards for the Committee on Public Works, U.S. Senate, September, 1974.
6. M. Shelef and J. T. Kummer, Chemical Engineering Progress, Symposium Series 67 (no. 115), 74 (1971).
7. Physical and Thermodynamic Properties of Elements and Compounds, booklet prepared by Chemetron Corp., 1969.
8. W. G. Hendrickson and F. Daniels, Ind. and Eng. Chem. 45, 2613 (1953).
9. H. Wise and M. F. Frech, J. Chem. Phys. 20, 22 (1952).
10. J. P. Fraser and F. Daniels, J. Phys. Chem. 62, 215 (1958).
11. E. L. Yuan, J. I. Slaughter, W. E. Koerner, and F. Daniels, J. Phys. Chem. 63, 952 (1959).
12. F. R. Taylor, Air Pollution Foundation Report No. 28, 1959.
13. R. R. Sakaida, R. G. Rinker, Y. L. Wang, and W. H. Corcoran, AIChE Journal 7, 658 (1961).
14. S. Sourirajan and J. L. Blumenthal, Proc. Second Inter. Congr. on Catalysis, Vol. II, p. 2521, Paris, 1960.
15. M. S. Peters, AEC Report No. TID-18423 (1963).
16. L. L. Wikstrom and K. Nobe, I&EC Proc. Des. and Devel. 4, 191 (1965).
17. T. M. Yur'eva, V. V. Popovskii, and G. K. Boreskov, Kinetika i Kataliz 6, 1041 (1965).
18. E. R. S. Winter, J. Catal. 22, 158 (1971).
19. K. J. Laidler, "Catalysis", Vol. I, P. H. Emmett, ed., p. 149, Reinhold, New York, 1954.
20. T. E. Green and C. N. Hinshelwood, J. Chem. Soc. 129, 1709 (1926).
21. P. W. Bachman and G. B. Taylor, J. Phys. Chem. 33, 447 (1929).
22. J. Zawadzki and G. Perlinski, Comptes Rend. 198, 260 (1934).
23. M. Shelef, K. Otto, and H. Gandhi, Atm. Environ. 3, 107 (1969).
24. A. Amirnazmi, J. E. Benson, and M. Boudart, J. Catal. 30, 55 (1973).
25. R. J. H. Voorhoeve, J. P. Remeika, and L. E. Trimble (submitted to Mat. Res. Bull.)
26. G. Mai, R. Siepmann, and F. Kummer, German Patent Appl. 2, 151, 958 (1973).
27. M. Shelef and H. S. Gandhi, Platinum Metals Rev. 18, 2 (1974).
28. T. P. Kobylinski and B. W. Taylor, J. Catal. 33, 376 (1974).
29. K. C. Taylor and R. L. Klimisch (submitted to J. Catal.).
30. S. W. Harris, E. F. Morello, and G. H. Peters, U.S. Patent 3,459,494 (1969).
31. R. A. Ogg and J. D. Ray, U.S. Patent 2,684,283 (1954).
32. D. M. Yost and H. Russell, "Systematic Inorg. Chem. of the Fifth and Sixth Group Nonmetallic Elements," Prentice-Hall, New York, 1944.

33. *C. N. Hinshelwood and C. R. Prichard, J. Chem. Soc. 127, 327 (1925).*
34. *H. Wise and M. F. Frech, J. Chem. Phys. 20, 1724 (1952).*
35. *F. Kaufman and J. R. Kelso, J. Chem. Phys. 21, 751 (1953).*
36. *M. Bodenstein, Z. Phys. Chem. 100, 106 (1922).*
37. *J. C. Treacy and F. Daniels, J. Am. Chem. Soc. 77, 2033 (1955).*
38. *J. F. Roth and R. C. Doerr, Ind. & Eng. Chem. 53, 293 (1961).*
39. *E. Muller and H. Barck, Z. Anorg. Chem. 129, 309 (1928).*
40. *A. Solbakken, Second Inter. Congr. on Catalysis, Vol. I, p. 341, Paris, 1960.*
41. *G. I. Golodets, Kinetika i Kataliz 6, 1123 (1965).*
42. *P. B. Weisz and C. D. Prater, Adv. in Catal. 4, 143 (1954).*
43. *H. C. Anderson, W. J. Green, and D. R. Steele, Ind. & Eng. Chem. 53, 199 (1961).*
44. *J. H. Jones, J. T. Kummer, K. Otto, M. Shelef, and E. E. Weaver, Env. Sci. and Tech. 5, 790 (1971).*
45. *H. C. Anderson, G. Cohn, W. J. Green, and D. R. Steele, French Patent 1,205,311 (1960).*
46. *H. C. Anderson and C. D. Keith, French Patent 1,233,712 (1960).*
47. *H. Schachner, Cobalt 9, 12 (1960).*
48. *G. K. Boreskov, Disc. Faraday Soc. 41, 263 (1966).*
49. *G. K. Boreskov, Adv. in Catal. 15, 285 (1964).*
50. *A. Terenin and L. Roev, Second Inter. Congr. on Catalysis, Vol. II, p. 2183, Paris, 1960.*
51. *G. Blyholder and M. G. Allen, J. Phys. Chem. 69, 3998 (1965).*
52. *M. Onchi and H. E. Farnsworth, Surf. Sci. 13, 425 (1969).*
53. *W. M. H. Sachtler, Second Inter. Congr. on Catalysis, Vol. II, p. 2197 (1960).*
54. *A. Alexeyev and A. Terenin, J. Catal. 4, 440 (1965).*
55. *Y.-H. Hu and J. W. Hightower (to be published).*
56. *G. W. Poling and R. P. Eischens, J. Electrochem. Soc. 113, 218 (1966).*
57. *C. V. Sternling, Seminar at Rice University, September 7, 1972.*
58. *A. Lawson, J. Catal. 24, 297 (1972).*
59. *V. Haensel, Fifth Inter. Congr. on Catalysis, Vol. I, p. xx, Palm Beach, 1972.*
60. *R. J. Slone, private communication, October 1, 1974.*

DISCUSSION

J. J. Carberry *(University of Notre Dame)*

I fail to see the relevance of any NO_2 species in the high temperature mechanisms since it is well known that its formation has a negative activation energy.

Hightower

Yes. I would agree with that and also NO_2 is simply thermodynamically unstable at high temperature, but you're not going to be able to desorb it from the surface. So I think the only time the NO_2 mechanism would be important probably would be in a low temperature reaction as Lawson has found, for example, at less than 100°C.

P. Emmett *(Portland State University)*

Joe, what about the silver-silver-oxide system? You said it wouldn't decompose NO and now you say it decomposes NO at 100°?

Hightower

No, I intended to say that the silver-silver oxide will decompose nitrous acid that is formed as an intermediate in the gas phase. I think Kokes suggested the formation of such species a few years ago. There is a possibility that nitrous acid can be decomposed with the silver catalyst. It's not really known what the mechanism of that reaction is, but whatever the mechanism, the decomposition does not proceed above 100°C. That's the observation.

R. L. Klimisch *(General Motors Research Laboratories)*

You could also interpret that silver work to involve the formation of silver nitrate or silver nitrite. I don't think you can dismiss the possibility of this kind of stoichiometric reaction.

Hightower

That is true. Would Dr. Lawson like to make a comment about this system?

A. Lawson *(Ontario Research Foundation)*

I think you have to be pretty careful about this. More work is needed to be done to try to distinguish between stoichiometric reactions and pure decomposition. We have carried out other studies since then on similar systems involving supported urea which appears to be both catalytic and stoichiometric, but it is clear that the catalytic activity eventually disappears. I think that the use of this type of system would be for extended scrubbing of NO, perhaps in stationary source control, but you cannot get long life time from these systems in the true catalytic sense.

G. J. K. Acres *(Johnson Matthey and Co., Ltd.)*

You mentioned tungsten at the end of your talk. We have done some work on various tungsten compounds and did find NO decomposition. We say "decomposition" because we used nitrides and carbides and after about 6 hours the "decomposition" reaction ceased. During the reaction, the carbides or nitrides are oxidized to tungsten oxide which is inactive. So there are some unusual properties of tungsten, but exactly what is happening isn't yet clear.

Also, at the beginning of your lecture, you mentioned that results obtained on NO decomposition in the presence of CO and hydrocarbons should be ignored. We agree entirely with you that the main problem of NO decomposition is not in decomposing the NO but in removing the oxygen so formed from the catalyst surface. We investigated a number of systems where by introducing CO and hydrocarbons into the reaction mixture we can get selective reduction of NO. A variety of catalysts will "decompose" NO in the presence of excess oxygen but in the main the reaction is temperature dependent.

Hightower

Yes. I'm glad you brought that out because this reaction is utilized in a very important commercial application today. Specifically, for tail gas control in nitric acid plants a small amount of ammonia is introduced in the presence of a large excess of oxygen, and the oxygen ignores the ammonia. So that's the reason I think anyone studying NO decompositon must eliminate all the reductants from this system in order to avoid being fooled. One can check this by using isotropic tracer techniques; by labeling the nitrogen with ^{15}N, one can tell from whence it came.

T. P. Kobylinski (Gulf R & D)

During your extensive review of the literature on the decomposition of nitric oxide, you omitted neodymium oxide. I'd like to know your reason for this. I remember Ethyl Corporation some time ago reported the use of this as a continuous system for the nitric oxide decomposition under rather mild conditions (using neodymium oxide)*. We tried to duplicate this work but didn't have much luck. Would anybody here have a comment on this.

Hightower

I think neodymium oxide was included in the group of catalysts tested by Winter.

M. Boudart (Stanford University)

In the Stanford dissertation of Ali Amirnazmi which was not mentioned, a mechanism was proposed which resembles one that you had on your slide which you attributed to Dr. Voorhoeve, but with one variation. This variation is that we thought that not only do you need anionic vacancies to get the NO down with the oxygen in the hole, but you also need an electron which has to be caught by the NO molecule according to the principles discussed earlier this morning. So the defect we need is then properly defined as an F-center. Now then, I think there is some confirmation for this view in work that was presented by my colleagues, Mason and Huggins, at the May meeting of the Electrochemical Society in San Francisco. In this work what was done was to remove the oxygen as was suggested a moment ago. This is of course what we all want to do in NO decomposition. And the way to get rid of oxygen is to pump it through your catalyst. This was done by means of a zirconium dioxide electrolyte. On the surface of the zirconium dioxide a gold electrode was deposited. The electrode is patchy so that there are holes. The gold itself is absolutely inactive for the decomposition of NO under the conditions of the experiment. Then when you start pumping the rate of NO decomposition increases by more than two orders of magnitude over what it is on the zirconium dioxide alone as measured in our own laboratory. What happens is that vacancies are pumped to the surface and, with electrons, they decompose NO as was postulated by Ali Amirnazmi.

*R. E. Stephens, U.S. Patent No. 3,524,721 (Aug. 18, 1970).

Hightower

This is a most interesting result. I would like to raise one question. I do not understand how all of the explanations that we have seen here can imply oxygen coming down to the surface. And yet, there is overwhelming evidence suggesting that NO must be bound in complexes through the nitrogen end. I would like to know if anyone has any explanation for that.

M. Shelef *(Ford Motor Company)*

In the paper by Yates and Madey* there is a scheme for surface rearrangement. The NO comes down with the nitrogen facing the surface, then it slides down and splits up.

R. J. H. Voorhoeve *(Bell Laboratories)*

I think it's not so difficult to see. In infrared you see NO adsorbed with N down and to reconcile the reactivity you see O down. If you look at the paper by Shelef on NO decomposition, the portion of sites active for NO decomposition is very small. The activation energy as well as the activity is low. The infrared measures the adsorbed NO that is not active. The active sites probably are oxygen vacancies with or without trapped electrons. These are low in concentration, so I don't think it's difficult to reconcile.

Hightower

In other words maybe the infrared measurements that we're looking at don't have anything to do with the kinetics.

*J. T. Yates and T. E. Madey, J. Chem. Physics 45, 1623 (1966).

CATALYTIC SYNTHESIS AND DECOMPOSITION OF AMMONIA

J. A. DUMESIC and M. BOUDART

Stanford University, Stanford, California

ABSTRACT

It is generally accepted that the rate determining process in ammonia synthesis is the chemisorption of dinitrogen, at least under the most widely studied process conditions. It is probable that also under these conditions chemisorbed nitrogen atoms are the most abundant surface intermediates. From our recent work it appears that surface iron atoms with a coordination number of 7 are more active than other surface atoms in nitrogen fixation. An explanation of this finding is sought in terms of molecular orbital symmetry and in terms of generally accepted views on the reaction mechanism.

INTRODUCTION

To make or break nitrogen-nitrogen bonds has long been recognized to be the rate determining step in ammonia decomposition and synthesis to and from dinitrogen and dihydrogen (1). Many studies of this reaction, carried out over iron catalysts in the vicinity of 700K, at atmospheric or higher pressures suggest that the most abundant surface intermediate is chemisorbed nitrogen, which is in quasi-equilibrium with dihydrogen and ammonia in the gas phase so that the mechanism can be summarized in two steps:

$$2* + N_2 \ \rightleftharpoons \ 2*N \tag{1}$$

$$*N + \frac{3}{2}H_2 \ \leftrightarrow \ NH_3 + * \tag{2}$$

of which the first is rate determining and the second is in equilibrium (2). Besides, the iron catalyst surface behaves as if it were broadly non-uniform, consisting of sites, denoted as *, possessing substantially different affinities and reactivities. This

References p. 104.

behavior leads to a kinetic treatment of the reaction largely associated with the name of Temkin, whose first paper on the subject, thirty-five years ago, has withstood the many attempts made at confirming or refuting the basic tenets summarized above.

More recently, suggestions have been made by Brill and coworkers concerning the larger affinity for nitrogen and the higher ammonia synthesis activity of (111) planes of iron, as compared to other planes (3,4). On the other hand, the decomposition of ammonia on tungsten, which has the same body-centered cubic (bcc) structure as iron, proceeds about ten times faster on (111) planes than on (100) or (110) planes of that metal (5).

With small metal particles, between 1 and 10 nm in size, surface anisotropy is not well described in terms of poorly developed crystallographic planes. Rather, it is more convenient to talk about surface atoms characterized by the number of nearest neighbors. Surface anisotropy of small particles is then defined by the relative number of atoms C_i, where C_i denotes a surface atom, which we shall call a site, with i nearest neighbors. Now, the surface anisotropy of small particles changes with particle size. In particular, for bcc structures, the relative proportion of C_7 sites increases by an order of magnitude as particle size grows from 1 to 5 nm (6). It must be noted that C_7 sites are found on (111) planes, but not on (100) or (110) planes of a bcc structure.

The higher ammonia synthesis activity of (111) planes or C_7 sites of iron, suggested by earlier work, has recently been confirmed by a study of the effect of particle size of supported iron catalysts on rate of ammonia synthesis (7,8,9). The turnover number, or number of molecules of ammonia made per site per second was found to increase with particle size in the range from 1 to 10 nm. Moreover, changes in Mössbauer spectra as well as a decrease in the amount of chemisorption of carbon monoxide as a result of a pretreatment of the surface with nitrogen which also increased turnover number in ammonia synthesis, were interpreted as caused by the increase in the relative amount of C_7 sites as a result of nitrogen pretreatment. These results have been summarized elsewhere (9).

If then C_7 sites possess a higher catalytic activity in ammonia synthesis and decomposition, why is it so? In this paper, we attempt to answer this question on the basis of molecular orbital symmetry. Since our considerations are quite general, they may well find an application in the catalytic chemistry of nitrogen oxides, when nitrogen-nitrogen bonds are made in decomposition or reduction of nitric oxide.

SYMMETRY OF SURFACE SITES ON A BCC LATTICE

The appropriate "symmetry operations" for C_4, C_5, C_6, and C_7 sites are shown in Fig. 1. A symmetry operation is a transformation of the site (e.g., a rotation about an axis or a reflection through a plane) that takes the site into itself. For example, with reference to Fig. 1, a rotation of the C_4 site by $\pi/2$ radians about the surface normal transforms the site into one indistinguishable from the unrotated site. Thus, the $\overline{C}_4$

rotation (a $\overline{C}_i$ rotation is a rotation by $2\pi/i$ radians) is a symmetry operation of the C_4 site, in the Schönfliess notation with overhead bars to avoid confusion with the notation of C_i sites.

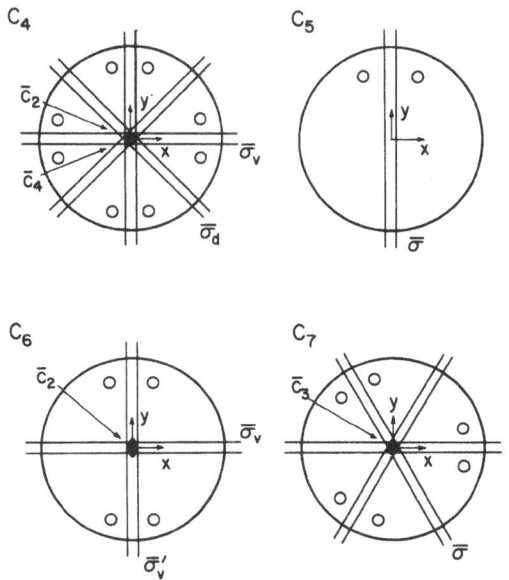

Fig. 1. Symmetry Operations for B.C.C. Surface Sites, o - represent symmetry related positions.

The symmetry operations of a particular site form a *point group*, and the point groups of the C_4, C_5, C_6, and C_7 sites are $\overline{C}_{4v}$, $\overline{C}_s$, $\overline{C}_{2v}$, and $\overline{C}_{3v}$, respectively. The symmetries of the various orbitals on a particular site are specified by the "irreducible representations" for which they form bases. For example, consider the p_x orbital for the C_6 site. The identity operation, $\overline{E}$, (which leaves the site unmoved) takes p_x into itself, by definition; the $\overline{C}_2$ operation takes p_x into $-p_x$; $\overline{\sigma}_v'$ (reflection through the yz plane) takes p_x into $-p_x$; and $\overline{\sigma}_v$ (reflection through the xz plane) takes p_x into itself. Thus, the symmetry of p_x can be *represented* by the coefficients of the various transformations:

Symmetry operations:	$\overline{E}$	$\overline{C}_2$	$\overline{\sigma}_v$	$\overline{\sigma}_v'$
Transformation coefficients:	1	-1	1	-1

In general, the symmetry operations may *mix* different orbitals. For example, a $\overline{C}_4$ operation on p_x for the C_4 site takes this orbital into p_y. Since the physics of the problem is unchanged by a $\overline{C}_4$ rotation (the site and its transform are indistinguish-

References p. 104.

able), the p_x and p_y orbitals must be degenerate. In this case, the symmetry operations are matrices instead of the previously discussed scalars. That is,

$$\overline{C}_4 \begin{pmatrix} p_x \\ p_y \end{pmatrix} \text{ transforms to } \begin{pmatrix} p_y \\ -p_x \end{pmatrix}$$

indicating that,

$$\overline{C}_4 = \begin{pmatrix} 0 & 1 \\ -1 & 0 \end{pmatrix}$$

TABLE 1

Symmetry Representations for Surface and Adsorbate Orbitals

Point Group	Representation Symbol	Transformation Traces					Orbitals
$\overline{C}_1$		$\overline{E}$					
	A	1					All Orbitals
$\overline{C}_2$		$\overline{E}$ $\overline{C}_2$					
	A	1 1					p_z, d_{z^2}, $d_{x^2-y^2}$, d_{xy}
	B	1 -1					p_x, p_y, d_{xz}, d_{yz}
*$\overline{C}_3$		$\overline{E}$ $\overline{C}_3$ $\overline{C}_3^2$					
	A	1 1 1					p_z, d_{z^2}
	E	$\begin{bmatrix} 1 & \epsilon & \epsilon^* \\ 1 & \epsilon^* & \epsilon \end{bmatrix}$					$\left[\begin{array}{c}(d_{x^2-y^2}, d_{xy}), (p_x, p_y) \\ (d_{yz}, d_{xz})\end{array}\right]$
$\overline{C}_{2v}$		$\overline{E}$ $\overline{C}_2$ $\overline{\sigma}_v$ $\overline{\sigma}_v'$					
	A_1	1 1 1 1					p_z, d_{z^2}, $d_{x^2-y^2}$
	A_2	1 1 -1 -1					d_{xy}
	B_1	1 -1 1 -1					p_x, d_{xz}
	B_2	1 -1 -1 1					p_y, d_{yz}
$\overline{C}_{3v}$		$\overline{E}$ $\overline{C}_3$ $\overline{\sigma}_v$					
	A_1	1 1 1					p_z, d_{z^2}
	A_2	1 1 -1					
	E	2 -1 0					$\left[\begin{array}{c}(p_x,p_y), (d_{xz}, d_{yz}) \\ (d_{x^2-y^2}, d_{xy})\end{array}\right]$
$\overline{C}_{4v}$		$\overline{E}$ $\overline{C}_4$ $\overline{C}_2$ $\overline{\sigma}_v$ $\overline{\sigma}_d$					
	A_1	1 1 1 1 1					p_z, d_{z^2}
	A_2	1 1 1 -1 -1					
	B_1	1 -1 1 1 -1					$d_{x^2-y^2}$
	B_2	1 -1 1 -1 1					d_{xy}
	E	2 0 -2 0 0					p_x, p_y, d_{xz}, d_{yz}
$\overline{C}_s$		$\overline{E}$					
	A'	1 1					p_x, p_y, d_{z^2}, $d_{x^2-y^2}$, d_{xy}
	A''	1 -1					p_z, d_{yz}, d_{xz}

*$\epsilon = \exp(2\pi i/3)$

(Adapted from Cotton, Ref. 10)

The matrices for the other symmetry operations on the C_4 site are the following:

$$\overline{E} = \begin{pmatrix} 1 & 0 \\ 0 & 1 \end{pmatrix} \qquad \overline{C}_2 = \begin{pmatrix} -1 & 0 \\ 0 & -1 \end{pmatrix}$$

$$\overline{\sigma}_v = \begin{pmatrix} 1 & 0 \\ 0 & -1 \end{pmatrix} \qquad \overline{\sigma}_d = \begin{pmatrix} 0 & -1 \\ -1 & 0 \end{pmatrix}$$

The symmetry of the p_x and p_y orbitals is now represented by the traces of the appropriate matrices:

Symmetry operations:	$\overline{E}$	$\overline{C}_4$	$\overline{C}_2$	$\overline{\sigma}_v$	$\overline{\sigma}_d$
Transform traces:	2	0	-2	0	0

Analogous arguments can also be made for the d-orbitals, and the results of such arguments (10) are summarized in Table 1 for various point group symmetries. The symmetries of the π-bonding and $\pi*$-antibonding orbitals of N_2 are shown on Fig. 2. It is seen that the π-orbitals transform as p-orbitals and the $\pi*$-orbitals transform as d-orbitals. It should be noted, however, that those lobes of the $\pi*$-orbitals on one atom, as viewed looking down the molecular axis, transform as p-orbitals.

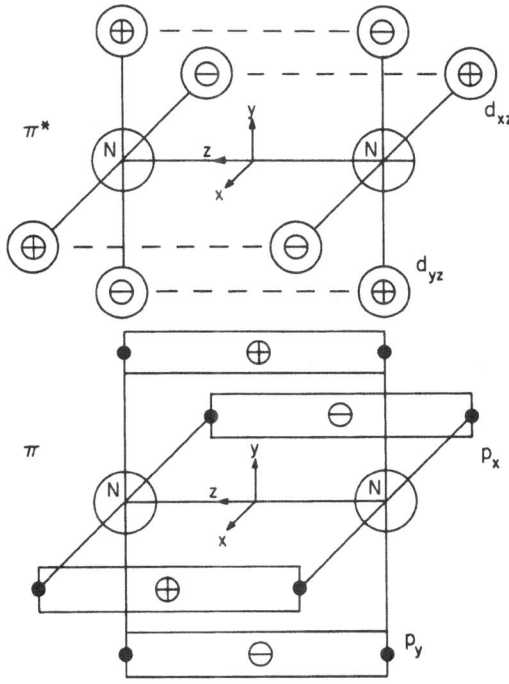

Fig. 2. Symmetry of $\pi*$- and π-Orbitals of Nitrogen.

References p. 104.

In Fig. 3 are shown two modes of nitrogen chemisorption of dinitrogen with associated coordinate systems. For chemisorption of dinitrogen with its molecular axis parallel to the surface normal, the symmetries of the C_4, C_5, C_6, and C_7 sites remain $\overline{C}_{4v}$, $\overline{C}_s$, $\overline{C}_{2v}$, and $\overline{C}_{3v}$, respectively; the π-orbitals of N_2 transform as p_x and p_y; and the $\pi*$-orbitals of N_2 also transform as p_x and p_y since only those lobes on one atom overlap with the surface orbitals of iron. It should be noted, that if the $\pi*$-orbitals are represented by the d_{xz}- and d_{yz}-orbitals (considering the $\pi*$ lobes on both N atoms) instead of the p_x- and p_y-orbitals (considering the $\pi*$ lobes on one N atom), the conclusions reached are not significantly different. For chemisorption of dinitrogen with its molecular axis perpendicular to the surface normal, the symmetries of the C_4, C_5, C_6, and C_7 sites are lowered to $\overline{C}_2$, $\overline{C}_1$, $\overline{C}_2$, and $\overline{C}_1$, respectively, for an arbitrary rotational orientation of the molecular axis about the surface normal; the π-orbitals of N_2 transform as p_x and p_z; and the $\pi*$-orbitals of N_2 transform as d_{xy} and d_{yz}.

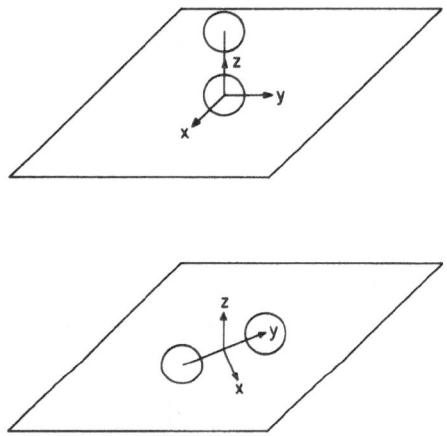

Fig. 3. Surface Modes of Nitrogen Chemisorption.

The matrix element between two orbitals, $\langle \psi_i | H | \psi_j \rangle$, is non-vanishing only when ψ_i and ψ_j belong to (or span) the same irreducible representation (10). For example, with reference to Table 1, the π-orbitals of N_2 (transforming as p_x and p_y) for chemisorption with the molecular axis parallel to the surface normal can only interact with the d_{xz}- and d_{yz}-orbitals of iron for the C_4 site, since these orbitals all belong to the E representation. All other matrix elements between the π-orbitals of N_2 and the d-orbitals of iron are identically zero. In Table 2 are summarized the allowed matrix elements between the π- and $\pi*$-orbitals of N_2 with the d-orbitals of iron for the two modes of chemisorption of dinitrogen.

The case of σ-bonding is now easily considered. A σ-orbital is totally symmetric with respect to rotations about the surface normal, and hence it transforms as the

TABLE 2

Nonvanishing Matrix Elements for Chemisorption of Dinitrogen
on Iron Sites, $\langle \psi_{N_2} | H | \psi_{site} \rangle$

Site	N$_2$ axis parallel to surface normal		N$_2$ axis perpendicular to surface normal	
	ψ_{N_2}	ψ_{site}	ψ_{N_2}	ψ_{site}
C$_4$	π p$_x$	d_{xz}, d_{yz}	π p$_x$	d_{yz}, d_{xz}
	π p$_y$	d_{xz}, d_{yz}	π p$_z$	$d_{z^2}, d_{x^2-y^2}, d_{xy}$
	π^* p$_x$	d_{xz}, d_{yz}	π^* d$_{yz}$	d_{yz}, d_{xz}
	π^* p$_y$	d_{xz}, d_{yz}	π^* d$_{xy}$	$d_{x^2}, d_{x^2-y^2}, d_{xy}$
C$_5$	π p$_x$	$d_{z^2}, d_{x^2-y^2}, d_{xy}$	π p$_x$	All d-orbitals
	π p$_y$	$d_{z^2}, d_{x^2-y^2}, d_{xy}$	π p$_z$	All d-orbitals
	π^* p$_x$	$d_{z^2}, d_{x^2-y^2}, d_{xy}$	π^* d$_{yz}$	All d-orbitals
	π^* p$_y$	$d_{z^2}, d_{x^2-y^2}, d_{xy}$	π^* d$_{xy}$	All d-orbitals
C$_6$	π p$_x$	d_{xz}	π p$_x$	d_{yz}, d_{xz}
	π p$_y$	d_{yz}	π p$_z$	$d_{z^2}, d_{x^2-y^2}, d_{xy}$
	π^* p$_x$	d_{xz}	π^* d$_{yz}$	d_{yz}, d_{xz}
	π^* p$_y$	d_{yz}	π^* d$_{xy}$	$d_{z^2}, d_{x^2-y^2}, d_{xy}$
C$_7$	π p$_x$	$d_{x^2-y^2}, d_{xy}, d_{xz}, d_{yz}$	π p$_x$	All d-orbitals
	π p$_y$	$d_{x^2-y^2}, d_{xy}, d_{xz}, d_{yz}$	π p$_z$	All d-orbitals
	π^* p$_x$	$d_{x^2-y^2}, d_{xy}, d_{xz}, d_{yz}$	π^* d$_{yz}$	All d-orbitals
	π^* p$_y$	$d_{x^2-y^2}, d_{xy}, d_{xz}, d_{yz}$	π^* d$_{xy}$	All d-orbitals

p$_z$-orbital. If the dominant interaction between the adsorbed species and the surface is provided by the σ-bond, then the symmetries of the C$_4$, C$_5$, C$_6$, and C$_7$ sites are not significantly changed from $\overline{C}_{4v}$, $\overline{C}_s$, $\overline{C}_{2v}$, and $\overline{C}_{3v}$, respectively. For this case, the nonzero matrix elements between this σ-orbital and the d-orbitals of iron are summarized in Table 3. For ammonia adsorption, if the full symmetry of the NH$_3$ must be considered, then the adsorption process lowers the symmetries of the C$_4$, C$_5$, C$_6$, and C$_7$ sites to $\overline{C}_1$, $\overline{C}_1$, $\overline{C}_1$, and $\overline{C}_3$, respectively, and the allowed matrix elements for this case are also summarized in Table 3.

TABLE 3

Nonvanishing Matrix Elements for σ-Bonding
to B.C.C. Surface Sites, $\langle \sigma | H | \psi_{site} \rangle$

Site	Symmetry	ψ_{site}	Symmetry	ψ_{site}
C$_4$	$\overline{C}_{4v}$	d_{z^2}	$\overline{C}_1$	All d-orbitals
C$_5$	$\overline{C}_s$	$d_{z^2}, d_{x^2-y^2}, d_{xy}$	$\overline{C}_1$	All d-orbitals
C$_6$	$\overline{C}_{2v}$	$d_{z^2}, d_{x^2-y^2}$	$\overline{C}_1$	All d-orbitals
C$_7$.	$\overline{C}_{3v}$	d_{z^2}	$\overline{C}_3$	d_{z^2}

APPLICATION TO AMMONIA SYNTHESIS ON A BCC LATTICE

The transfer of electrons from the surface to dinitrogen involves the overlap of the antibonding orbitals of the molecule with those of the surface. Consider first the case of adsorption with the molecular axis of dinitrogen oriented parallel to the iron surface normal. Adsorption on a surface atom in this manner does not change the symmetry of the surface site. Thus, the symmetries of the C_4, C_5, C_6, and C_7 sites remain $\overline{C}_{4v}$, $\overline{C}_s$, $\overline{C}_{2v}$, and $\overline{C}_{3v}$, respectively, upon chemisorption. The nitrogen antibonding orbitals are antisymmetric with respect to a two-fold rotation about the molecular axis, and the C_4 and C_6 sites possess a two-fold rotation axis parallel to the surface normal. Matrix elements, $\langle \psi_{N_2*} | H | \psi_{Fe} \rangle$, between the nitrogen antibonding orbitals, ψ_{N_2*}, and the orbitals of iron, ψ_{Fe}, are nonvanishing only when ψ_{Fe} belongs to a representation which includes the irreducible representation for ψ_{N_2*}. For example, if it is assumed that ψ_{Fe} can be approximated by a sum of appropriate atomic 3d and 4s orbitals centered on each surface atom, then $\langle \psi_{N_2*} | H | \psi_{Fe} \rangle$ is nonvanishing only when ψ_{Fe} is d_{xz} or d_{yz} for the C_4 and C_6 sites. On the C_5 site, the antibonding orbitals of dinitrogen interact with three d-orbitals and the s-orbital of iron, and on the C_7 site these orbitals of dinitrogen interact with four of the d-orbitals.

Secondly, consider the case of adsorption of dinitrogen on a surface atom with the molecular axis oriented perpendicular to the surface normal. In this case, the symmetries of the C_4, C_5, C_6, and C_7 sites are lowered to $\overline{C}_2$, $\overline{C}_1$, $\overline{C}_2$, and $\overline{C}_1$, respectively, for an arbitrary rotational orientation of the molecular axis about the surface normal. The presence of a two-fold symmetry axis for the C_4 and C_6 sites requires one of the antibonding orbitals of dinitrogen to be symmetric and the other to be antisymmetric with respect to this rotation. This requirement, not present for the C_5 and C_7 sites, gives rise to vanishing matrix elements between the antibonding orbitals of dinitrogen and the surface iron orbitals. Thus, as in the previous case, the nitrogen molecule may interact more strongly with the C_5 and C_7 sites than with the C_4 and C_6 sites.

The interaction between chemisorbed dinitrogen and the iron surface many also be partially the result of electron transfer from dinitrogen to iron, and this transfer will involve the π-bonding orbitals of nitrogen and vacant iron orbitals (or d-band holes). The symmetry considerations for the π-bonding orbitals are identical to those for the antibonding orbitals discussed previously. Thus, as with electron transfer from the iron to the nitrogen, the process of electron transfer from the nitrogen to the iron seems to be symmetry facilitated for the C_5 and C_7 sites.

Let us finally discuss the symmetry of σ-bonded species, such as nitrogen N or ammonia NH_3 adsorption on various surface sites. Adsorption of the ammonia will involve the overlap of one of the ammonia hybrid orbitals (approximately sp^3) with iron orbitals, this overlap forming a σ-bond between ammonia and the surface. In general, as shown in the previous section, the symmetry for σ-bonding is particularly favorable for C_5 sites, less so for C_6 sites, and least favorable for C_4 and C_7 sites; and in particular, for ammonia adsorption considering the full symmetry of the NH_3, all sites except the C_7 are favorable for σ-bonding.

CONCLUSIONS

The general symmetry arguments collected in Tables 2 and 3 are summarized in Table 4, which shows that C_7 sites of iron are the best for binding dinitrogen N_2 in π or $\pi*$ fashion and the worst for binding nitrogen N in σ fashion, if by best and worst we mean the largest and the smallest total number of orbitals respectively of the site that can be used for binding, whether the site keeps its full symmetry following adsorption or not.

TABLE 4

Total Number of Orbitals of Iron Sites
That Can Participate in Bonding

π or $\pi*$ bonding	Type of Site	σ-bonding
4	C_4	6
8	C_5	8
4	C_6	7
9	C_7	2

To apply these considerations to ammonia synthesis on iron, a potential energy diagram of the reaction recognizing both π(or $\pi*$) and σ-bonded species at the surface is drawn in Fig. 4 for a site (full line) and for the best site (dashed line) defined as the one with the strongest binding in the π (or $\pi*$) mode and the weakest in the σ mode. In this diagram, we represent adsorbed dinitrogen as a precursor to adsorbed nitrogen. Activation barriers to adsorption of the precursor and to back adsorption of ammonia are assumed to be negligible. The kinetic scheme pictured in Fig. 4 is therefore indistinguishable from the generally accepted one discussed in the introduction. In particular the rate determining step is still the dissociation of dinitrogen, although it is now specified that the precursor to dissociation is adsorbed dinitrogen.

As suggested by Fig. 4, the best site will be one with a higher rate for dissociation because of either a lower activation energy or a higher concentration of π (or $\pi*$) bonded precursors, and because of the fact that more sites will be available as the concentration of the most abundant σ bonded intermediates is decreased on the best site.

This tentative explanation of the higher ammonia synthesis activity assigned to C_7 sites of iron should be further tested by examination of the kinetic and equilibrium

References p. 104.

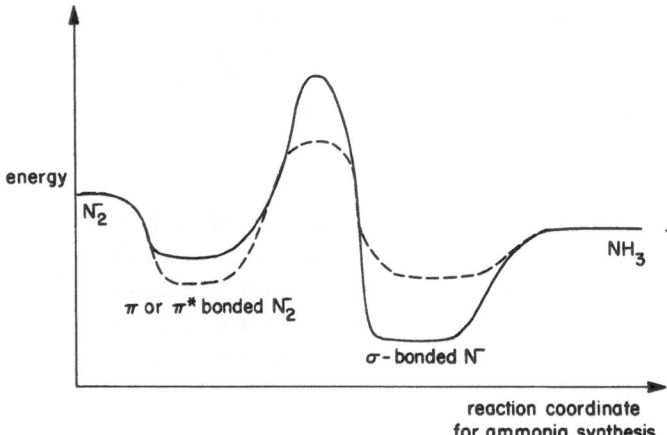

Fig. 4. Potential Energy Diagram for Ammonia Synthesis on Iron full line: any site, dashed line: optimum site.

constants extracted from kinetic data obtained on iron catalysts of varying particle size (11). Such an examination is underway and will be published elsewhere.

ACKNOWLEDGMENTS

Partial support by ARPA through the Center for Materials Research at Stanford University.

REFERENCES

1. P. H. Emmett and S. Brunauer, J. Am. Chem. Soc., 56, 35 (1934).
2. M. Boudart, AIChE Journal, 18, 465 (1972).
3. R. Brill, E. L. Richter and E. Ruch, Angew. Chem. Intern. Ed. 6, 882 (1967).
4. R. Brill and J. Kurzidim, Colloques Int. C.N.R.S. No. 187, 99 (1969).
5. J. McAllister and R. S. Hansen, J. Chem. Phys. 59, 414 (1973).
6. R. van Hardeveld and F. Hartog, Surf. Sci. 15, 189 (1969).
7. M. Boudart, A. Delbouille, J. A. Dumesic, S. Khammouma and H. Topsøe, J. Catal. in press.
8. J. A. Dumesic, H. Topsøe and M. Boudart, to appear in Surface Sci.
9. M. Boudart, H. Topsøe and J. A. Dumesic, The Physical Basis for Heterogeneous Catalysis, R. Jaffee et al., Eds. Plenum Press, in press.
10. F. A. Cotton, Chemical Applications of Group Theory, 2d ed., John Wiley and Sons, New York, 1971.
11. J. A. Dumesic, Ph.D. Dissertation, Stanford University (1974).

DISCUSSION

R. S. Hansen *(Iowa State University)*

I'm reluctant to make too many extrapolations from work on tungsten to iron, but I will comment in a few places. In the first place you have reported, and this is widely done, as to the rate limiting process in the synthesis of ammonia being dinitrogen adsorption. I think it's worth commenting that on tungsten, and this is probably also true on iron, there are a series of nitrogen structures so that, for example on the (100) face, the C(2 x 2) structure is quickly formed and the sticking coefficient is of the order one. The nitrogen atom density can be doubled going from the C(2 x 2) structure to the (1 x 1) structure by further nitrogen dosage, but the sticking coefficient for that is orders of magnitude lower. So when you say that the rate limiting step is the adsorption of dinitrogen, I think it's worthwhile commenting on what structures this is to occur. It is not, in my judgment, on bare surfaces, in the case of tungsten, and probably not in the case of iron either. Now, in our judgment, although this is perhaps less well documented, there are yet further states of nitrogen containing more than one nitrogen atom per tungsten atom which are relevant in the synthesis and decomposition. For that reason I am extremely concerned with a bonding model which binds the nitrogen that's going to do the decomposing directly to the tungsten in our case, iron in your case. I make these reservations without being sure that they are right. But I suspect they need to be looked into.

Boudart

I appreciate your comments but I would like to point out that the way I have represented it here, the molecularly bonded dinitrogen at the surface bound in a pi or pi-star fashion is one that is bound without an activation energy. So it is not a case of activated adsorption. The rate determining step is the process of going from that pi-bonded state to another sigma-bonded state.

P. Emmett *(Portland State University)*

I just want to point out that you carried the calculation with regard to the (111) surface one step farther than Brill in that you postulate when you go back to pure hydrogen, you can reverse the flow and go from (111) back to the less active (100) and (110) which hadn't been done before.

G. Parravano *(University of Michigan)*

Michel, if I understand you, the anisotropy effect ... the magnetic anisotropy effect, is a rather large one.

Is that right? Because if that is right, it seems to me the interesting step here would be to use polarized gamma rays, because the polarization factor is affected by the magnetic anisotropy. And it would be very interesting to do the adsorber effect in which you can rotate at different angles, one with respect to the other. You should find some very interesting results if I understand the effects correctly.

Boudart

We are very much interested in pursuing this but the effect is unfortunately not as large as all that. You have seen the details in the spectral area and it is of the order of 10 to 15%; now that's perfectly measurable, but it is not a lot.

Parravano

Yes, but the field of polarized gamma rays is a very interesting one. Just as in the case of the two nickels you can completely neglect all the background noise and concentrate on the rest.

J. Carberry *(University of Nortre Dame)*

Michel, at what particle size do you find a lack of accord between the Mössbauer and the bulk magnetic properties? In other words, what is the population of clusters so that you no longer find a correspondence between bulk and Mössbauer?

Boudart

The average particle size for the smallest particle size sample is 15 angstroms and at that size we do not see any discrepancy with bulk properties.

T. P. Kobylinski *(Gulf Research and Development Co.)*

Prof. Boudart, by changing the particle size in your experiments you have shown that nitrogen fixation is more or less a structure sensitive reaction. I have a comment here which proposes a process to you. We've been doing ammonia decomposition, especially over ruthenium, and we have strong indications that ammonia decomposition over ruthenium is very sensitive . . . is a very structure sensitive process. We did some work on changing the particle size of ruthenium and we observed a very strong effect.

I was just wondering whether you agree with me or whether you have observed anything similar on iron as far as the ammonia decomposition is concerned.

Boudart

No. We have not studied the ammonia decomposition, per se, although it takes place while we make ammonia because we approach equilibrium rather readily at atmospheric pressure.

CATALYTIC ACTIVATION OF NITROGEN OVER TRANSITION METALS

A. OZAKI, K. AIKA and K. URABE

Tokyo Institute of Technology, Tokyo, Japan

ABSTRACT

Ammonia synthesis rates over metals are strikingly increased by addition of potassium as a promoter so that significant activities are obtained for the ammonia synthesis over rhodium and iridium, with which no synthesis activity has been known. The mechanism of enhancement by potassium has been investigated by means of kinetics of isotopic mixing in molecular nitrogen and of ammonia synthesis both over ruthenium. The kinetics of isotopic mixing is in conformity with the equation

$$R_E = kP/(1 + KP)^2 \qquad (1)$$

which is consistent with the dissociative chemisorption. From the temperature dependence of k and K at 320 to 380°C, the activation energy and the heat of adsorption are found to be 13 and 47 kcal/mol respectively. This is a remarkably high value as a heat of nitrogen adsorption on ruthenium, for which no nitride is known. The kinetics of ammonia synthesis is also in conformity with the equation

$$R_s = kP_{N_2}/(1 + KP_A/P_{H_2}^x)^2 \qquad (2)$$

The inverse isotope effect as found on iron catalysts is observed on Ru-AC-K, whereas, on pure ruthenium, no isotope effect was observed. These results demonstrate that the adsorption bond of nitrogen is strengthened by the potassium addition, thus promoting the noble metals for the ammonia synthesis.

INTRODUCTION

The synthesis of ammonia is an old theme for catalysis ever since the discovery of the Haber Process. The conventional iron catalyst, $Fe-Al_2O_3-K_2O$, has been

References p. 116.

thoroughly investigated by many workers. On the basis of these studies, it has been understood that the potash promoter brings about beneficial effects on the intrinsic activity of iron, presumably because of its electron-donating nature. This presumption led us to the discovery of a novel catalyst system involving a transition metal on active carbon promoted by alkali metal. Ruthenium was found to be the most active transition metal. The role of the alkali metal in this catalyst seemed ascribable to its electron-donating nature since the catalytic activity increased with decrease in ionization potential of the alkali, as shown in Fig. 1 (1) i.e., in the order

$$Cs \rangle K \rangle Na$$

An immediate question occurs as to the effect of the alkali metal on the kinetic behavior of the nitrogen activation reaction, since nitrogen chemisorption is accepted as the rate determining step of ammonia synthesis. The present status of investigations into this problem is introduced in the following.

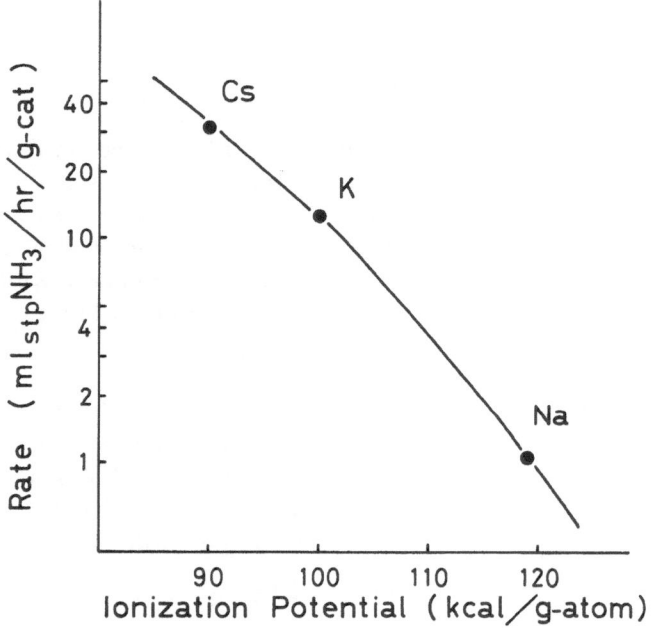

Fig. 1. The rate of ammonia synthesis over Ru-Ac-alkali at 290°C as a function of ionization potential of alkali metal. (4 mg-atom, g cat alkali).

RELATIVE EFFICIENCIES OF METALS

Although a large number of investigations have been made on ammonia synthesis, no comparable data for the ammonia synthesis rate is available with metals. Instead, the relative values of effluent ammonia concentration under a fixed condition (2 g catalyst, 550°C, 100 atm. and 60 l/hr flow rate) are taken from the classical work by

Mittasch (2), and plotted against the parameter ΔH: (heat of formation of highest oxide per gram atom of metal), with which the initial heat of chemisorption of gases, including nitrogen, is linearly correlated (3). It is obvious from Fig. 2(a) that Os and Fe are the most active elements. No activity had been known for Rh, Ir and Pd for the synthesis reaction.

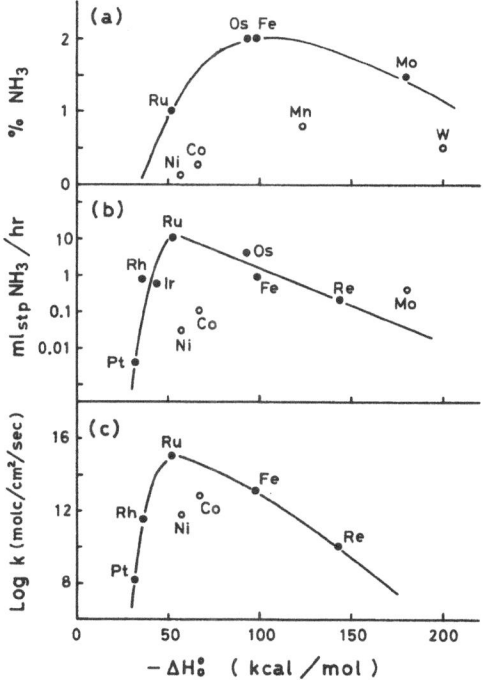

Fig. 2. Activity patterns of transition metals: (a) Effluent ammonia concentration in high pressure operation at 550°C, (b) Rate of ammonia synthesis over metals on AC-K at 250°C, (c) Rate constants of ammonia decomposition over metal films at 400°C.

On the other hand, the relative rate constants over metal films are available for the decomposition reaction at 400°C as shown in Fig. 2-(c) (4). It is to be noted that the maximum is shifted to Ru and that Rh also exhibits a considerable activity. Fig. 2-(b) shows the relative rates of ammonia synthesis at 250°C over the transition metal-AC-K catalysts (5). There is a striking similarity between the activity patterns of Fig. 2-b and -c. It appears that the potassium metal promoter which is characteristic for the new catalyst system brings about a preferential activation of noble metals. Although the reaction temperatures are different (250°C vs. 400°C), the activity pattern for the synthesis at 400°C would be essentially unchanged because the apparent activation energy of synthesis over Ru (23 ∿ 27 kcal/mol) is higher than the others except for Co (6).

It is to be noted in Fig. 2-(b) that the iron family of metals is always the most active in the various transition series.

References p. 116.

KINETICS OF NITROGEN ACTIVATION

It has been accepted that nitrogen chemisorption is rate-determining in ammonia synthesis over metals. This is confirmed by the inhibition of isotopic equilibration during the synthesis reaction over Ru-AC-K and Ru-K as shown in Fig. 3 (7). It is obvious that although the isotopic equilibration proceeds in the presence of hydrogen with an equilibrium amount of ammonia, it is suddenly slowed down by trapping ammonia. This result demonstrates that, once nitrogen is chemisorbed, it is quickly hydrogenated to ammonia. It is also observed that the isotopic equilibration does proceed even with trapping ammonia if the hydrogen pressure is lowered, suggesting that the hydrogenation of chemisorbed nitrogen is not extraordinarily fast. The kinetics of isotopic equilibration over Ru-AC-K at 280°C was one half order with respect to nitrogen pressure (8). There have been three other determinations of the reaction order for isotopic equilibration, i.e., on pure iron at 350°C (9) and on doubly promoted iron at 433°C (10a), and 500°C (10b). Incidentally they are one-half order, also. Such kinetics are usually interpreted in terms of the adsorption constant. But the adsorption constant of N_2 on Ru seemed unlikely to be large enough because the kinetics of ammonia synthesis was first order with respect to N_2 in conformity with the absence of a deuterium isotope effect (11). Thus one might imagine that there is a mechanistic reason to give the one-half order of kinetics, while no rationalization could be made.

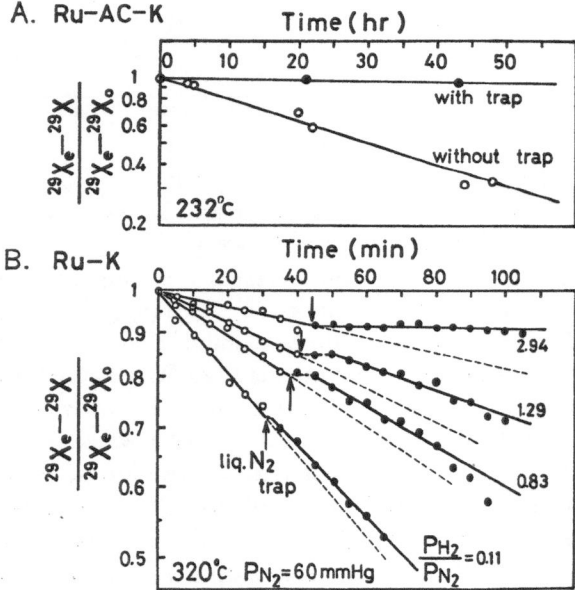

Fig. 3. Isotopic equilibration of N_2 during ammonia synthesis over (A) Ru-AC-K at 232°C and (B) Ru-K at 320°C.

It was accordingly decided to examine the half-order kinetics by the temperature effect (12), where Ru-K was adopted. Unexpectedly it turned out that the reaction order increased from 0.4 at 320°C to 0.8 at 380°C, as shown in Fig. 4, suggesting the feasibility of interpretation in terms of the adsorption constant. Since those measurements were made under conditions of adsorption equilibrium, the rate of adsorption, Va, is equal to that of desorption, Vd. If the Langmuir equation for the dissociative adsorption is applied, the rate of exchange, R, is given by

$$R = Va = Vd = kP/(1 + \sqrt{KP})^2 \tag{1}$$

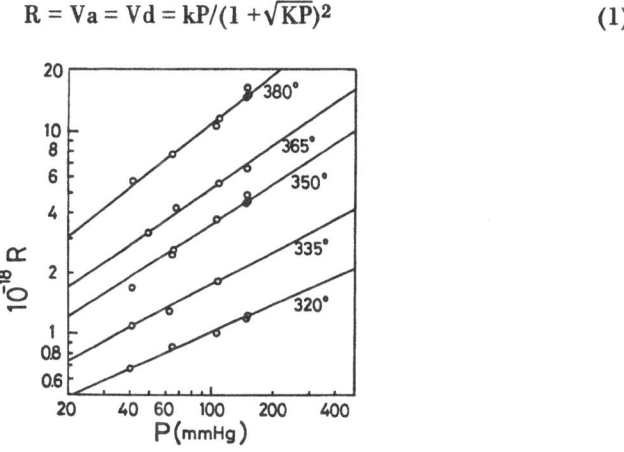

Fig. 4. Pressure dependence of isotopic equilibration over Ru-K.

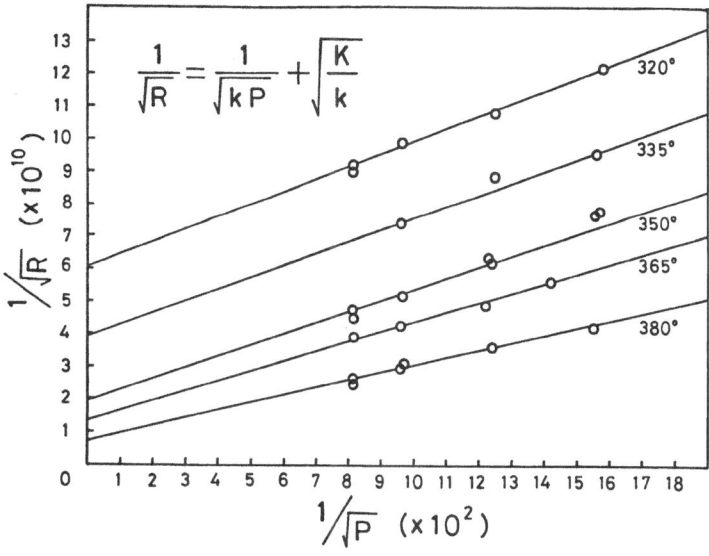

$$\frac{1}{\sqrt{R}} = \frac{1}{\sqrt{kP}} + \sqrt{\frac{K}{k}}$$

Fig. 5. Test of equation (1).

The results in Fig. 4 fit this equation as shown in Fig. 5 where $1/\sqrt{R}$ is plotted against $1/\sqrt{P}$. From the straight lines of Fig. 5, the values of rate constant, k, and adsorption constant, K, are determined. Their Arrhenius plots are shown in Fig. 6, which gives an activation energy of 13 kcal/mol and a heat of adsorption of 47 kcal/mol. An alternative plot of 1/R versus 1/P resulted in a curved Arrhenius plot.

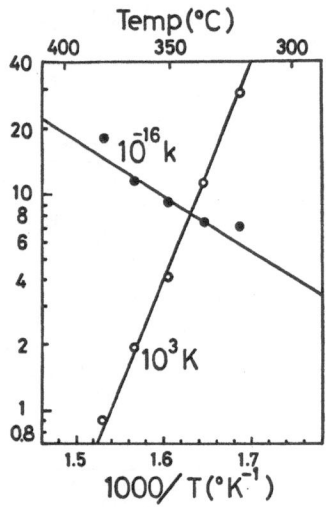

Fig. 6. Arrhenius plots of rate constant and adsorption constant.

The pure ruthenium catalyst without added potassium was considerably less active so that the reaction order could not be obtained even at 470°C, whereas the apparent activation energy was estimated to be 28 kcal/mol. In order to clarify the effect of potassium, another series of kinetic measurements was made on a more active catalyst, $Ru-Al_2O_3$. The results are summarized in Table 1.

TABLE 1

Kinetic Constants of Isotopic Equilibration

Catalyst	Reaction order	Temp. range	E_{app}	E_o kcal/mol	Q
Ru	—	430-480°C	28	—	—
Ru-K	0.4-0.8	320-380	32	13	47
Ru-Al₂O₃	1.0	380-440	12	12	—
Ru-Al₂O₃-K	0.3-0.45	280-320	27	9	28

It is obvious that the addition of potassium gives rise to the fractional reaction order caused by strong adsorption of nitrogen, while the heat of adsorption seems to be affected by the alumina support. Since the first order kinetics suggest an immeasurably low value of the heat of adsorption, it may be concluded that the addition of potassium intensifies the adsorption strength of nitrogen on Ru. Indeed, no nitride is known for noble metals, and no data is available for the heat of nitrogen chemisorption over those metals, presumably because the interaction is too weak. However it is now disclosed that the inactive surface of noble metal acquires a high affinity towards nitrogen. This must be an important reason for the preferential activation of noble metals as shown in Fig. 2-b.

As a result of the above mentioned effect of added potassium, an I.R. spectrum of chemisorbed nitrogen was observed on Ru- and Rh-Al$_2$O$_3$-K at around 2000-2050 cm^{-1}(13). It did not disappear on evacuation and was transformed into ammonia with hydrogen treatment at 350°C.

EFFECT OF HYDROGEN

The effect of hydrogen on the rate of isotopic equilibration is illustrated in Fig. 7. It is seen that the equilibration rate is increased on pure Ru but is decreased on the Ru-K catalyst by the addition of hydrogen. An analogous effect is observed for the relative rates of ammonia synthesis and isotopic equilibration, as shown in Fig. 8. That is, the rate of ammonia synthesis is faster than equilibration on pure Ru but is slower on Ru-K.

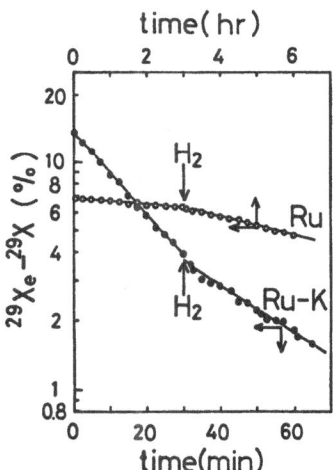

Fig. 7. The effect of hydrogen on the rate of isotopic equilibration over Ru at 450°C and Ru-K at 380°C.

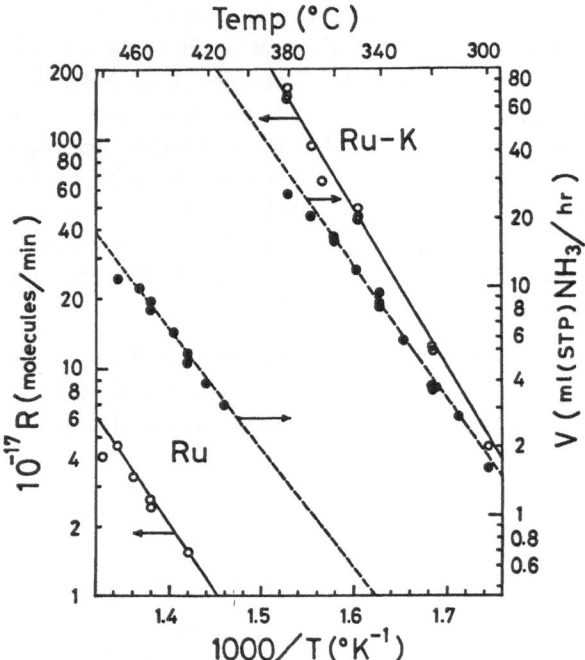

Fig. 8. Arrhenius plots of rate of equilibration and synthesis over Ru and Ru-K.
(P_{N_2} = 150 torr)

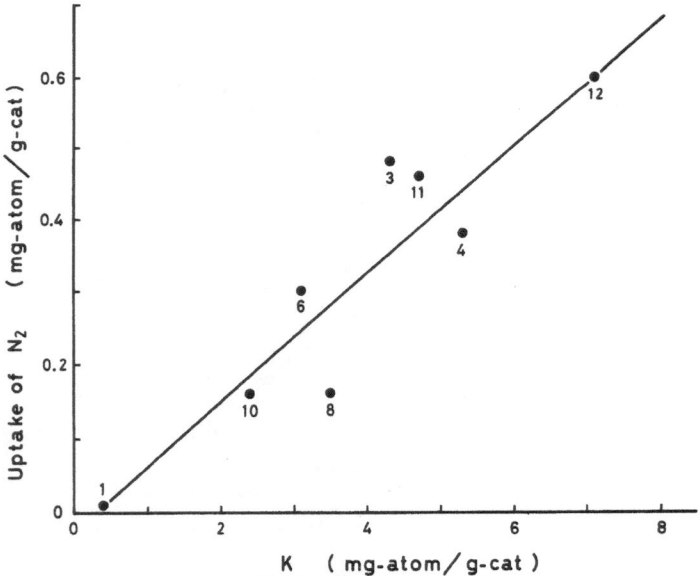

Fig. 9. Amount of adsorbed N_2 as a function of added K on Ru-AC.

The retarding effect of hydrogen may be explained by partial transformation of potassium into an amide. It was observed that the amount of chemisorbed nitrogen linearly increases with the potassium concentration in Ru-AC-K as shown in Fig. 9 (14). Since the amount of chemisorbed nitrogen was too high to be held by ruthenium, the nitrogen must be partly held by potassium. Such nitrogen seems to be transformed into amide in the presence of hydrogen (14), reducing the electron-donating nature of potassium. On the other hand in the case of pure Ru, adsorbed hydrogen or NH_x would act as an electron donor.

KINETICS OF AMMONIA SYNTHESIS

It was previously reported that the kinetics of ammonia synthesis over Ru-celite are first order with respect to N_2, in agreement with the absence of a deuterium isotope effect (11). In the case of Ru-AC-K, however, the synthesis rate was affected by the flow rate despite the very low conversion to ammonia, demonstrating that the reaction rate is a function of ammonia partial pressure. Thus the rate analysis gave equation (2) for the synthesis rate, V,

$$V = kP_{N_2} / (1 + K_A P_{NH_3}/P_{H_2}^x)^2 \qquad (2)$$

where K_A denotes the equilibrium constant for the dissociative adsorption of NH_3;

$$NH_3 = NH_{(3-2x)}(a) + xH_2 \qquad (3)$$

The activation energy and the heat of adsorption of ammonia were obtained from k and K_A to be 21.5 ± 1.0 and 27.5 ± 2.0 kcal/mol respectively (15), whereas the latter value was likely to depend on the catalyst condition. The corresponding value of heat of adsorption of ammonia during the synthesis reaction over iron catalysts was around 5 kcal/mol (16).

In addition to this, the synthesis rate over Ru-AC-K at 230°C showed the deuterium isotope effect as follows;

f_D	0	0.63	0.84	0.95
V mlNH_3/hr.gcat	3.7	5.4	6.8	7.1

which is in conformity with the rate equation (2) as previously shown for iron catalysts (10).

In this way both the kinetics and isotope effect of ammonia synthesis are consistent with the intensification of Ru-N bond by the addition of K metal. Such stabilization of the adsorbed state should reduce the activation energy for nitrogen chemisorption. This is seen for Ru as shown above; the same effect of K metal would operate on other noble metals, giving rise to the observed activation.

References p. 116.

REFERENCES

1. A. Ozaki, K. Aika and Y. Morikawa, *Vth Int. Congr. (Palm Beach) Paper 90 (1970)*.
2. A. Mittasch, *Adv. Catalysis 2, 81 (1950)*.
3. K. Tanaka and K. Tamaru, *J. Catalysis 2, 366 (1963)*.
4. S. R. Logan and C. Kemball, *Trans. Faraday Soc., 56, 144 (1960)*.
    Estimated by G. C. Bond in *Catalysis by Metals*. p. 381 Academic Press (1962). Correlated by
    K. Tanaka and K. Tamaru, *Kinetika i Kataliz, 7, 242 (1966)*.
5. K. Aika, J. Yamaguchi and A. Ozaki, *Chem. Lett., 1973, 161*.
6. A. Ozaki, K. Aika and F. Hori, *Bull. Chem. Soc. Jap., 44, 3216 (1971)*.
7. K. Urabe, K. Aika and A. Ozaki, to be published.
8. K. Urabe, K. Aika and A. Ozaki, *J. Catalysis, 32, 108 (1974)*.
9. G. Shulz and H. Schaefer, *Z. Phys. Chem. N.F. 64, 333 (1969)*.
10. a) N. Takezawa and I. Toyoshima, *J. Catalysis 19, 271 (1970)*, b) G. K. Boreskov, A. I.
    Gorbunov and O. L. Masanov, *Doklady Akad. Nauk USSR, 123, 90 (1958)*.
11. K. Aika and A. Ozaki, *J. Catalysis 16, 97 (1970)*.
12. K. Urabe, K. Aika and A. Ozaki, to be published.
13. M. Ohkita, K. Aika and A. Ozaki, to be published.
14. K. Aika and A. Ozaki, *J. Catalysis* in press.
15. K. Urabe, K. Koyama, K. Aika and A. Ozaki, *35th Discussion of Catalysis Soc. Jap., Paper
    #1D08 (1974)*.
16. A. Ozaki, H. S. Taylor and M. Boudart, *Proc. Roy. Soc. (London), A258, 47 (1960)*, von
    R. Krabetz and Cl. Peters, *Ber. Bunsen. Phys. Chem. 67, 381 (1963)*.

DISCUSSION

P. Emmett *(Portland State University)*

I just wanted to ask you how your frequency of chemisorbed nitrogen compared to that which Eischens found for nitrogen on nickel. Are they about the same?

Ozaki

Eischens observed 2200-2220 cm^{-1} while ours was at 2020 cm^{-1}.

C. C. Chang *(General Motors Research Laboratories)*

I would like to know how strong is the nitrogen chemisorbed on the ruthenium surface you detected by infrared spectrometry. And also I would like to know how hydrogen influences the bands you found in the infrared.

Ozaki

Evacuation at room temperature for one day doesn't change the band. The IR spectra were taken at room temperature and at the highest, 100°C. We are planning to study heating effects on the spectra, but haven't done it. If you treat with hydrogen ... at high temperature (350°C) the band disappears. Hydrogen treatment at room temperature causes no shift.

Chang

Another question. When you say you find an amide on the surface, can you see this in the infrared too?

Ozaki

This work was of course intentionally aimed at finding the amide. But we haven't found it yet, because the alumina support gave us trouble. I'm now trying to use another support.

R. Klein *(National Bureau of Standards)*

We looked at the thermal desorption of nitrogen from ruthenium, and it comes off at 160 K. Also, it comes off in first order kinetics. The binding is, of course, very weak.

MICRO AND MACRO CHANGES IN A STAINLESS STEEL CATALYST DURING REDUCTION OF NO

E. E. PETERSEN, J. LANDAU and E. SAUCEDO

University of California, Berkeley, California

ABSTRACT

A newly constructed 304 stainless steel recycle batch reactor was found to be significantly reactive to NO above 250°C. The extent of reactivity depended upon pretreatment schedules used and whether the reactor surfaces were in an oxidized or reduced state. This work suggests that a breakdown of the protective chromium oxide surface caused by cyclic oxidizing and reducing atmospheres leads to the development of a large surface roughness. Further work with supported iron and chromium oxides substantiates this mechanism.

INTRODUCTION

The walls of a newly constructed 304 stainless steel recycle batch reactor were found to be significantly active toward the reactions of nitric oxide above 250°C. To ascertain what contribution these wall reactions may have made to previous studies performed in stainless steel reactors and to assess the feasibility of using stainless steel as a converter catalyst, a preliminary investigation of the catalytic potential of stainless steel was undertaken. Attention was directed primarily to the catalytic potential of stainless steel for promoting the reduction of NO.

The reactivity of alloys and stainless steel had been documented in reports (1) and patents (2) prior to its recent appearance in the published literature (3,4,5). Yet compared to the extensive work performed on supported and unsupported metal oxide catalysts as well as noble metal catalysts, little is known about the effects of the reaction environment on the activity and selectivity of the reactions important in nitric oxide abatement systems.

References pp. 127-128.

Three reactions appear to be most important. They are the reduction of NO by H_2, the reduction of NO by NH_3 and the decomposition of NH_3 (6). Ammonia forms primarily by the reduction of NO by hydrogen sources. It has been suggested that NH_3 decomposition may be a major path to N_2 at temperatures above 600°C (7), whereas at lower temperatures the NO-NH_3 reaction becomes competitive (8).

Consequently, preliminary experiments in this study have concentrated on the nature of the above reactions on active stainless steel, supported iron oxide and chromium oxide catalysts.

EXPERIMENTAL METHOD

Reactor — The reactions were studied in a stainless steel recycle batch reactor. The major components of the system are the heating zone where reaction takes place, the heat exchange section which acts as a preheater for the gas entering the heating zone, a well mixed reservoir and an 8-port sample valve. Construction details of the reaction system are discussed in reference 9, and the analytical system is described in reference 10.

Pretreatment of the Surface — The heated stainless steel surface was pretreated prior to some runs, but not all of them. When the pretreatment consisted of 100 torr H_2 and 900 torr He recycled for 12-18 hours at subsequent reaction temperature, the surface was considered activated in the reduced state and will be referred to as an RAS surface. If the pretreatment was 760 torr air plus 240 torr He at reaction temperature for 12-18 hours, the surface was considered activated in the oxidized state and will be referred to as an OAS surface.

EXPERIMENTAL RESULTS

Reduction of Nitric Oxide on Stainless Steel Catalyst — The reaction between NO and an RAS surface at 480°C was very rapid, initially, as shown on Fig. 1, Run 29. The curve through the NO points represents the material balance line according to Equation 1 and the N_2 measurements. Then the reaction rate fell very rapidly and remained slow. When a new charge of NO was added, the reaction rate continued at approximately the same small value. The lack of O_2 in the product stream suggested that the reduced surface was being "titrated" with O atoms from NO. The line on this figure represents the material balance based upon the appearance of nitrogen.

The corresponding reactivity of an OAS surface is shown by Run 30. No initial fast reaction is exhibited by this surface.

Nitric Oxide — Hydrogen Reactions on Stainless Steel Catalyst — Four runs were performed using almost identical concentrations of NO and H_2. The activities of the RAS and OAS surfaces for these reactions were measured at 480°C and 340°C and are shown on Figs. 2, 3, 4 and 5. Rapid initial disappearance of NO is characteristic of

RAS surfaces similar to that shown on Fig. 1. The nitrous oxide appeared as a product at 340°C but not at 480°C.

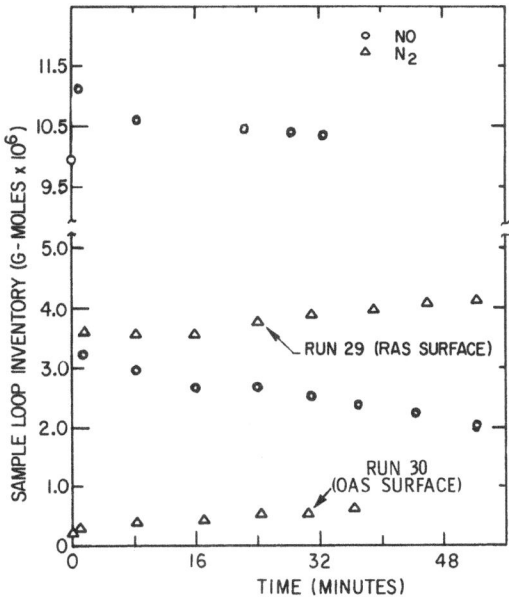

Fig. 1. Reduction of nitric oxide by a RAS and "oxidized" stainless steel surface (Runs 29 and 30).

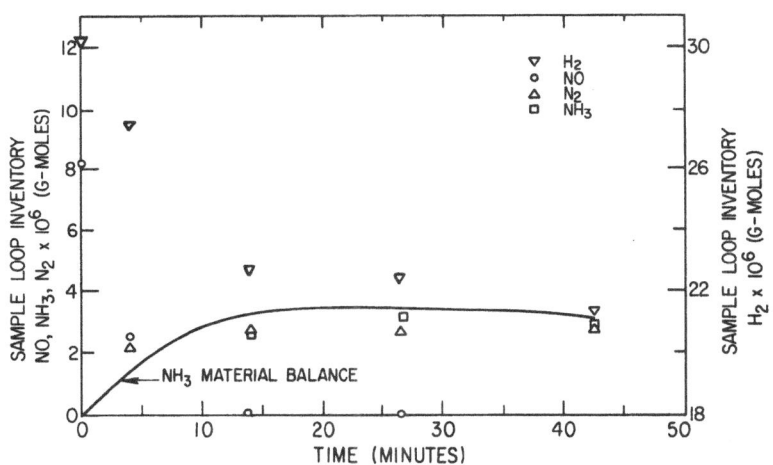

Fig. 2. Reduction of nitric oxide by hydrogen on a RAS stainless steel surface at 478°C (Run 7).

References pp. 127-128.

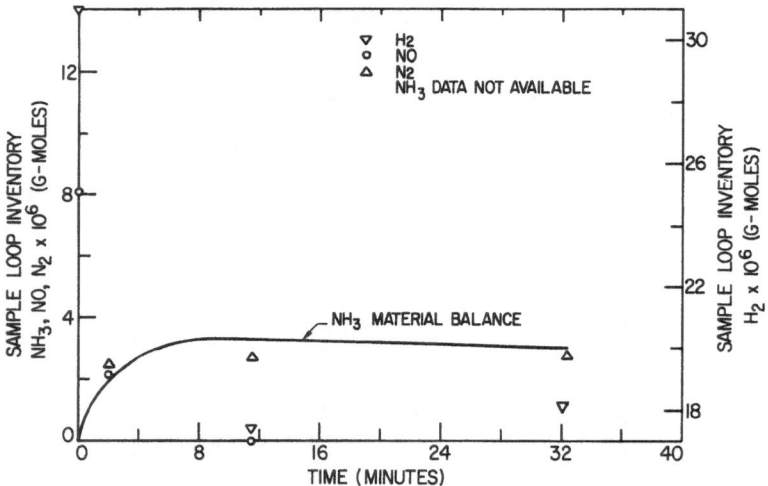

Fig. 3. Reduction of nitric oxide by hydrogen on an OAS stainless steel surface at 482°C (Run 11).

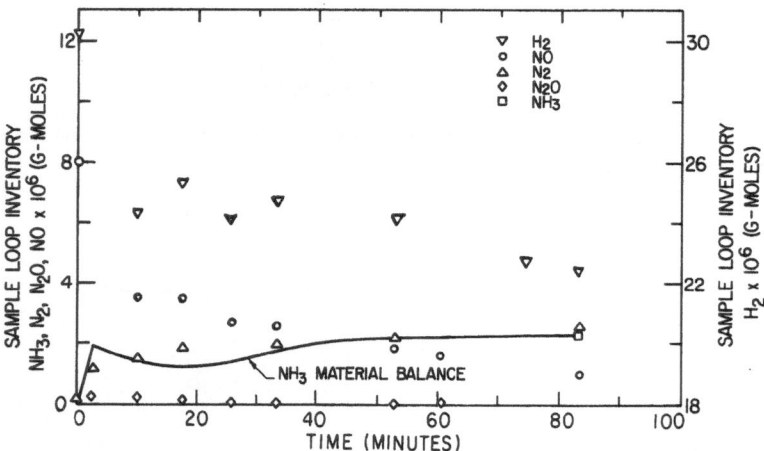

Fig. 4. Reduction of nitric oxide by hydrogen on an RAS stainless steel surface at 348°C (Run 14).

The lines on Fig. 2 and 3 are obtained from the following material balance equations:

$$2NO \rightarrow N_2 + 2O_{ads} \qquad \text{(1)}$$

$$H_2 + O_{ads} \rightarrow H_2O \qquad \text{(2)}$$

$$NO + \frac{5}{2} H_2 \rightarrow NH_3 + H_2O \qquad \text{(3)}$$

These equations lead to the result that initially oxygen goes on the surface of the RAS catalyst of Fig. 2, and that oxygen is removed from the OAS catalyst of Fig. 3.

Figs. 4 and 5 are interpreted by adding another equation for the formation of N_2O to the set above.

$$2NO + H_2 \rightarrow N_2O + H_2O \tag{4}$$

This catalyst was also active for the NH_3-NO reaction.

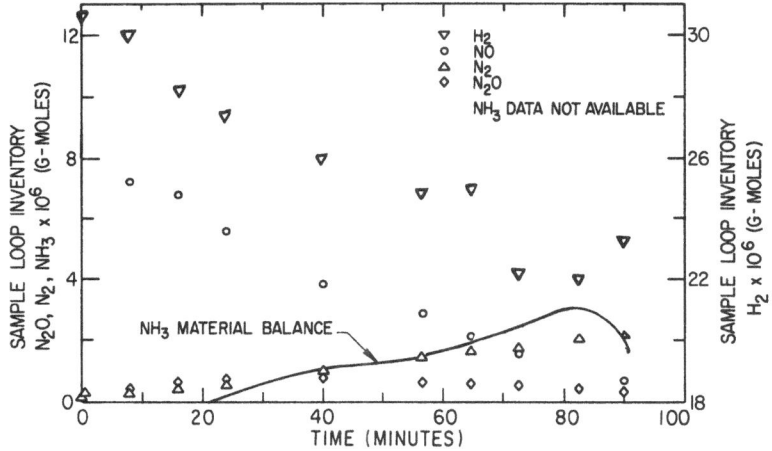

Fig. 5. Reduction of nitric oxide by hydrogen on an OAS stainless steel surface at 336°C (Run 13).

Surface Roughness of the Stainless Steel Catalyst — Figs. 6 and 7 show the roughness of the stainless steel before and after exposure to activation procedures. The photographs are magnifications of about 1000 X and clearly show increased surface roughness. It is difficult to assess, however, how this would translate to area because the micropore structure is not detectable.

Experiments with Supported Iron Oxide and Chromium Oxide Catalysts — A large number of experiments using supported chromium oxide and iron oxide catalysts show similar behavior to the stainless steel catalyst. The significant finding in this work is that N_2O formed on chromium and stainless steel catalysts but very little formed on the iron catalyst. No experiments were made on supported NiO catalysts, although the literature cited below indicates that NiO is active.

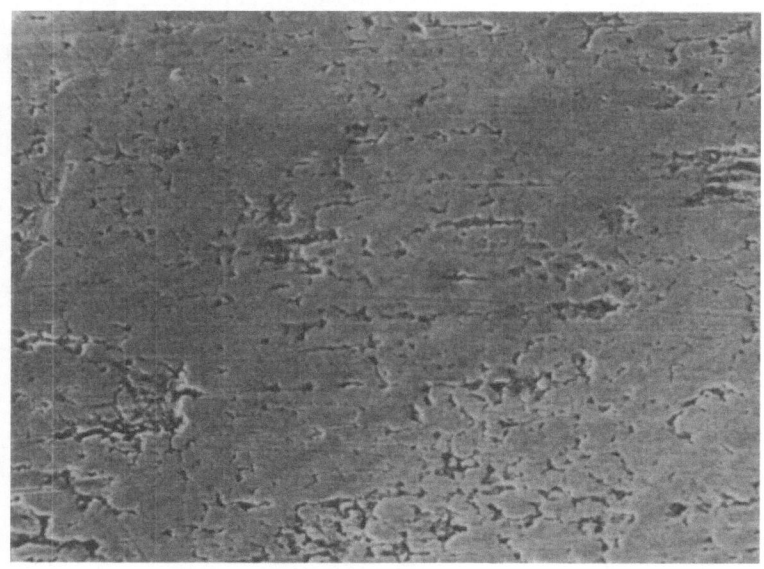

Fig. 6. Photomicrograph of 304 stainless steel before surface development.

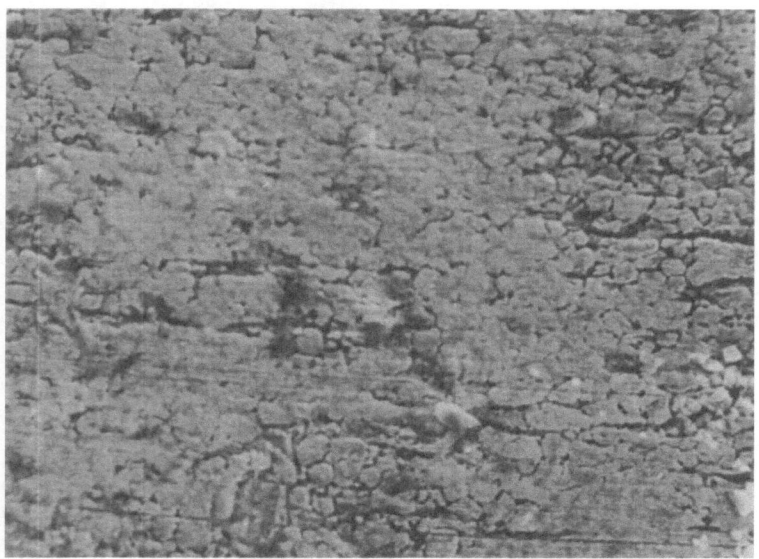

Fig. 7. Photomicrograph of 304 stainless steel after surface development.

DISCUSSION

These results imply changes of a stainless steel catalyst at three different levels during the course of the experiments:

a. Macroscopic development of surface roughness of the catalyst.

b. Addition and depletion of oxygen from RAS and OAS surfaces respectively during the initial reaction period.

c. Suggestion that the mechanism of the catalyst operation is an oxidation-reduction scheme.

The increase in surface roughness is inferred from the increase in catalytic activity with successive oxidation and reduction of the surface.

Stainless steel has been found to be catalytically active for the reduction of nitric oxide by H_2 and NH_3 and for NH_3 decomposition. The preliminary evidence presented suggests that the extent and condition of the surface governs to a great extent the path of a reaction; therefore, it is important to understand what physical and chemical changes might be occurring to the stainless steel surface during these studies.

The literature indicates that the concentration of species on a s.s. surface that has been heated bears little resemblance to the initial bulk composition. When an 18:8 stainless (18% Cr, 8% Ni, rest Fe) is heated to 800°C in an inert atmosphere and then cooled to 500°C, chromium carbide precipitates near the grain boundaries dropping these regions below 12% Cr concentration. The alloy is then no longer "stainless," and the grain boundaries are active for corrosive attack (11). Undoubtedly this was taking place in the heating zone of the s.s. reactor, since the heating zone was raised near 800°C during the initial check out phase of the system.

Wood and coworkers have examined the segregation that takes place during the oxidation of iron chromium and nickel chromium alloys (16). These studies would indicate that once the almost pure Cr_2O_3 protective surface scale (formed during initial oxidation) is cracked, local castastrophic scaling is likely to occur exposing increasingly larger areas of the bulk to attack. The studies on the Ni - Cr alloys also showed that it was possible for NiO to migrate preferentially to the surface.

To some extent segregation must have been occurring on the stainless steel surfaces of this reactor during oxidation. Visible inspection found that the surface did not retain its metallic characteristics. Generally the surface was bluish-black and very dull. Scanning electron micrographs indicated that there was extensive cracking and channeling on the surface. This seems to indicate that an oxide scale had indeed formed at one time. The work on Fe - Cr alloys suggests that the stainless steel surface is enriched with Cr greatly above its bulk composition of 18%. Whether a NiO scale forms is not known, but the possibility is apparently present.

References pp. 127-128.

The oxidation of the metal surface is just one half of the many cycles that the surface went through. Reduction of the surface after oxidation probably caused the surface structures to become even more porous and less corrosion resistant as evidenced dramatically by the scanning electron micrographs.

Oxidation State of the Surface During Reaction — The experimental reaction data indicate that the surface undergoes changes in oxidation state during certain reactions. Although no experiments were performed attempting to quantify these changes, information in the literature strongly indicates its occurrence under similar conditions.

Shelef and coworkers have used gravimetric methods when studying the adsorption of NO to determine the oxidation states of oxides in their reduced and oxidized form. Of particular interest to the study of stainless steel are the studies performed on iron oxide (15), chromium oxide (16, 17), and nickel oxide (18). Oxidation of the oxides in O_2 at 450°C gave respectively Fe_2O_3, $Cr_2O_{5.5}$, and NiO. Since conditions during activated oxidation of the stainless steel were similar for the current experiments, it is assumed that this is a close description of the oxidized species on the OAS surface.

Using 20% CO in CO_2 at 450°C to reduce the oxides, Shelef found reduced states of Fe_3O_4, Cr_2O_3, and NiO, respectively. Although the temperature was similar, H_2 was used in the present reduction procedures rather than CO. Weller and Voltz (19, 20) indicate that the reduced state of chromia obtained using H_2 at 500°C depends on the dryness of the system. Thermodynamics predicts that bulk chromia may be reduced to chromous oxide at atmospheric pressure of H_2 if the H_2O partial pressure is kept below 0.017 torr. Although the thermodynamics are presumably different for the surface chromia, the water content also plays a role in the oxidation state of the surface. During activation, the partial pressure of H_2 was 100 torr and probably the H_2O partial pressure was greater than 0.017 torr. Therefore the RAS surface of chromia in the present system is assumed to be Cr_2O_3.

An analogous situation applies to the reduced iron oxide present on the surface of the stainless steel. The reduction is performed in a *closed* system, and the dynamic equilibrium between the oxygen on the surface and the H_2O and H_2 in the circulating gas does not allow a reduction of the iron oxide to elemental iron. The RAS surface of the iron oxide is postulated to be FeO or a combination of FeO and Fe_3O_4. At 600°C, Shelef was able to reduce his iron oxide sample to $Fe_{0.8}O_{0.2}$ in H_2 (20). Yet more significantly, he concluded that the ability of iron oxide to adsorb NO did not increase when it was reduced below Fe_3O_4.

Reduction of NO by a Reduced Stainless Steel Surface — The reduction of NO by an initially reduced surface plays an important role in the reactions examined above. This reaction results in N_2 on a RAS surface at 480°C. When the surface is reduced at lower temperatures, near 340°C, a reaction with the surface also occurs but seems to yield a gaseous product of N_2O. In both cases the nitric oxide oxidizes a RAS surface even in the presence of a highly reducing gaseous environment.

The fast initial reaction with the surface observed here on reduced stainless steel also was observed by Otto and Shelef on reduced chromia (16, 17) and iron oxide (15). At lower temperatures (-78°C), the adsorption of NO was reversible, but at higher temperatures (150°C) an appreciable amount of the adsorption was irreversible (that is, there was a reaction between the NO and the surface that resulted in chemical changes). Only N_2O was observed as a product of the irreversible adsorption (reaction with the surface). The observations made on stainless steel with the present system seem to be a logical extension of that work. This may be the major reason why N_2O is seen as a gas phase intermediate at moderate temperatures (150-300°C) and is not present at higher temperatures ()400°C) in most heterogeneous catalytic reactions of NO.

The observations in the present stainless steel system support the basic kinetic conclusions presented by Otto and Shelef for NO adsorption on iron oxide and chromium oxide. A sharp break in the Elovitch plot was seen for NO adsorption on their reduced surfaces. There was no such break for NO adsorption on their oxidized surfaces. This corresponds to the observations made earlier concerning Runs 29 and 30. The conclusion in both cases must be the same. The surface is initially in a reduced state, yet after introduction of NO it rapidly becomes "oxidized." Once it becomes "oxidized," it continues to adsorb NO, but at a much slower rate. Since this rate and the rate for the OAS surface are similar, one can conclude that the active adsorption sites are probably similar in both cases. Presumably the RAS and OAS surfaces have greatly different intrinsic rate constants for the reaction with nitric oxide.

Shelef and Gandhi have shown that nickel oxide behaves very differently from iron oxide and chromia oxide (18). Since the stainless steel observations are similar to the iron oxide and chromium oxide results, it can be concluded that at least on reduced surfaces NiO does not strongly affect the observations of NO adsorption on stainless steel. However, this does not indicate whether or not nickel oxide is actually on the surface in large or small quantities or what effect it may have in promoting the reactions that involve NH_3 (particularly NH_3 decomposition).

REFERENCES

1. A. Lamb and E. L. Tollefson, Preprint, 22nd CSChE Conference, "Catalysis and Sorption in Air Pollution Control," Symposium, Toronto, Sept. 1972.
2. G. H. Meguerian and C. R. Lang, Preprint Automotive Engineering Conference, Jan. 1971.
3. J. R. Jenkins and M. A. Voisey, Atm. Environment 7, 177 (1973).
4. J. R. Jenkins and M. A. Voisey, Atm. Environment 7, 187 (1973).
5. J. London, Ph.D. Dissertation, Univ. of California, Berkeley (1972).
6. J. E. Hunter, General Motors Research Publication GMR-1061, March 23, 1971.
7. R. L. Klimisch and G. J. Barnes, Env. Sci. and Tech. 6 (6), 543 (1972).

8. K. Otto and M. Shelef, J. Phys. Chem. 76 (1), 37 (1972).
9. J. I. Landau, M. S. Thesis, Univ. of California, Berkeley (1973).
10. J. I. Landau and E. E. Petersen, J. Chromat. Sci. 12, 362 (1974).
11. J. H. Braphy, R. Rose, and J. Wulff, The Structure and Properties of Materials, Vol. II, Thermodynamics of Structure, John Wiley & Sons, Inc., New York 1964.
12. D. P. Whittle and G. C. Wood, J. Electrochemical Soc. 114, 986 (1967).
13. G. C. Wood and D. P. Whittle, J. Electrochem. Soc. 115, 126 (1968).
14. D. P. Whittle and G. C. Wood, J. Electrochem. Soc. 115, 133 (1968).
15. K. Otto and M. Shelef, J. Catalysis 18, 184 (1970).
16. K. Otto and M. Shelef, J. Catalysis 14, 223 (1969).
17. M. Shelf, J. Catalysis 15, 289 (1969).
18. H. Gandhi and M. Shelef, J. Catalysis 24, 241 (1972).
19. S. W. Weller and S. E. Voltz, J. Am. Chem. Soc. 76, 4695 (1954).
20. S. E. Voltz and S. W. Weller, J. Am. Chem. Soc. 76, 4701 (1954).

DISCUSSION

K. H. Ludlum *(Texaco)*

I'd like to say, in support of what Prof. Petersen said, Eischens and I have noted that when carbon monoxide is exposed to stainless steel we've seen nickel-carbonyl, perhaps manganese and/or iron carbonyl formed. We published the results of this work in Surface Science* some time last year. We think that many times people have mistaken these gaseous carbonyls for surface-derived species that were easily chemisorbed and pumped away. I refer you to Surface Science for the details.

M. S. Peters *(University of Colorado)*

Gene, I have a graduate student sitting in his laboratory back home who has run into exactly the same problem. We were doing some work on NO and CO with rare earths such as cerium oxide. We only intended to go to about 250 or 300°C. We got excellent results. We ran all of our blanks; no conversions at all. Then we decided to cover the whole range of the rare earths so we had to go to higher temperature. The first time we took our stainless steel equipment, which is 304 and 316, and had it at 500°C, we ran our normal blank and got about 50% conversion, immediately. This was about a month ago and we spent about 3 weeks figuring out exactly what you have said today . . . that it is the stainless steel. The only thing to do of course was to switch to another kind of reactor. What I would like to bring out here is that you mentioned last night that you would bet my graduate student let it go up to 800°. The oven we are using has a top limit of about 580°C, so I don't think it's been heated above this. This tells me that, when you get the stainless steel up to around 500°C, it may do it right now which means that some of you folks who are using stainless steel probes to take samples don't have to let the probes get heated way up before they become active. What do you think? Do you think it wouldn't work unless my graduate student let it get up to 700 or 800°C?

*K. H. Ludlum and R. P. Eischens, Surface Science 40, 397 (1973).

Petersen

Max, I don't want to be quoted that 800°C is necessary to activate the stainless steel. I suspect that it's a time-temperature phenomenon. The information I referred to suggested the figure of 800°C. It seems quite a bit higher than what your graduate student has told you.

K. H. Ludlum

With reference to the temperature at which this takes place with carbon monoxide, when we reduced overnight in flowing hydrogen at 350°C and cooled to room temperature and admitted carbon monoxide at several torr pressure, we developed nickel-carbonyl bands within 5 minutes, and slowly thereafter the other bands developed which I figured were either manganese or iron carbonyl. So we found that the stainless steel was susceptible to attack by carbon monoxide even at pretreatment temperatures, much lower than what you're talking about.

Petersen

The initial carbon content of the stainless steel appears to be important to get this phenomenon. If one is working in a hydrogen atmosphere it may take a rather high temperature for a low carbon stainless steel to precipitate these carbides. But if you're using carbon monoxide, you probably have the carbon that you need to form these carbides at much lower temperature.

G. H. Meguerian *(Amoco)*

Nitrogen oxide will do the same thing. We first discovered the effect on 304 stainless steel when we were studying nitrogen oxide formation in flames. We used stainless steel thermocouples in the reactors. The first day we got very high nitrogen oxide. The second day we got less nitrogen oxide. As we kept using the thermocouple, we got less and less nitrogen oxide. And then we discovered it was a thermocouple effect at 350°C. Nitrogen oxide will do the job also.

Petersen

But why? Do you know why? What's going on at the surface?

Meguerian

It corrodes.

Petersen

But what . . . the nitrides instead of the carbides . . .

J. W. Hightower *(Rice University)*

We heard a lot about what not to do. But I would like to know if you have any recommendations about what to do. First of all we can't have a reactor and we can't measure the temperature, and I was just wondering what we can do?

M. Shelef *(Ford)*

A quartz reactor is the solution.

Petersen

I hope that he's right when he says quartz. Because that's what we're now using.

Shelef

A quartz sheathed thermocouple will help also.

J. J. Mooney *(Englehard Industries)*

A replacement material is Inconel 601. We've used this.

J. B. Butt *(Northwestern)*

You said that you think that perhaps the formation of this increased surface roughness is due to carbide formation. I was wondering if it might also be possible under oxidizing conditions to have interpenetration of the grain boundaries by chromium oxide. I think that the metallurgists call this green rot.

Petersen

I don't think these are unrelated. My information is that at these grain boundaries the precipitation of the carbide means that the chromium content is down and the material is no longer stainless. And it's subject to all of these rots that you're talking about. It does come in at the grain boundary and begins to penetrate the grains themselves.

P. Emmett *(Portland State University)*

I just wanted to add that you can measure the surface area of these materials very easily by Xenon as you probably know. Since you're not going to use stainless steel any more it probably doesn't make any difference, but you can use Xenon with good accuracy.

V. Haensel *(UOP)*

Gene, one quick question. As far as I can see, you use an empty reactor and you're getting 50% conversion, you must be getting extremely good contact somehow. It is an empty reactor, isn't that right?

Petersen

It is an empty reactor . . . a recirculation reactor. The time you see is of the order of 10 to 20 minutes. Otherwise I'd have a very good catalyst.

N. A. Gjostein *(Ford)*

As probably the only metallurgist in the audience, let me say that I don't think you have to invoke the carbide to explain it, but if you want to test that idea there are stainless steel stabilizers for carbide precipitations of which titanium is an example. So you could check that.

Petersen

I think I'll use quartz and avoid the problem.

KINETICS OF NO REDUCTION BY NH_3 OVER SUPPORTED Pt AND THE EFFECT OF SO_2

J. R. KATZER

University of Delaware, Newark, Delaware

ABSTRACT

Kinetics of the catalytic reduction of NO by NH_3 over supported Pt were determined in a steady-state flow microreactor. The rate of reaction is satisfactorily described by a single-site Langmuir-Hinshelwood kinetic model involving associative NO adsorption and dissociative NH_3 adsorption, with reaction between adsorbed NO and an adsorbed nitrogen fragment appearing to be rate limiting. O_2 markedly enhances the rate of reaction. Auger studies indicate that H_2S and SO_2 form different compounds upon reacting with Pt, Pd and Ni, and the susceptibility of attack by H_2S and SO_2 increases with Pt $\langle$ Pd $\langle\langle$ Ni. The rate of NO reduction by NH_3 is markedly reduced by very low concentrations of SO_2. Pt, Pd, Ru and Ni which show the following activity sequence Pt $\rangle$ Pd $\rangle\rangle$ Ru $\rangle$ Ni, in the absence of SO_2, decrease in activity according to Ru $\rangle\rangle$ (Pt, Pd, Ni) in the presence of SO_2. O_2 markedly enhances the rate of NO reduction over Pt in the presence of SO_2.

BACKGROUND

Selective Reduction — Control of NO_x ($NO + NO_2$) emissions is an important, complex problem. In combustion processes approaches such as staging or generating a reducing agent *in situ* are being widely investigated. A potentially important process for the control of NO_x emissions from chemical plants, and stationary and mobile combustion sources, is catalytic reduction. It has been commercially applied to the reduction of NO_x from nitric acid plants, on a pilot-plant scale to the reduction of NO_x in flue gas from utility boilers and other combustion sources, and on a prototype basis to NO_x reduction in automobile exhaust. Reducing agents include CO, H_2, NH_3, CH_4 and other hydrocarbons; but they generally react more rapidly

References pp. 165-166.

with O_2 than with NO_x. Important exceptions are H_2 under certain limited conditions and NH_3, possibly hydroxyl amine, hydrazine and primary amines, under considerably broader conditions. Thus, the use of CH_4 to abate NO_x from nitric acid plant tail gas requires considerable fuel and two catalyst beds, with interbed cooling to remove heat generated by the reduction of several percent O_2, before significant fractions of the NO_x are reduced.

Using a selective reducing agent such as NH_3 to reduce NO in the presence of O_2, without reacting with a significant fraction of the O_2, eliminates both the large reductant consumption and heat generation problems. Selective reduction should allow the operation of combustion processes at maximum efficiency and the removal of NO_x with minimum amounts of reductant. For example, Andersen, Green and Steele (1) showed that with a Pt catalyst 99% NO removal could be achieved with twice the stoichiometric amount of NH_3 required to reduce the NO in the presence of 10 moles of O_2 per mole NH_3 (Fig. 1). Selectivity remained high to 523 K. There is little quantitative data on the kinetics of selective NO reduction, however.

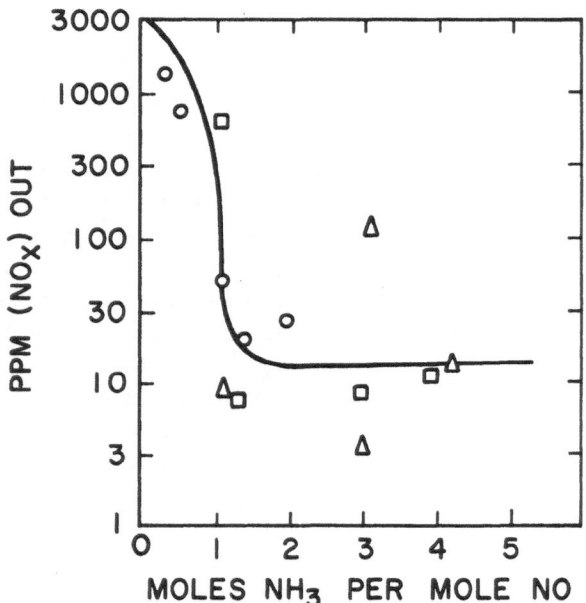

Fig. 1. Selective reduction of NO_x by NH_3 over supported Pt;
       3.4% O_2, 0.28% NO, 0% H_2O, 10 ml cat., inlet temp. 422-439 K, 0 psig
       3.4% O_2, 0.28% NO, 1.6% H_2O, 10 ml cat., inlet temp. 422-439 K, 0 psig
       3.0% O_2, 0.30% NO, 0.66% H_2O, 400 ml cat., inlet temp. 436-453 K, 100 psig
       (catalyst: 0.5% Pt/Al_2O_3, balance of gas: N_2).

Kinetics — In 1939 Michailova (2) studied the kinetics of the NO-NH_3 reaction over a platinum wire in a static reaction system between 500 and 530 K and 2 cm Hg initial total pressure. The reaction rate did not depend on the absolute value of the initial pressure of NO or NH_3 but depended only on their ratio. An observed rate maximum occurred at $P_{NH_3}/P_{NO} = 2.1$, and the calculated activation energy was 24.8 kcal/gmol. No mention was made of N_2O formation; the rate equation for N_2 formation was

$$\frac{d\,P_{N_2}}{dt} = \frac{P_{NH_3}/P_{NO}}{\left(P_{NH_3}/P_{NO} + B\right)^2} . \tag{1}$$

Michailova proposed that oxygen inhibited the reaction since she found that heating the platinum wire in oxygen reduced the observed rate (2).

More recently, Markvart and Pour (3, 4) reported that in flow reactor studies molecular oxygen markedly enhanced the rate of NO reduction by NH_3. They gave no data on N_2O formation.

Gupta (5) studied the kinetics of NO reduction by NH_3 over a 3% Ni − 0.1% Pt on alumina catalyst. The rate of N_2 formation was fitted to a Langmuir-Hinshelwood model involving non-dissociative adsorption of both reactants;

$$r_{N_2} = \frac{k\,K_{NO}\,K_{NH_3}\,P_{NO}\,P_{NH_3}}{\left(1 + K_{NO}\,P_{NO} + K_{NH_3}\,P_{NH_3}\right)^2} . \tag{2}$$

Below 673 K the rate of N_2O formation was fitted to

$$r_{N_2O} = \frac{k\,P_{NO}}{\left(1 + K_{NH_3}\,P_{NH_3}\right)^2} . \tag{3}$$

Although the data were represented fairly well by the rate equations given, the behavior of the constants in the rate expression, for example, indicating endothermic adsorption of reactants, was not consistent with theoretically expected behavior. No attempt was made to relate the behavior to the catalytic chemistry.

The first major insights into the catalytic chemistry of the NO-NH_3 reaction system came from ^{15}N isotope work of Otto, Shelef and Kummer. By tagging N in

References pp. 165-166.

NH_3 and then in NO with the stable isotope ^{15}N, they clearly demonstrated that the catalytic reduction of NO by NH_3 involved the following reaction paths (6-8):

$$NH_3 \rightleftharpoons NH_3 \,_{(ads)} \tag{4}$$

$$NH_3 \,_{(ads)} \rightleftharpoons NH_2 \,_{(ads)} + H_{(ads)} \tag{5}$$

$$NO \rightleftharpoons NO_{(ads)} \tag{6}$$

$$NO_{(ads)} + NH_2 \,_{(ads)}$$
$$\longrightarrow [\text{surface complex I}] \rightarrow N_2 + H_2O \tag{7}$$
$$\dashrightarrow N_2O + 2H_{(ads)}$$

$$H_{(ads)} + NO_{(ads)} \rightarrow HNO_{(ads)} \tag{8}$$

$$2HNO \xrightarrow{(ads)} [\text{surface complex II}] \rightarrow N_2O + H_2O$$
$$+ 2H_{(ads)} \tag{9}$$
$$\dashrightarrow N_2 + 2H_2O$$

Bold arrows in steps (7) and (9) represent major reaction paths, dashed arrows minor reaction paths. Product desorption steps are implicit, surface intermediates were considered plausible, but reaction steps or surface complexes other than those implied here are possible.

NH_3 adsorption is thought to be mostly associative at room temperature (9, 10) with the extent of dissociation to $NH_{2\,(ads)} + H_{(ads)}$ increasing with temperature (10-13).

If the minor reaction pathways in steps (7) and (9) are neglected, the selectivity to N_2 is determined by the material balance on adsorbed NH_2 and H, and r_{N_2}/r_{N_2O} must equal 2.0. Otto, Shelef and Kummer (6, 8, 14) clearly recognized the importance of N_2O and demonstrated that both N_2 and N_2O were primary products by showing that the N_2/N_2O ratio and isotopic distribution in the products were independent of conversion in batch experiments. Because of the involvement of the minor reaction pathways the r_{N_2}/r_{N_2O} ratio was less than 2.0. Similar reaction pathways were also observed for NO-NH_3 reaction over copper (8) and ruthenium (7) indicating a degree of generality to the proposed reaction mechanism.

Otto, Shelef and Kummer (6) showed that the rate of H_2 (as adsorbed H) reaction with NO over Pt was considerably faster than that of NH_3, and in accordance with their surface mechanism reduction with H_2 produced copious quantities of N_2O. They also demonstrated that the N_2O reduction rate was about an order of magnitude slower than the NO reduction rate using NH_3. Apparent diffusional limitations prevented further kinetic analysis of the data.

In a further study using NH_3 and ND_3 a kinetic isotope effect was demonstrated, and the reaction rate showed close to zero-order dependence with respect to both reactants and products (14). However, neither reactant partial pressure nor reactant ratio was varied sufficiently to establish the kinetics. If the catalytic surface is essentially saturated with reactants in a given pressure range, varying total reactant pressure at a constant reactant ratio will have no effect on the observed rate. Zero order indicated that the rate limiting step occurred in the adsorbed layer, and the kinetic isotope effect indicated that it involved breaking N-H [N-D] bonds.

Markvart and Pour (3) give conversion vs. temperature data for Pt, Pd, Ru and Ru/Pd catalysts and reported that the relative order of activities was Pt⟩Pd⟩⟩Ru. The Ru/Pd activity was between that of Pd and Ru. Using H_2 as a reductant Kobylinski and Taylor (15) observed that the order of catalytic activities was Pd⟩Pt⟩Rh⟩Ru. When CO or a CO/H_2 mixture was used, these latter researchers observed that the activity sequence was reversed because of the very strong adsorption of CO on Pt and Pd.

Markvart and Pour (3) observed that for Pt the NO reduction rate increased rapidly with increasing molecular oxygen concentration to about 0.6% O_2 and remained unchanged from 0.6 to 3.2% O_2. For Pd, however, O_2 caused a decrease in the activity. The effect of oxygen on the Pt catalyst was explained on the basis of O_2 removing ammonia fragments from the surface, thus allowing NO to chemisorb and react with reactive adsorbed NH_3 (or NH_2) species.

Sulfur Poisoning — Sulfur has been widely considered a poison for NO reduction catalysts, yet Pt has been a commercial SO_2 oxidation catalyst (16). A review of the literature leads to the conclusion that fundamental and practical work on sulfur poisoning in NO reduction reactions have yet to be fully reconciled. Saleh and coworkers (17-20) have shown by adsorption measurements that H_2S adsorbed dissociatively at low temperatures on Pt apparently forming surface metal sulfides. Above room temperature sulfur was incorporated into Pd and Ni apparently forming bulk sulfides, whereas sulfur was not incorporated into the Pt lattice to a significant extent at temperatures to 523 K. Saleh in 1968 found that sulfur from SO_2 was incorporated into Ni above 373 K presumably forming a bulk sulfide but that sulfur from SO_2 was not incorporated into either palladium (18) or platinum (17) at temperatures to 523 K.

French researchers working with Benard and Oudar (21-27) have shown for a number of metals including Ni and Pt that under reducing conditions (H_2-H_2S) reversible two-dimensional surface sulfides, which exist in a number of substates representing different coverages and symmetries, are formed with increasing H_2S pressure. At still higher H_2S pressures a stable bulk sulfide is formed.

The high stability of metal sulfides under strongly reducing conditions in NO_x removal service suggests that sulfur compounds should poison metal catalysts rather severely. Meguerian et al. (28) found that as little as 0.05 wt % sulfur in the fuel

References pp. 165-166.

resulted in almost complete deactivation of a nickel oxide NO_x removal catalyst at 821 K, and that 0.003 wt % sulfur in the fuel required a 100 K increase in the catalyst temperature to maintain the same catalytic activity found for sulfur-free fuel. High CO levels reduced the effect of sulfur. Above 973 K activity reduction by sulfur is typically not observed (28, 29). Campau *et al.* (30) and Gagliardi, Smith and Weaver (31) also observed poisoning of noble metal catalysts. Gagliardi *et al.* (31) also showed that the extent of deactivation increased with sulfur level and that activity almost completely recovered when sulfur was removed from the fuel. McArthur (32) and Jackson, McArthur and Simpson (33) have shown that 45 ppm SO_2 had little effect on the activity of a Pt-Ni catalyst for NO_x reduction in synthetic exhaust gas mixtures, but SO_2 in concert with lead was exceedingly detrimental. SO_2 deactivated a Pt catalyst much more severely than it did the Pt-Ni catalyst. Similarly, SO_2 severely deactivated a Cu-Ni catalyst; yet, the Pt-Ni catalyst showed little sensitivity to SO_2. Such a positive synergistic effect would not be predicted.

Platinum, an excellent catalyst for converting SO_2 to SO_3 under *oxidizing* conditions (16), is not affected by SO_2 in CO and hydrocarbon oxidation of automotive exhaust (30, 31, 34, 35). Base metal catalysts in the same service are typically severely poisoned by sulfur compounds probably because of the stability of the surface sulfates formed (29, 36).

In 1970 Romeo reported that platinum-group metal catalysts used in the catalytic reduction of NO_x in nitric acid plant tail gas are deactivated by sulfur compounds present in natural gas, naptha and other hydrocarbon fuels (37). The reduction process considered was non-selective with a considerable O_2 concentration in the feed gas, and the poisoning problems alluded to were promotion of Pt crystallite size growth (loss of catalytic surface area) and destruction of the alumina support through $Al_2(SO_4)_3$ formation. No mention was made of the poisoning of surface catalytic activity by the sulfur compounds. Jones *et al.* (38) found that for selective reduction of NO by H_2 in an engine exhaust containing about 1.0% O_2, complete deactivation of both Pt- and Pd-containing catalysts by sulfur (0.02 wt % in fuel) occurred in about 35 hours for a catalyst temperature of 449 K. Meguerian *et al.* (28), however, reported for a supported nickel catalyst that O_2 was a more severe poison than sulfur.

NO-NH$_3$ REDUCTION KINETICS

Objectives — The literature provides sufficient information to develop several qualitative models of the kinetics of the reduction of NO by NH_3 supported by fundamental information on the reaction mechanism. The task remaining is to synthesize this information into appropriate models; to generate sufficiently good rate data to discriminate between models; and then to determine the kinetic parameters in the appropriate model. Similarly, the literature provides significant information on the interaction of sulfur compounds with metals and on the NO_x reduction behavior of these metals in the presence of SO_2, but integration of the two types of information is required.

The objectives of this work were to quantitatively measure the rate of reduction of NO by NH_3, to model the data with realistic kinetic equations and to determine the effect of O_2 and SO_2 on the rate of reduction. This has been done in detail for supported platinum, and other precious metals have been compared with respect to their activity and the effect of SO_2 on them. Auger studies of H_2S and SO_2 interactions with Pt, Pd and Ni were carried out to determine if metal properties measured in this way could be correlated with catalytic properties in the presence of sulfur compounds. Of necessity this work has involved relatively short term reaction studies, but we believe that the findings should be useful in the real world of abatement catalysts with its requirement of long catalyst life.

Apparatus and Procedure — The apparatus consisted of a manifold system for measuring and controlling the flow of premixed gases to the reactor; a packed bed, flow, microreactor operated at atmospheric pressure and immersed in a fluidized sand bath for temperature control ($\pm$ 1°C) (39). A thermocouple in the catalyst bed demonstrated equality with sand-bath temperature. The system was entirely of 316 stainless steel. A dual-column gas chromatograph with dual gas sample valves permitted samples of effluent gas to be analyzed with either of the two columns allowing complete separation of components and determination of point rates of reaction.

Because oxygen enhances the rate of $NO-NH_3$ reaction, extreme care was taken to eliminate it from the system, and tests indicated less than one ppm O_2 under reaction conditions (3, 4, 40). Freshly prepared catalyst as 30-40 mesh range particles was reduced under flowing hydrogen by heating at 1 K/min to 623 and 773 K for the unsintered and sintered catalyst (described below) and holding for 12 hr before placing in the reactor. Rereduction in the reactor did not affect the steady-state activity of the catalyst although it did affect initial activity. Typically several tenths of a gram of catalyst were used. Space velocities based on the feed stream at 298 K were varied from 170,000 to 390,000 hr^1 for the unsintered catalyst and from 13,500 to 154,000 hr^1 for the sintered catalyst. Conversion of limiting reactant was maintained below 10% in all cases, and differential reactor operation was demonstrated in this region by showing that rate was independent of space velocity. Because the catalysts underwent deactivation to steady-state activity in about 12 hours, they were always deactivated under 0.25% NO and 2.0% NH_3 for 18 hr. Kinetic data were obtained with one-half hour allowed for equilibration after each concentration change. The rate at the conditions of the initial deactivation was shown to reproduce after all kinetic data were obtained. Concentration ranges of 0.25 to 1.0% NO and 0.1 to 2.0% NH_3 were studied at 423, 448, and 473 K.

For the $NO-NH_3-O_2$ system the stream containing O_2 was mixed with other reactants just prior to the catalyst bed to prevent formation of reactive NO_2 prior to the catalyst, and a liquid nitrogen cold trap placed immediately behind the reactor trapped out NH_3 and NO preventing the formation of NO_2 and the subsequent formation of significant amounts of N_2 from gas-phase reaction between NH_3 and NO_2. A run employing an equivalent weight of platinum-free alumina in the reactor quantitatively demonstrated that the amount of N_2 formed independent of platinum was small. N_2O was not produced without the platinum catalysts. Space velocities of 680,000 and 857,000 hr^{-1} were used for the unsintered and sintered catalysts. The feed composition was 0.25% NO, 2% NH_3 and 0.5% O_2 in helium, and the temperature was 423 K.

All analyses were by gas chromatography. Nitrogen and nitric oxide were separated by a 4 m Chromasorb 104 column, and SO_2 was separated from the other species by a 2 m Chromasorb 104 column. Both columns were operated at room temperature. The oxygen-nitrogen separation was achieved with a 2 m 5A molecular sieve column operated at 313 K.

Rates are given as turnover numbers which are defined as the number of g moles of N_2 or N_2O produced per second per gram mole of calculated surface sites. Each surface site was assumed to consist of two surface platinum atoms where surface platinum atoms were determined by hydrogen chemisorption. For catalysts that we have not fully characterized the rates are per gram of catalyst.

Materials — High purity helium ($\langle$ 20 ppm O_2), and 10% NO ($\langle$ 40 ppm air), 6% NH_3 ($\langle$ 25 ppm N_2) and 25% high purity O_2 in high purity helium were used. Helium was passed through a heated (473 K), reduced copper oxide catalyst bed followed by a 4A molecular sieve column to remove O_2 and H_2O. H_2 was electrolytic grade, further purified with a reduced copper catalyst and molecular sieve unit.

Three Pt catalysts were studied. Two were 1 wt % Pt on alumina prepared and characterized by Ostermaier (41). The unsintered catalyst had a platinum dispersion of 0.38 corresponding to a surface average platinum crystallite size of 2.7 nm assuming a cube with five sides available for hydrogen chemisorption (42) and a 1:1 hydrogen atom to surface platinum atom stoichiometry (43, 44). It chemisorbed 9.83 x 10^{-6} g mole H_2/g catalyst. Electron micrographs indicated that essentially all the platinum crystallites fell between 2.0 and 3.0 nm. A portion of this catalyst was sintered in air at 923 K for 24 hr causing a decrease in H_2 chemisorption to 1.71 x 10^{-6} g mole H_2/g catalyst and in platinum dispersion to 0.065 corresponding to a surface average platinum crystallite size of 15.5 nm. X-ray indicated a volume average crystallite size of 14 nm and electron micrographs a crystallite size range 8.0 to 20 nm. The third catalyst was a commercial 0.5% Pt on alumina catalyst (referred to as type M). It chemisorbed 2.78 x 10^{-6} g mole H_2/g catalyst and had a dispersion of 0.236 corresponding to a crystallite size of 4.3 nm.

Results and Discussion — Nitric Oxide-Ammonia System

Deactivation — At 473 K no measurable reaction occurred in either the empty reactor or the reactor filled with alumina. Reproducibility of rate across the entire concentration range was typically better than 15% for independent runs carried out several weeks apart. Diffusion calculations indicated mass transfer limitations were not present.

A marked decline in activity occurred during the first few hours on stream (39). Catalyst activity typically decreased by a factor of about 3.3 independent of temperature when steady-state activity was reached. Fig. 2 shows typical deactivation behavior for the unsintered 1% Pt catalyst after *in situ* reduction and illustrates the effect of initial surface condition on behavior. The rate of N_2 formation was independent of the presence of hydrogen on the catalyst surface. N_2O formation underwent a large maximum for a hydrogen covered surface. Initial rates of N_2 and N_2O formation were about twice as high when the surface was contacted with oxygen at room temperature after reduction, giving a monolayer of chemisorbed oxygen.

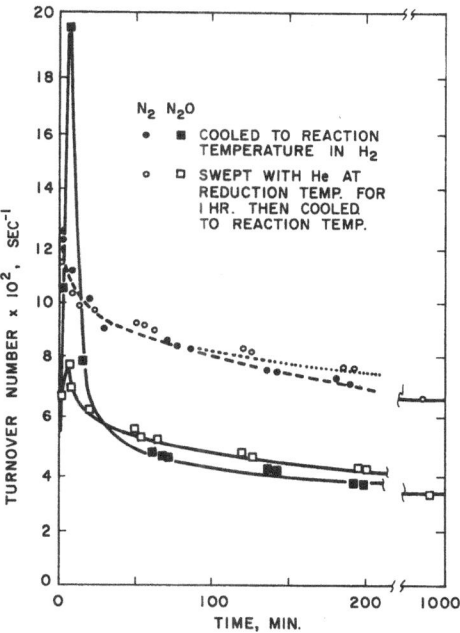

Fig. 2. Effect of reduction conditions on initial rate behavior; unsintered 1% Pt/Al_2O_3, 0.25% NO, 2% NH_3, 473 K, reduced for 14 hr at 623 K.

The enhanced N_2O production is consistent with the mechanism of Otto, Shelef and Kummer (6) (Equations (8) and (9)) and their observation of enhanced N_2O production in reduction of NO with H_2 over Pt at 473 K. In contrast, Otto and Shelef (7) reported that at 195 and 173 K adsorbed hydrogen on platinum reduced NO almost exclusively to N_2.

Deactivation occurred only under reaction conditions; the presence of oxygen on the surface resulted in enhancement of initial activity, and flowing NH_3 or NO

References pp. 165-166.

separately over the catalyst did not significantly reduce the initial activity. Passing NH_3 or H_2 in He over the deactivated catalyst at reaction temperature resulted in complete regeneration of catalytic activity, and deactivation under the standard reaction feed gas thereafter followed the same curve as for fresh catalyst. Regeneration of activity could also be achieved by heating above 400°C in flowing He. Activity was not restored by treatment under flowing NO. This behavior suggests that the deactivation resulted from the formation of an oxide layer on the platinum.

The deactivation observed apparently results from the formation of a platinum oxide during reaction; this oxide is different from chemisorbed oxygen on the surface of the metal. Further, this oxide formation occurs in a system which is overall reducing (0.25% NO – 25% NH_3).

The rapid initial catalyst deactivation observed and the time required to reach steady-state activity imply great difficulty in obtaining quantitative kinetic data from batchwise experiments such as those reported previously since such experiments give the superposition of two transient events. Continuous flow experiments should provide more quantitative kinetics.

Rates and Selectivity – Table 1 gives initial specific catalytic activities, obtained by extrapolation of semi-log plots of rate to zero time, and the extents of deactivation observed. Initial specific catalytic activity is independent of platinum crystallite size within the estimated error of the zero-time extrapolation of ±25%.

TABLE 1

Initial Specific Catalytic Activities and Extents
of Deactivation for Nitrogen Formation*

Temp., K	Catalyst	Initial Activity, Turnover no., s^{-1}	r_I/r_{ss}
423	Sintered 1% Pt	0.015	3.4
448	Unsintered 1% Pt	0.08	3.5
448	Sintered 1% Pt	0.07	3.0
473	Unsintered 1% Pt	0.22	3.3
473	Sintered 1% Pt	0.23	5.3
473	Type M	0.33	3.1

Feed composition 0.25% NO, 2.0% NH_3 in helium. Estimated error in initial rate was ±25%.

Fig. 3 summarizes the effect of temperature and platinum crystallite size on the steady-state specific catalytic activity of the three catalysts studied for the NO-NH_3 reaction. At 473 K the initial specific catalytic activity of the two catalysts is the same, but the sintered catalyst appeared to deactivate more severely. Since the data for the unsintered catalyst were duplicated and reproduced well, while those for the sintered catalyst were not repeated, the steady-state results for the sintered catalyst are considered less reliable. Type M catalyst, which has an average platinum crystallite size 1.6 times that of the unsintered catalyst, gave the same specific catalytic activity as the unsintered catalyst. We conclude from these data that specific catalytic activity is independent of crystallite size in the $NO-NH_3$ system.

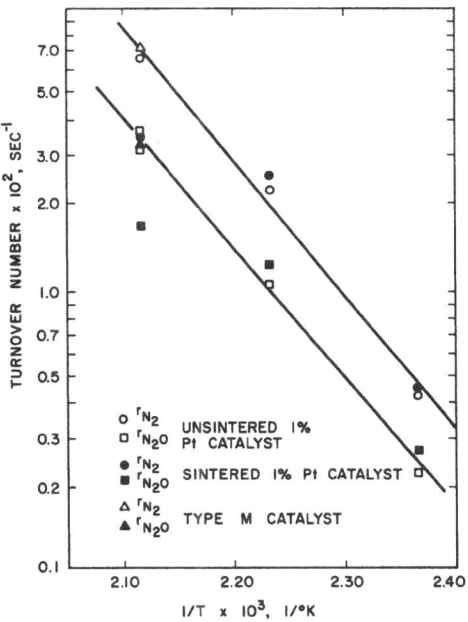

Fig. 3. Arrhenius temperature dependence of observed steady-state rate of reaction for 0.25% NO and 2.0% NH_3 in helium.

The specific catalytic activity in the reduction of NO by NH_3 over Pt shows no dependence on crystallite size and thus should be considered a facile reaction (45). The absence of a crystallite size effect indicates that reaction involves rather simple sites on the metal surface and is thus insensitive to changes in surface detail which occur with changes in crystallite size. This behavior is to be contrasted with the large crystallite size effect observed in the low-temperature oxidation of NH_3 to N_2 and N_2O over these same catalysts (41).

The activation energies calculated from Fig. 3 are 21.6 ± 2 kcal/mol for N_2 formation and $20.8 \pm$ kcal/mol for N_2O formation excluding the 473 K data for the sintered catalyst. The difference is not significant, and the value calculated from the initial rate data is essentially the same. The value reported by Michialova (2) (24.6 kcal per g mol) is remarkably close to our values.

Steady-state selectivity to N_2, defined as the ratio r_{N_2}/r_{N_2O}, is independent of platinum crystallite size, feed gas composition and temperature (see Table 2). Our

References pp. 165-166.

selectivity data show better agreement with the simplified version of the mechanism proposed by Otto *et al.* (6) than do their own data, probably partly because of the lower temperature and lower concentrations used here and because our results are under steady-state conditions. Our data during deactivation for *in situ* reduced surfaces showed selectivities characteristic of theirs.

TABLE 2

Steady-State Selectivity to Nitrogen for
NO-NH$_3$ Reaction over Platinum

Temp., K	Catalyst	Average Selectivity*, r_{N_2}/r_{N_2O}	Standard Deviation
423	Unsintered 1% Pt	1.99	0.14
423	Sintered 1% Pt	1.70	0.14
448	Unsintered 1% Pt	1.78	0.04
448	Sintered 1% Pt	1.92	0.04
473	Unsintered 1% Pt	1.81	0.13
473	Sintered 1% Pt	1.79	0.27
473	Type M	1.86	0.17

Average over 12 feed compositions covering 0.25 to 1% NO and 0.1 to 2% NH$_3$ in helium. No trend with feed composition was observed.

Kinetic Model — The rate increased with both NH$_3$ and NO partial pressure at low partial pressures and became weakly dependent on partial pressure at higher pressures indicating that the rate limiting step was a surface step as suggested by Otto *et al.* (14). A number of kinetic models can be derived from the reaction pathways and mechanism proposed by Otto *et al.* (6) depending on which steps are assumed to be rate limiting. If it is assumed that reaction steps (7) and (8) are rate limiting, a Langmuir-Hinshelwood type kinetic model [Equation (10)] can be readily derived by setting $r_{N_2} = k_7 \, \theta_{NH_2} \, \theta_{NO}$, writing a steady-state balance on θ_H, and expressing surface coverages in terms of gas-phase partial pressures using a Langmuir isotherm (46). The resultant equation is

$$r_{N_2} = \frac{k_{s,N_2} K_{NO} K_{NH_3}^{1/2} P_{NO} P_{NH_3}^{1/2}}{\left(1 + K_{NO}P_{NO} + K_{NH_3}^{1/2} P_{NH_3}^{1/2}\right)^2} . \tag{10}$$

Since the r_{N_2}/r_{N_2O} ratio was constant at steady state, the rate of N$_2$O formation should be given by a similar equation:

$$r_{N_2O} = \frac{k_{s,N_2O} K_{NO} K_{NH_3}^{1/2} P_{NO} P_{NH_3}^{1/2}}{\left[1 + K_{NO}P_{NO} + K_{NH_3}^{1/2} P_{NH_3}^{1/2}\right]^2} . \tag{11}$$

K_{NH_3} is a product of K'_{NH_3}, K_{s4} and the ratio of surface rate constants for steps (7) and (8). Other rate equations may be derived based on other assumptions of rate limiting steps or other mechanisms.

The NO-NH$_3$ rate data, which consisted of a 12 point rate-concentration grid for each catalyst and each temperature, were fitted to this and several other Langmuir-Hinshelwood kinetic models using a non-linear, least squares fitting technique. The above model and one derived assuming that NO and NH$_3$ adsorbed on separate sites were the only models which gave good fits as judged by standard error between observed and predicted rate;

$$\text{standard error} = \left\{ \sum_{i=1}^{n} (r_{obs,i} - r_{pre,i})^2/(n-k) \right\}^{1/2} .$$

The single-site model, represented by Equations (10) and (11), was chosen because the absence of crystallite size effect indicated the reaction involved only simple, unspecific sites and thus the requirement of two different sites seemed unlikely. The standard error was less than 10% of the rate for all conditions except with the sintered catalyst at 473 K and the unsintered catalyst at 423 K where the error was $\sim 20\%$. Plots of rate of N$_2$ formation vs. the square root of NH$_3$ partial pressure gave straight lines passing through the origin, whereas plots of the rate of N$_2$ formation vs. NH$_3$ partial pressure were curved and had a large positive intercept. The rate clearly showed half-order dependence on NH$_3$ partial pressure at low NH$_3$ partial pressures.

The goodness of fit does not verify the assumed mechanism and rate limiting step but only suggests plausibility; however, mechanisms and rate limiting steps which give rate models that did not fit the data are eliminated.

Additional support for the model based on reaction step (7) as rate limiting comes from a reevaluation of the data of Michailova (2) by Wanner (40) in our laboratory. Her data fit Equation (10) about as well as our data did. Also in favor of reaction step (7) as a rate limiting step is the observation that the reaction of H$_2$ with NO is more rapid than that of NH$_3$ (6) suggesting that reaction steps (8) and (9) are not rate limiting. The apparently satisfactory agreement of the observed kinetics with a model based on reaction step (7) as rate limiting suggests that this step may indeed be the important step kinetically.

Fig. 4 shows the temperature dependence of the surface rate constants for N$_2$ and N$_2$O formation for the unsintered and sintered catalysts. The results are correlated by the equations:

$$k_{s,N_2} = 1.37 \times 10^5 \ e^{-11,800 \ (\pm 3000)/RT} \ _s\text{-}1 \tag{12}$$

$$k_{s,N_2O} = 1.56 \times 10^4 \ e^{-10,300 \ (\pm 1000)/RT} \ _s\text{-}1 \tag{13}$$

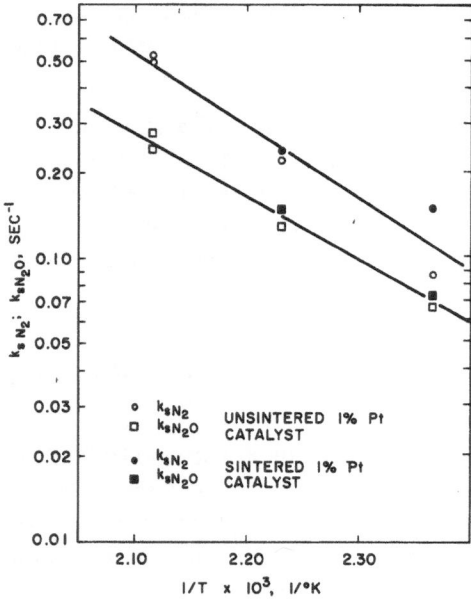

Fig. 4. Arrhenius temperature dependence of surface reaction rate constants for single-site Langmuir-Hinshelwood kinetic model fit to NO-NH$_3$ data for 1% Pt catalysts.

Fig. 5 shows the temperature dependence of the adsorption constants from both Equation (10) and (11) for NO and NH$_3$ for the unsintered and sintered catalysts. The NO adsorption coefficients for the unsintered and sintered catalysts agree very well, indicating independence of crystallite size. The NH$_3$ adsorption parameter agrees reasonably well with the exception of 423 K. The NH$_3$ adsorption parameter is the square of the computed parameter so that scatter in the computed parameter is magnified. Further, because the contribution of the NH$_3$ adsorption term in the denominator is small relative to that for NO, the term does not become dominant at low temperatures even at the highest NH$_3$ concentrations. This results in a coupling of k_s and $(K_{NH_3})^{1/2}$ and an inverse correlation between these parameters. This caused some difficulty in converging on realistic values of k_s and K_{NH_3} at 423 K. The data are correlated by the equations:

$$K_{NO} \quad 4.89 \times 10^7 \, e^{-11,300 \, (\pm \, 1,000)/RT} \, _{atm} \cdot 1 \tag{14}$$

$$K_{NH_3} = 1.04 \times 10^{13} \, e^{-23,500 \, (\pm \, 4,000)/RT} \, _{atm} \cdot 1 \tag{15}$$

A run at 548 K and 0.25% NO exhibited a rate maximum at about 0.1% NH$_3$ and weak dependence on NH$_3$ partial pressure between 0.5 and 0.07% NH$_3$, indicating that K_{NH_3} is quite large at this temperature.

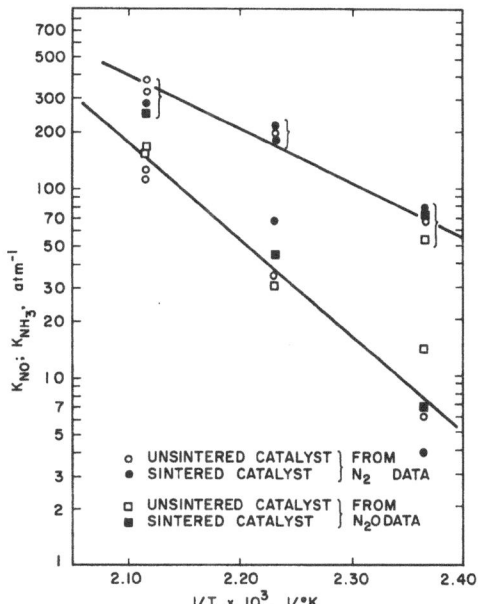

Fig. 5. Arrhenius plot of NH₃ and NO adsorption parameters for single-site Langmuir-Hinshelwood kinetic model fit to NO-NH₃ data for 1% Pt catalysts. Bracketed data represent K_{NO} values.

Since K_{NH_3} in Equation (10) contains a ratio of rate constants, a positive temperature coefficient (23,500 cal/mol) is possible. The positive temperature coefficient for K_{NO} is consistent with positive values reported by others (5, 47) although the reason for this is not clear.

If the rate of surface dissociation of NH₃ were rate limiting as suggested by Otto *et al.* (1971), the observed rate of reaction should be first order in NH₃, zero order in NO, and should tend to minus first order in NO at high NO pressure. This was not observed. The apparent zero order in both reactants observed by Otto *et al.* (14) is explained by the fact that the rate of reaction is near its maximum as observed by us and thus the surface is essentially covered by reactant molecules at the reactant concentrations used in their study. Thus, the rate is quite insensitive to changes in reactant concentration over their range of concentrations and even less sensitive to changes in concentration at constant ratio of reactants. Other surface reaction steps can give a kinetic isotope effect, and thus, this need not mean that NH₃ dissociation is rate limiting.

Results and Discussion — Nitric Oxide — Ammonia-Oxygen System

Kinetics and Selectivity — The NO-NH₃-O₂ system was studied at 523 K with a varying feed composition (40) and at 423 K with a constant feed composition of

References pp. 165-166.

0.25% NO, 2.0% NH_3, and 0.5% O_2 (39). No deactivation occurred with O_2 in the feed gas. The rate enhancement by O_2 was similar to that reported by Markvart and Pour (3, 4).

The higher temperature studies involved only the type M catalyst. The rates of N_2 and N_2O formation both increased with one-half power of O_2 partial pressure with no further rate increases occurring for O_2 partial pressures above about 1% O_2. One-half order dependence indicates dissociative adsorption of O_2 which is consistent with the expected mode of O_2 chemisorption on Pt. The data of Markvart and Pour (3) are also indicative of one-half order dependence of NO removal on O_2 partial pressure with no further increase being observed above 0.5% O_2.

At 423 K when 0.5% O_2 was added to the feed, the rate of N_2 formation for the unsintered catalyst increased by a factor of 12.5 and that of N_2O by a factor of 22. For the sintered catalyst the rates increased by factors of 66 and 117. The data are summarized in Table 3. The ratio of steady-state specific catalytic activity of the sintered catalyst to that for the unsintered catalyst is 5.8 and 6.6 for N_2 and N_2O formation respectively. Reproducibility was demonstrated with a separate catalyst charge. Thus a crystallite size effect is present indicating that the reaction becomes demanding in the presence of O_2. The selectivity to N_2 (r_{N_2}/r_{N_2O}) was 0.96 for the sintered catalyst and 1.11 for the unsintered catalyst with 0.5% O_2 in the feed. This value is about one-half that without O_2.

Table 4 gives the initial specific rate of formation of unfixed nitrogen (r_{N_2} + r_{N_2O}) at 423 K for feeds containing 2% NH_3 and either 0.5% O_2, or 0.25% NO, or 0.5% O_2 and 0.25% NO in helium. Any effect of deactivation has been eliminated by comparing initial specific rates. The NH_3-O_2 data are from Ostermaier et al. (41). NH_3 reacts with O_2 and NO separately at about the same rate, particularly over the

TABLE 3

Effect of Oxygen on Reaction Between
Nitric Oxide and Ammonia at 423 K

Catalyst	Feed Composition, mole %	Steady-state Specific Rate of Reaction, Turnover No., s^{-1}		Selectivity r_{N_2}/r_{N_2O}
		N_2	N_2O	
Unsintered 1% Pt	2.0% NH_3 0.25% NO 0.0% O_2	0.0042	0.0022	1.9
Sintered 1% Pt		0.0045	0.0027	1.7
Unsintered 1% Pt	2.0% NH_3 0.25% NO 0.5% O_2	0.052	0.047	1.11
Sintered 1% Pt		0.30	0.31	0.96

unsintered catalyst. When O_2 and NO are both present in the feed, the rate is markedly higher. This could not be predicted from the two-reactant feed data. Further, under these conditions NH_3 should be reacting selectively with NO. The sum of rates of formation of unfixed nitrogen for the unsintered catalyst in the NH_3-O_2 and NH_3-NO systems is 42% of the rate for the same catalyst in the NH_3-NO-O_2 system. For the sintered catalyst the sum of rates for the first two systems is 12% of the rate in the NH_3-NO-O_2 system.

TABLE 4

Effect of Different Reaction Components on Rate of
Formation of Unfixed Nitrogen at 423 K

System	Feed Composition, mole %		Catalyst	Initial Specific Rate*, Turnover No., s^{-1}
NH_3-O_2**	{	2% NH_3 0.5% O_2	Unsintered 1% Pt Sintered 1% Pt	0.014 0.049
NH_3-NO	{	2% NH_3 0.25% NO	Unsintered 1% Pt Sintered 1% Pt	0.022 0.022
NH_3-NO-O_2	{	2% NH_3 0.25% NO 0.5% O_2	Unsintered 1% Pt Sintered 1% Pt	0.099 0.61

*Rate equals $(r_{N_2} + r_{N_2O})$, initial rates were obtained by extrapolation to zero time. For NH_3-NO-O_2 system no deactivation was observed.
**From Ostermaier et al. (41).

The rate enhancement caused by O_2 is consistent with that reported by Markvart and Pour (3, 4); our report of the effect of O_2 on selectivity is new. The literature offers no satisfactory explanation for the rate enhancement by the presence of O_2. Selectivities with O_2 in the feed (Table 3) are not due to gas-phase reactions since these produce only N_2 below 473 K (48) and only small amounts of N_2O above 473 K (49). Only small amounts of N_2 were produced in the absence of the platinum catalyst. Homogeneous N_2 production was subtracted from the observed rates of N_2 formation.

In the presence of O_2 the following additional reaction steps are proposed (41, 50):

$$O_2 \rightleftharpoons 2O \text{ (ads)} \tag{16}$$

$$O \text{ (ads)} + NH_2 \text{ (ads)} \rightarrow NH_2O \text{ (ads)} \tag{17}$$

$$NH_2O \text{ (ads)} + O \text{ (ads)} \rightarrow NO \text{ (ads)} + H_2O \tag{18}$$

References pp. 165-166.

Surface species are speculative, but since NO is observed to be the first product of interaction between NH_3 and O_2 on platinum, more complex reaction mechanisms are not indicated (12, 50). Half-order dependence on O_2 partial pressure observed in low-temperature NH_3 oxidation (41) and here in the NO-NH_3-O_2 reaction supports step (16).

The decreased selectivity to N_2 in the presence of gas-phase O_2 may result from the intervention of oxygen in the reaction via steps (17) and (18) to increase the amount of H(ads) available on the surface for steps (8) and (9) eliminating the material balance requirement that $r_{N_2}/r_{N_2O} = 2.0$ as observed in the absence of O_2. All additional H(ads), resulting via step (5) from steps (17) and (18), which is not oxidized to H_2O by adsorbed oxygen should result in additional N_2O. Thus r_{N_2}/r_{N_2O} can be considerably less than 2.0. This is consistent with the observations that H_2 reacts selectively with NO in the presence of O_2 (38, 51); that selectivities to N_2 which are much less than one were observed when gas-phase H_2 is added to the NO-NH_3 system (6); and that starting a run with the platinum surface covered with chemisorbed hydrogen results in high rates of N_2O formation relative to N_2 formation initially. If the O_2 had reacted with a large fraction of the H(ads), the selectivity to N_2 would have increased.

Crystallite Size Effect — Since the rate limiting step in the NH_3-NO system appears to be step (7) and since this system does not show a crystallite size effect, the crystallite size effect in the presence of O_2 appears to enter earlier in the reaction network. Steps (17) and (18) are the most logical candidates for kinetically important steps which may exhibit crystallite size dependence. Since steps (8) and (9) appear rapid relative to step (7), steps (17) and (18) may provide another mechanism of reducing the concentration of NH_2(ads) and altering the relative concentrations of adsorbed NH_2 and NO-like species. This shift in relative surface concentrations could explain the rate enhancement by O_2 and also explain the crystallite size effect in a manner consistent with that in the NH_3-O_2 system (41). Another explanation involves the claimed formation of reactive NO_2 from O_2 and NO over platinum (52). The mechanism of the O_2 effect is not clear, and its magnitude suggests that steps (16) through (18) may be simplistic.

INTERACTION OF SULFUR COMPOUNDS WITH METAL FOILS

Apparatus and Procedure — The interaction of H_2S and SO_2 with polycrystalline foils of Pt, Pd and Ni was studied to provide comparison between metals. Surface analysis and component depth profiles, particularly of sulfur, were determined with a Physical Electronics Auger Electron Spectrometer with an electron gun for argon ion sputtering. The maximum escape depth of Auger electrons is 3-8Å (53).

Auger spectra were taken at low beam current to reduce surface effects caused by the beam (54, 55), and during sputtering a multiplex system allowed almost continuous monitoring of the several desired peaks. The sputtering current was 10 milliamps. The spectra were analyzed by determining peak to peak heights for the desired peaks and dividing these by the peak to peak height of the best situated (usually most intense) metal peak giving S/M ratios, thus normalizing the data for each metal. This removed variations in absolute sulfur peak intensity due to differences in focusing and in Auger spectrometer settings. However, since the intensity of the metal peak was different for each metal, the same sulfur coverage gave quite different S/M ratios on each metal.

Foils of the metals were treated in a pyrex tube placed in a muffle furnace. The treating tube was connected via high-vacuum stopcocks to a gas manifold system and vacuum pump to allow evacuation and contacting with desired gas environments in a flow system.

Foils were held in a porcelain boat in the treating tube and were first cleaned by treatment under flowing O_2 at 773 K for one hour. Samples were then reduced under flowing hydrogen at 773 K for one hour and then flushed with He. Experiments demonstrated that this treatment resulted in substantially complete removal of sulfur and carbon originally present on the foils. The foils were then contacted with H_2S or SO_2 at a partial pressure of 4mm Hg in flowing He at the desired temperature for one hour. The tube was then flushed with He and removed from the oven to cool. All three metals were treated simultaneously to ensure that observed differences were due to metal behavior and not to different conditions experienced during contacting.

Samples were then quickly transferred to the Auger spectrometer, and the unit was evacuated. The transfer procedure did not result in significant surface contamination by sulfur or carbon as shown by our experiments with cleaned, untreated foils and as shown by Pignet, Schmidt and Jarvis (56). Cleaned, untreated foils after transfer had a relatively strong oxygen peak probably resulting from oxygen picked up during transfer. This peak could be reduced to zero in less than one minute of sputtering at 10 mA, indicating that it was only surface oxygen. After treating with H_2S and transferring, an oxygen signal, which was approximately zero for the maximum sulfur signals observed and larger for smaller sulfur signals, was present. For Pt and Pd, sputtering typically reduced the oxygen signal to zero very rapidly and did not result in an increase in the sulfur signal indicating that oxygen had not altered the surface sulfur concentration. Ni was particularly sensitive to oxygen and unless extreme care was taken during the transfer the surface sulfur signal was significantly reduced by oxygen as indicated by sputtering.

We, therefore, cannot rule out the possibility that contact with the atmosphere during transfer resulted in changes in surface composition, and future work will be done entirely within the Auger spectrometer to avoid this problem. However the emphasis of this study was to compare the metals rather than to obtain quantitative numbers, and we believe that our observations are valid in this context.

Results and Discussion — The effect of H_2S and SO_2 on the various metals was different enough that it is useful to discuss the results for each gas separately.

References pp. 165-166.

H₂S Surface Composition — The surface composition of Pt, Pd and Ni as a function of temperature of contact with H_2S varied significantly as shown in Fig. 6. The data at 523 K represent apparent complete coverage of the surface with sulfur as indicated by sulfur peak-to-peak values of 69, 62 and 67 for Pt, Pd and Ni. The data points fall at different values because of the difference in the metal signal. The results for Ni are the same at all temperatures although oxygen picked-up during transfer occasionally reduced the observed surface sulfur level. Other data suggest that a bulk sulfide was formed in all cases. With Pd the S/M ratio decreases below 373 K indicating less than monolayer coverage by sulfur at 310 K. With Pt there is a marked change in the S/M ratio with temperature. The value of 0.53 at 310 K was considerably higher than the typical background sulfur level at about 0.03 for "cleaned" foils. Between 310 and 523 K the S/M ratio increased about 11 fold for Pt.

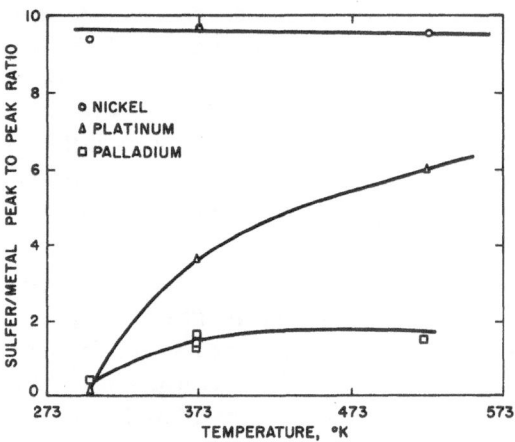

Fig. 6. Effect of treatment temperature on surface sulfur concentration for one hour contact time with 4 mm H_2S pressure in helium.

Our data for Ni and Pt appear to be consistent with results of other investigators. The work of Saleh (19) showing ready incorporation of sulfur into the Ni lattice at low temperatures suggests that Ni should have at least a monolayer of sulfur as a result of our H_2S treatments, whereas only surface coverage with some reversible adsorption would be expected with Pt (17). Our observation of reduced sulfur concentration at 310 K for Pd is unexpected in view of Saleh's finding (18) that sulfur incorporation into Pd lattice occurred at and above room temperature. We speculate that a kinetic effect was operative here. Since the Auger samples somewhat more than one surface layer, a low sulfur signal could result if the H_2S treatment time for Pd at 310 K was too short for more than a surface sulfide layer to form. Higher temperatures would mean more penetration of sulfur and a stronger signal.

Our observation that sulfided nickel samples were easily covered with oxygen, while this did not appear to be a problem with platinum, also is consistent with the literature. Holloway and Hudson (57) have shown that sulfur appeared to be more easily removed from nickel by oxygen than from platinum. Bonzel and Ku (58) showed that the removal of sulfur from Pt by oxygen under CO oxidation conditions was very slow below 473 K.

H_2S *Component Depth Profiles* — The extent (depth) of sulfide formation was determined by argon ion sputtering at a rate estimated to be between 25 and 50Å per minute. Fig. 7 shows the time required to sputter to a sufficient depth that the sulfur peak is essentially zero.

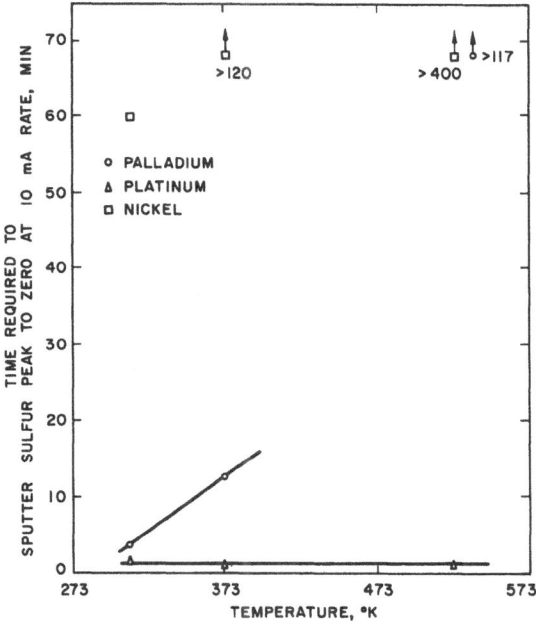

Fig. 7. Effect of temperature of one hour contact with H_2S on time required to sputter sulfur peak to zero at 10 mA rate.

Ni formed extensive sulfide layers from H_2S contact; the sample treated at 310 K required 60 min to sputter the sulfur peak to zero (Fig. 7). Samples treated at 375 K and 523 K showed only slightly diminished sulfur peaks after sputtering for 120 and 400 min respectively. The extent of sulfiding of Pd at 310 and 373 K was much less (Fig. 7); the sulfur peak was reduced to zero in 12 min. However, sulfiding at 523 K was so extensive that the sulfur signal was still present after sputtering for 117 minutes. If the H_2S treatment time at 373 K was increased from 1 to 3 hr, the sulfur removal time about doubled (Fig. 8).

Pt did not undergo significant incorporation of sulfur under any of our treatment conditions (Figs. 7 and 8).

References pp. 165-166.

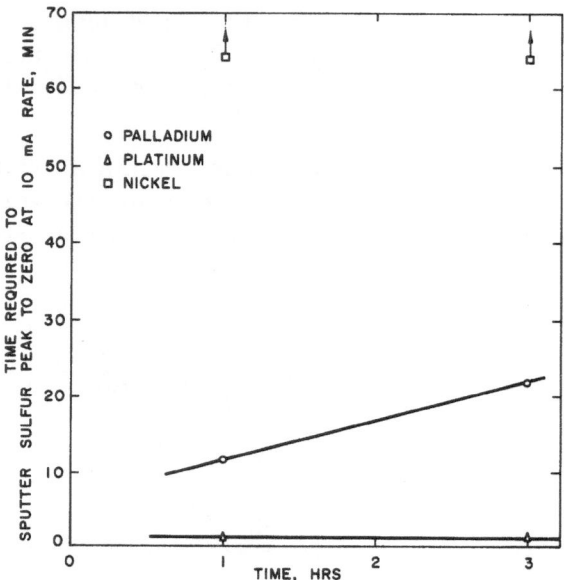

Fig. 8. Effect of H_2S contact time on time required to sputter the sulfur peak to zero for 4 mm H_2S at 273 K.

Fig. 9 shows typical component depth profiles in terms of sputtering time for Pd and Pt after contact with 4mm H_2S for one hour. Sensitivities represent signal at full

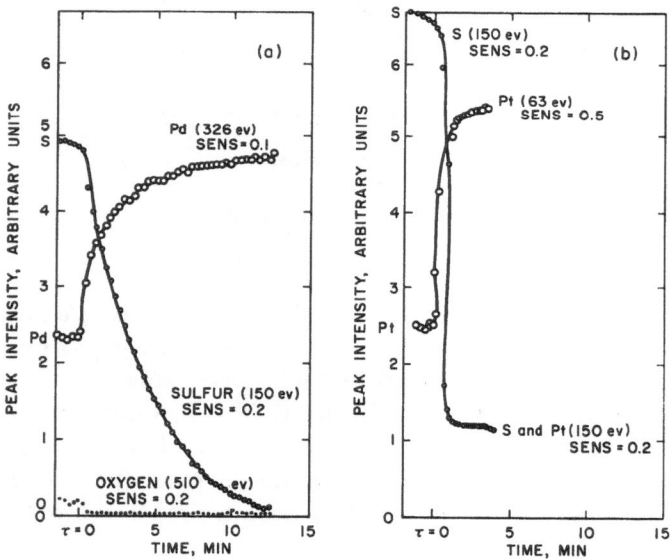

Fig. 9. Component depth profiles for palladium and platinum after one hour treatment with 4 mm H_2S; (a) Pd contacted at 373 K, (b) Pt contacted at 523 K.

scale. Pd showed considerable incorporation of sulfur into the lattice, whereas Pt did not as indicated by the very sharp drop in the sulfur signal upon sputtering. A weak Pt peak occurs at the same location as the sulfur peak (150 ev), and thus, this peak does not sputter to zero as with Pd.

These results clearly show the ease of sulfiding with H_2S is in the order

$$Ni \rangle Pd \rangle\rangle Pt$$

and are in agreement with results of Saleh and coworkers (17-20).

Treatment with SO_2 — Contact with SO_2 had some markedly different effects from that of H_2S. Ni showed oxygen incorporation rather than sulfur incorporation (Fig. 10), but the results were somewhat eratic. Two of the five samples treated at 523 K showed a sulfur concentration peak immediately below the surface; the remainder were similar to Fig. 10.

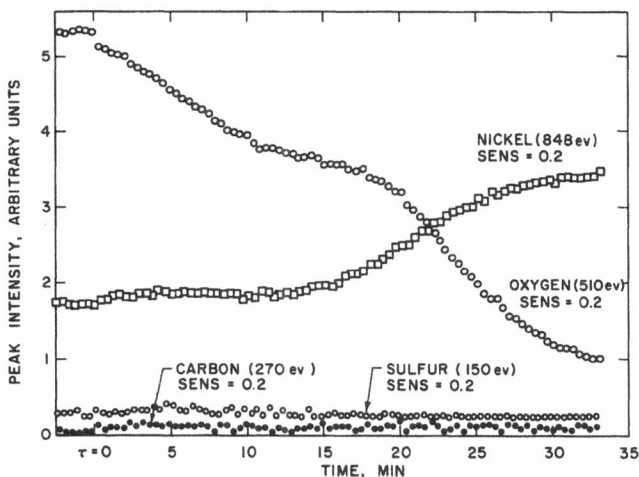

Fig. 10. Component depth profiles for nickel after one hour treatment with 4 mm SO_2 at 523 K.

Fig. 11 shows the surface S/M ratio for Pt and Pd after SO_2 treatment at the various temperatures studied. For both metals sulfur peak intensity increased between 310 and 373 K and remained constant from 373 to 573 K. As with H_2S treatment, in SO_2 Pd readily and extensively underwent sulfur incorporation into the lattice, whereas Pt did not. Fig. 12 demonstrates this in terms of the time required to sputter the sulfur peak to zero as a function of contact temperature. For Pd, the sputter time rises sharply with contact temperature as with H_2S (Fig. 7).

Fig. 13 shows the component depth profiles for Pd and Pt in terms of sputtering time after one hour SO_2 contact at 523 K. Pd and Pt exhibited a substantial surface oxygen peak for all contacting conditions. No relative peak intensity changes (O/M or

References pp. 165-166.

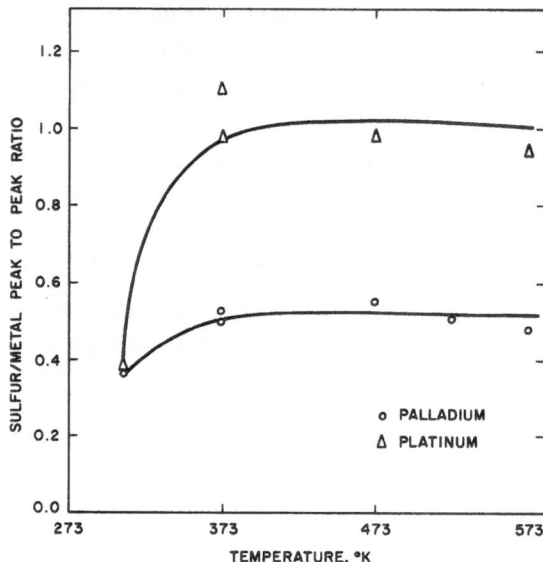

Fig. 11. Effect of treatment temperature on surface sulfur concentration for one hour treatment with 4 mm SO_2 pressure in helium.

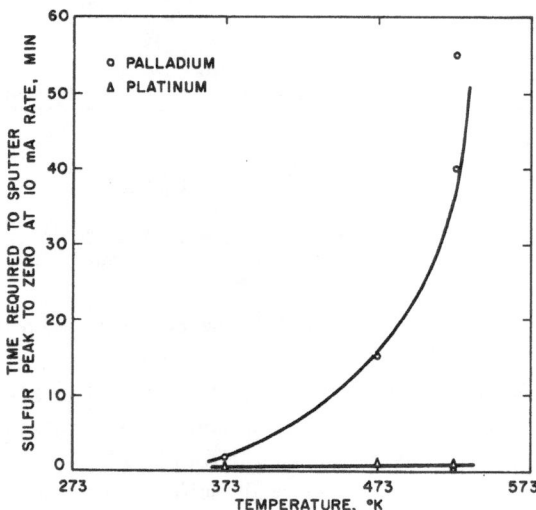

Fig. 12. Effect of temperature of one hour SO_2 treatment on time required to sputter sulfur peak to zero at 10 mA rate.

S/M) were observed in increasing from zero beam current to typical operating values indicating that in short periods of time there was neither decomposition or desorption of the adsorbed species. On sputtering the sulfur signal typically increased 15-20% as

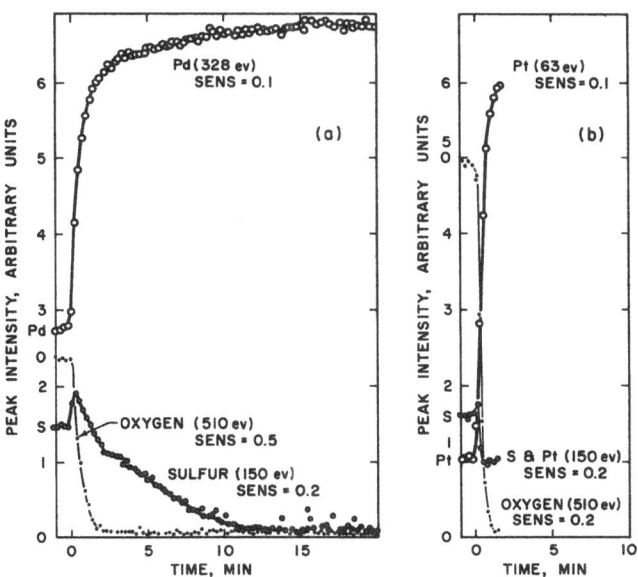

Fig. 13. Component depth profiles for (a) palladium and (b) platinum after one hour treatment with SO_2 at 523 K.

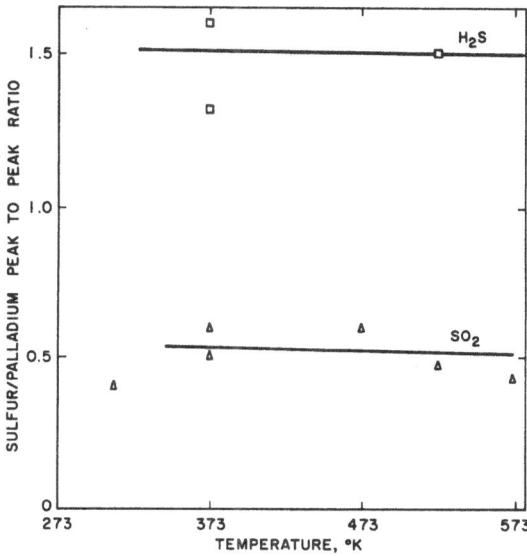

Fig. 14. Difference in sulfur concentration resulting from treatment with H_2S and SO_2 with palladium for one hour.

References pp. 165-166.

the oxygen signal dropped rapidly to zero (Fig. 13). This suggests desorption of either chemisorbed SO_2 or oxygen but does not indicate which. Even with this increase, the sulfur concentration, as measured by peak-to-peak intensity, was only about 60 percent of that resulting from contact with H_2S. This is shown in Fig. 14 for Pd; results for Pt were similar. Although we have not identified the metal-sulfur compounds formed, it appears that a different sulfur compound (metal sulfide) was formed by contact with SO_2 than by contact with H_2S. Nickel formed an oxide rather than a sulfide in the presence of SO_2. The metal-sulfur compound formed is dependent on the environment, including the metal, and thus the manner in which sulfur compounds interact with working catalyst surfaces could be sensitive to the reaction conditions.

For Ni, Saleh *et al.* (19) reported that at temperatures greater than 373 K sulfur was incorporated into the nickel lattice. We typically see oxygen incorporation rather than sulfur incorporation. They had no analysis of either the metal surface or bulk but relied solely on pressure measurements to infer what had occurred. Incorporation of oxygen rather than sulfur may have occurred. This clearly indicates the need for information on both the gas and solid phases.

Saleh also reported that there was no evidence of incorporation of sulfur from SO_2 into Pd (18) or Pt (17) foils at temperatures to 523 K. SO_2 adsorption on Pt was reported to be completely reversible to 523 K. This was clearly not observed here; the differences may be due to the lower pressures used by Saleh.

Further work is needed to more completely define the interactions of sulfur compounds with metals separately and then to define their behavior in the presence of other species which may well be reactants or products in a catalyzed reaction. Reaction conditions may well alter the mode of interaction with the metal. Similarly, partial pressures of sulfur compounds in the range typically found in reaction systems requires further attention.

EFFECT OF SO_2 ON NO REDUCTION KINETICS

Procedure — Preliminary results of the effect of SO_2 on the reduction of NO by NH_3 over several supported metal catalysts are reported below. This work will be published in greater detail when it has reached a more complete stage. The work utilized the reactor system described earlier. Space velocity was 24,000 hr^{-1} for most runs, and temperatures to 733 K were investigated. Catalysts studied included 0.5% Pt/Al$_2$O$_3$ (type M catalyst), 0.3% Pt/SiO$_2$, 0.5% Pd/Al$_2$O$_3$, 0.5% Ru/Al$_2$O$_3$ (all from Engelhard) and 11% Ni as NiO on Al$_2$O$_3$ (Harshaw). All catalysts were crushed and screened to 28-35 mesh and reduced under flowing H$_2$ at 773 K for 2 hr. Feed NO and NH$_3$ concentrations of 1.0% were maintained constant throughout. Since the majority of these catalysts have not been fully characterized including chemisorption surface area, rates are reported on a per gram basis.

To determine if a given support might have definite advantages in terms of interactions with SO_2, the adsorption of SO_2 from helium on Al_2O_3 and SiO_2 at 453 K was determined. Al_2O_3 adsorbed enough SO_2 to represent approximately 40% of a monolayer of $SO_4^=$ on the surface; SiO_2 adsorbed sufficient SO_2 to account for about 5% of a monolayer.

Results at 473 K — Table 5 gives the relative activities of Pt, Pd, Ru and Ni for the reduction of NO by NH_3 at 473 K. Without SO_2 Pt is slightly more active than Pd. Ru and Ni are about two orders of magnitude less active. The 20 times higher Ni content indicates that the Ni is less active than Ru since X-ray line broadening showed that the Ni was quite highly dispersed.

TABLE 5

Steady-State Rates of Reduction in the Low-Temperature
Reduction of NO by NH_3

Catalysts	r_{N_2} [a,b]	r_{N_2O} [a,b]	r_{N_2}/r_{N_2O}
0.5% Pt/Al_2O_3	1.1×10^{-3}	0.66×10^{-3}	1.7
0.5% Pd/Al_2O_3	1.0×10^{-3}	0.62×10^{-3}	1.8
0.5% Ru/Al_2O_3	1.1×10^{-5}	1.3×10^{-5}	0.84
11% Ni/Al_2O_3	4.1×10^{-5}	0.98×10^{-5}	4.2
0.5% Pt/Al_2O_3 [c]	No Conversion Observed		—
0.5% Pt/Al_2O_3 [d]	4.6×10^{-4}	9.5×10^{-4}	0.48

[a] *473 K, 1% NO, 1% NH_3, 24,000 hr^{-1}, except Pd catalyst 476 K*
[b] *rate in g moles formed per gram catalyst per hr*
[c] *473 K, 1% NO, 1% NH_3, 24,000 hr^{-1}, 50 ppm SO_2*
[d] *473 K, 1% NO, 1% NH_3, 24,000 hr^{-1}, 50 ppm SO_2, 0.5% O_2*

The selectivity of Pd to N_2 formation is equivalent to that of Pt. Ru seems much less selective although the low N_2O concentration observed here may indicate significant error in the N_2O rate even though it was unchanged over a 13 hr period. Ni was much more selective to N_2 formation possibly because of the incorporation of oxygen from NO (N_2O) into the Ni lattice. Oxidation of the reduced Ni to NiO was observed under these reaction conditions and complete oxidation of the Ni has possibly not occurred by the time steady-state activity was observed.

Our results are consistent with those reported by Markvart and Pour (3) for the order of reactivity in NO reduction by NH_3,

$$Pt > Pd >> Ru$$

For NO reduction by H_2 the order of reactivity observed by Kobylinski and Taylor (14) was

$$Pd \rangle Pt \rangle Rh \rangle Ru$$

For the 0.5% Pt/Al_2O_3 catalyst at 473 K the presence of 50 ppm SO_2 in the feed gas reduced the activity to below our limits of detectibility (Table 5). However, the addition of 0.005 atm O_2 in the presence of 50 ppm SO_2 resulted in a marked enhancement in the rate, sufficiently that the rates of N_2 and N_2O formation were of the same magnitude as those observed without O_2 and SO_2 (Table 5). The selectivity is much lower in the presence of O_2. An increase in the SO_2 concentration to 250 ppm at the same O_2 level resulted in a significant decrease in both the N_2 and N_2O formation rates with the N_2 selectivity decreasing even further.

The reason for the positive effect of O_2 in the presence of SO_2 is not clear, just as it is not clear how O_2 enhances the rate in the absence of SO_2. However, we speculate that it may be due to the creation of an oxidizing rather than a mildly reducing micro-environment at the metal surface. Thus, SO_2 might be "oxidized" rather than "reduced." Pt is not affected by SO_2 in the oxidation of CO and hydrocarbons and is a good catalyst for the commercial oxidation of SO_2 to SO_3. The fate of SO_2 under our conditions was not determined. However, the effect of O_2 on activity is most intriguing and is the subject of further study. The low selectivity to N_2 in the presence of O_2 is consistent with, although greater than, the observed selectivity reduction caused by O_2 in the absence of SO_2.

Results at 673 K — The deactivation behavior of 0.5% Pt/Al_2O_3 catalyst with 50 ppm SO_2 in the feed at 673 K is shown in Fig. 15. Initially conversion was 100%, and essentially only N_2 was observed. The observed increase in N_2O formation with decreasing N_2 conversion is consistent with the parallel-consecutive nature of the reaction network (6, 59). N_2 selectivity passes through a minimum during the latter part of the deactivation indicating effects beyond the parallel-consecutive nature of the reaction. Time required to deactivate the catalyst to steady-state activity was inversely proportional to the SO_2 concentration in the feed. Steady-state activities showed no further change over several days. Table 6 summarizes the results of the 673 K runs with SO_2 in the feed. For 50 ppm SO_2 the Pt, Pd and Ni catalysts show approximately the same activity, and the activity is always less than that measured for pure Al_2O_3 in the reactor. Without Al_2O_3 reaction was not observed at 673 K. This indicates that the presence of 50 ppm SO_2 essentially eliminates the activity of the metal component on these catalysts in the $NO-NH_3$ reaction. This is further supported by the observation that at 250 ppm SO_2 a 60 K higher temperature was required to achieve the same observed rate of reaction for 0.3% Pt/SiO_2 as for 0.5% Pt/Al_2O_3 (Table 6). SiO_2 is a much more inert support for the adsorption of the reactants involved here as illustrated by the lower SO_2 adsorption. At 673 K and 250 ppm SO_2 about 20% of the SO_2 was reduced to elemental sulfur; no H_2S was observed nor was there any evidence of SO_3 formation. The surface micro-environment appears to be mildly reducing.

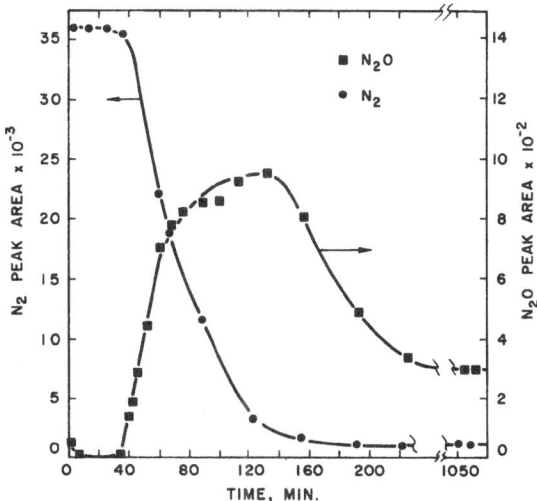

Fig. 15. Transient behavior of 0.5% Pt/Al$_2$O$_3$ during deactivation by 50 ppm SO$_2$ at 673 K; 1% NO, 1% NH$_3$ in He, 24,000 hr^{-1}, conversion is 100% at zero time.

TABLE 6

Steady-State Rates of Reaction in NO Reduction
by NH$_3$ in the Presence of SO$_2$

Catalyst	Temp, K	SO$_2$ conc, ppm	r_{N_2}[a]	r_{N_2O}[a]	r_{N_2}/r_{N_2O}
0.5% Pt/Al$_2$O$_3$	673	50	1.5 X 10^{-4}	0.69 X 10^{-4}	2.1
0.5% Pd/Al$_2$O$_3$	673	50	5.4 X 10^{-4}	0.61 X 10^{-4}	8.8
11% Ni/Al$_2$O$_3$	673	50	5.4 X 10^{-4}	0.19 X 10^{-4}	28.6
0.5% Ru/Al$_2$O$_3$	673	50	Complete Conversion of NO		
0.5% Ru/Al$_2$O$_3$	523	50	2.0 X 10^{-4}	1.5 X 10^{-3}	1.4
Al$_2$O$_3$	673	50	6.0 X 10^{-4}	0.6 X 10^{-4}	10
0.33% Pt/SiO$_2$	733	250	1.7 X 10^{-4}	0.48 X 10^{-4}	3.5
0.5% Pt/Al$_2$O$_3$	673	250	1.8 X 10^{-4}	0.96 X 10^{-4}	1.9

[a] *1% NO, 1% NH$_3$, space velocity = 24,000 hr^{-1} (except for Ni catalyst), rate in g moles formed per gram catalyst per hour*

References pp. 165-166.

The Ru catalyst activity was sufficiently high that complete conversion of NO occurred at 673 K. Complete conversion was also observed with 2000 ppm SO_2 in the feed. At 523 K the Ru catalyst underwent a significant deactivation upon the introduction of 50 ppm SO_2, but the steady-state rates were almost equivalent to those observed for Pt at 473 K without SO_2 present (Table 6). Thus, Ru shows uniquely high tolerance to SO_2 in the reduction of NO by NH_3, a property requiring further study.

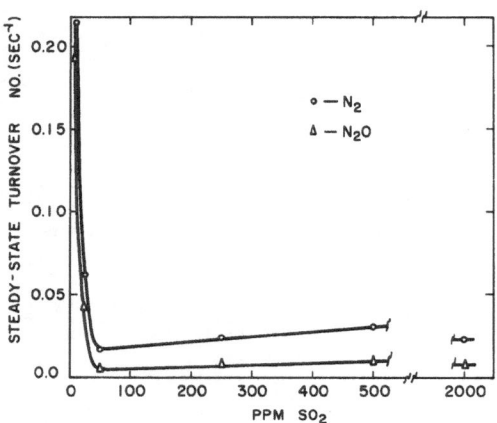

Fig. 16. Effect of SO_2 concentration on N_2 and N_2O formation rate over 0.5% Pt/Al$_2$O$_3$ at 673 K; 1% NO, 1% NH_3.

Fig. 16 shows the extreme sensitivity of the NO reduction reaction over Pt to SO_2. Turnover numbers can be converted to rate per gram cat. per hr by multiplying by 0.0106. The rate decreased by a factor of about 5 when the SO_2 concentration was increased from 8 to 25 ppm and at the 50 ppm level the activity of the Pt was reduced so severely that essentially only support activity remained. A slight, reproducible increase occurred at higher SO_2 levels, but the trend did not continue to 2000 ppm SO_2 (Fig. 16).

For 250 ppm SO_2 the observed activation energy for N_2 formation was 25 kcal/mol (Fig. 17). That for N_2O formation was about the same. This is only slightly higher than that observed for NO reduction by NH_3 over supported Pt in the absence of SO_2. The similarity is surprising in view of the apparent lack of metal activity at 250 ppm SO_2.

The Ni catalyst showed a very high selectivity to N_2 at steady-state activity over a period of several days. We speculate that this may again be related to the conversion of the reduced Ni catalyst to nickel oxide during reaction. After the run ($\sim$ 6 days) at 673 K and 50 ppm SO_2 the catalyst had approximately the same NiO peak intensity by x-ray as did the original unreduced NiO/Al$_2$O$_3$ catalyst indicating reoxidation of

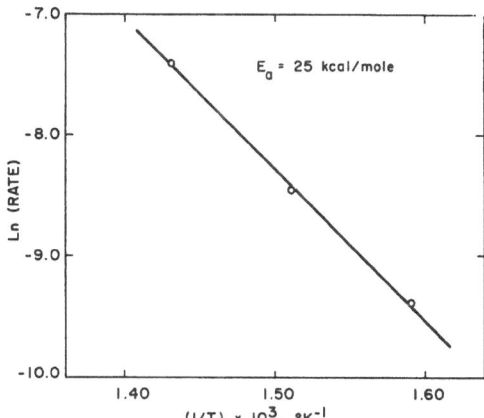

Fig. 17. Arrhenius plot of N_2 formation rate for 0.5% Pt/Al_2O_3 in the presence of 250 ppm SO_2; 1% NH_3 in He, 24,000 hr^{-1}.

the reduced Ni during reaction. There were no observable nickel sulfide or sulfate x-ray peaks present. This does not rule out sulfide formation but suggests that if it occurred it did so to a lesser extent than NiO formation. This oxidation rather than sulfiding is consistent with the behavior observed in our Auger studies for Ni in the presence of SO_2.

Low activity of Ni in NO reduction is often ascribed to the oxidation of the surface. Oxygen is observed to be a worse poison than sulfur (28,32). If the surface can be reduced and kept reduced, possibly by adding a small amount of Pt, Ni catalysts show high activity (32).

We might speculate that the high SO_2 tolerance observed by McArthur (32,33) for a 10% NiO-0.14% Pt catalyst may relate to the formation of a nickel oxide rather than a sulfide in the presence of SO_2 with the Pt contributing to keeping the nickel oxide surface partially reduced. The kinetics remained those of NiO in the presence of SO_2 (32). Without Pt present we observed a very severe reduction in the activity of a Ni catalyst caused by SO_2. Our Pt catalyst underwent a similarly severe deactivation caused by SO_2. These data would not predict tolerance to SO_2 for a Pt-Ni catalyst.

Our study of sulfur deactivation in these systems is in its preliminary stages, but the following results appear significant. Sulfur as SO_2 severely poisons the reduction of NO by NH_3 over Pt, Pd and Ni in the absence of O_2. Very low levels of SO_2 are sufficient to apparently reduce the activity of the metal below that of the Al_2O_3. Ru shows marked tolerance to SO_2 and deserves further study. The presence of O_2 markedly counters the poisoning potential of SO_2 on Pt; the effect must be due to a change in the oxidizing-reducing nature of the micro-environment at the surface of the metal. This effect also deserves further study not only because of its potential importance in pollution control applications but also because sulfur compounds may

References pp. 165-166.

be useful as probes of the micro-environment (oxidizing-reducing) at the catalyst surface to determine the overall parameters, e.g., the gas-phase composition, of the system.

NOTATION

B	undefined constant
K_i	adsorption parameter in Langmuir-Hinshelwood kinetic model for species i, atm^{-1}
K'_{NH_3}	adsorption equilibrium coefficient for associative adsorption of ammonia, atm^{-1}
K_{s4}	surface dissociation equilibrium constant for reaction step 4
k	number of parameters to be fit in model
$k_{s,i}$	surface reaction rate constant in Langmuir-Hinshelwood kinetic model for formation of species i,s^{-1}
k_i	surface reaction rate constant for reaction step i,s^{-1}
n	number of data points used in fitting
P_i	partial pressure of species i, atm
R	gas constant, cal/(gmol)(K)
r_i	rate of formation of species i as turnover number, s^{-1}
$r_{obs,i}$	observed rate of formation of species i,s^{-1}
$r_{pre,i}$	rate of formation of species i predicted from the model, s^{-1}
T	temperature, K
t	time, s

Greek Letters

θ_i	fraction of metal surface covered by species i

ACKNOWLEDGMENTS

Financial support for the kinetics part of this work by the National Science Foundation through Grant GK-34612X1 and by the Environmental Protection Agency through Grant No. 1 RO1 APO1419-01 is gratefully acknowledged. Financial support for the sulfur portion of the work is from the General Motors Research Laboratory and is greatly appreciated. The review material from our laboratory has come largely from the kinetic studies of the NO-NH$_3$ reaction by J. D. Wanner and R.

J. Pusateri, from Auger studies of A. Huss and R. M. Gould, and from the sulfur poisoning studies of J. M. Foley and D. R. Sullivan. Cooperation of W. H. Manogue on all this work, and helpful discussions with G.C.A. Schuit are gratefully acknowledged.

REFERENCES

1. H. C. Andersen, W. J. Green, and D. R. Steele, *Ind. Eng. Chem.* **53**, 199 (1961).
2. E. A. Michailova, *Acta. Physico-Chemica USSR*, **10**, 653 (1939).
3. M. Markvart and V. Pour, *Chem. Prum.* **19** (1), 8 (1969).
4. M. Markvart and V. Pour, *J. Catal.* **7**, 279 (1967).
5. B. P. Gupta, "Study of the Catalytic Reaction Between Nitric Oxide and Ammonia," Ph. D. thesis, Univ. Cincinnati, Ohio (1970).
6. K. Otto, M. Shelef, and J. T. Kummer, *J. Phys. Chem.* **74**, 2690 (1970).
7. K. Otto and M. Shelef, "Studies of Surface Reaction of NO by Isotope Labeling. VI. The Reduction of Nitric Oxide by Ammonia and Hydrogen over Supported Ruthenium," paper presented at Am. Chem. Soc. Meeting, Chicago (1973).
8. K. Otto and M. Shelef, *J. Phys. Chem.* **76**, 37 (1972).
9. G. Blyholder and R. W. Sheets, *J. Catal.* **27**, 301 (1972).
10. C. E. Melton and P. H. Emmett, *J. Phys. Chem.* **68**, 3318 (1964).
11. R. E. Mardaleishvili, Hu Sin-chou, and Zh. Ya. Smorodinskaya, *Kinet. Katal.* **8**, 786 (1967).
12. C. W. Nutt and S. Kapur, *Nature* **224**, 169 (1969).
13. J. W. May, R. J. Szostak, and L. H. Germer, *Surface Sci.* **15**, 37 (1969).
14. K. Otto, M. Shelef, and J. T. Kummer, *J. Phys. Chem.* **75**, 875 (1971).
15. T. P. Kobylinski and B. W. Taylor, *J. Catal.* **33**, 376 (1974).
16. J. K. Dixon and J. E. Longfield, "Oxidation of Ammonia, Ammonia and Methane, Carbon Monoxide and Sulfur Dioxide," in "Catalysis," Vol. VII, P. H. Emmett, ed., p. 281, Reinhold, New York, 1960.
17. J. Saleh, *Trans. Faraday Soc.* **67**, 1830 (1971).
18. J. M. Saleh, *Trans. Faraday Soc.* **66**, 242 (1970).
19. J. M. Saleh, C. Kemball, and M. W. Roberts, *Trans. Faraday Soc.* **57**, 1771 (1961).
20. J. M. Saleh, *Trans. Faraday Soc.* **64**, 796 (1968).
21. J. Bénard, *Bull. Soc. Chim. France*, p. 203 (1960).
22. J. Bénard, "The Chemical Adsorption of Sulfur on Metals: Thermodynamics and Structure," in "Catalysis Reviews" ed., H. Heinemann, Vol. 3, p. 93, Marcel-Dekker, New York, 1970.
23. J. Bénard, J. Oudar, and F. Cabahé-Brouth, *Surface Sci.* **3**, 359 (1965).
24. Y. Berthier, M. Perdereau, and J. Oudar, *Surface Sci.* **36**, 225 ((1973).
25. J. L. Domange and J. Oudar, *Surface Sci.* **11**, 124 (1968).
26. J. L. Domange, J. Oudar, and J. Bénard, "Growth Mechanism and Structure of Adsorption Layers," in "Molecular Processes on Solid Surfaces," ed., E. Drauglis, R. D. Gretz, and R. I. Jaffee, p. 353, McGraw-Hill Book Co., New York, 1969.
27. M. Kostelitz, J. L. Domange, and J. Oudar, *Surface Sci.* **34**, 431 (1973).
28. G. H. Meguerian, F. W. Rakowsky, E. H. Hirschberg, C. R. Lang, and D. N. Schock, "NO_x Reduction Catalysts for Vehicle Emission Control," SAE Paper No. 720480 (May, 1972).
29. J. E. Hunter, Paper No. 720122, presented at National Meeting Society of Automotive Engineers, Detroit, Michigan, January, 1972.
30. R. M. Campau, A. Stefan, and E. E. Hancock, "Ford Durability Experience on Low Emission Concept Vehicles" SAE Paper 720488 (May, 1972).

31. J. C. Gagliardi, C. S. Smith, and E. E. Weaver, "*Effect of Fuel and Oil Additives on Catalytic Converters,*" Paper No. 63-72 at 37th Midyear A.P.I. Meeting, New York, May 8-11, 1972.

32. D. P. McArthur, "*Degradation and Deactivation of NO_x Catalysts,*" Progress Report Accompanying Personal Communication, Union Oil Company of California, Brea, California 92621, 1973.

33. H. R. Jackson, D. P. McArthur, and H. D. Simpson, "*Catalytic NO_x Reduction Studies,*" SAE Paper 730568 (May, 1973).

34. S. S. Hetrick, and F. J. Hills, "*Fuel Lead and Sulfur Effects on Aging of Exhaust Emission Control Catalysts,*" SAE paper 730596 (May, 1973).

35. R. A. Giacomazzi and M. F. Homfeld, "*The Effect of Lead, Sulfur and Phosphorus on the Deterioration of Two Oxidizing Bead-Type Catalysts,*" SAE Paper 730595 (May, 1973).

36. N. A. Fishel, R. K. Lee, and F. C. Wilheim, Environ. Sci. Technol. 8, 260 (1974).

37. P. L. Romeo, "*Catalysis and the Control of Oxides of Nitrogen,*" paper at Joint NAPCA/MUCOM Seminar, Joliet, IL, June 3-5, 1970.

38. J. H. Jones, J. T. Kummer, K. Otto, M. Shelef, and E. E. Weaver, Environ. Sci. Technol. 5, 790 (1971).

39. R. J. Pusateri, "*The Reduction of Nitric Oxide by Ammonia on Supported Platinum Catalysts,*" M.Ch.E. thesis, Univ. of Delaware, Newark (1973).

40. J. D. Wanner, "*Selective Reduction of Nitric Oxide by Ammonia on a Supported Platinum Catalyst,*" M.Ch.E. thesis, Univ. of Delaware, Newark (1972).

41. J. J. Ostermaier, J. R. Katzer, and W. H. Manogue, J. Catal. 33, 457 (1974).

42. C. R. Adams, H. A. Benesi, R. H. Curtis, and R. G. Meisenheimer, J. Catal. 1, 336 (1962).

43. G. R. Wilson and W. K. Hall, J. Catal. 17, 190 (1970).

44. G. R. Wilson and W. K. Hall, J. Catal. 24, 306 (1972).

45. M. Boudart, A. Aldag, J. E. Benson, N. A. Dougharty, and C. G. Harkins, J. Catal. 6, 92 (1966).

46 J. M. Smith, Chemical Engineering Kinetics, 2nd Edit., McGraw-Hill, New York (1970).

47. R. J. Ayen and M. S. Peters, Ind. Eng. Chem. Process Design Develop. 1, 204 (1962).

48. F. Falk and R. N. Pease, J. Am. Chem. Soc. 76, 4746 (1954).

49. G. Bedford and J. H. Thomas, Trans. Faraday Soc. 68, 2163 (1972).

50. Ya. M. Fogel, B. T. Nadykto, V. F. Rybalko, V. I. Shvachko, and I. E. Korobchanskaya, Kinet. Katal. 5, 431 (1964).

51. M. Shelef and H. S. Gandhi, Ind. Eng. Chem. Prod. Res. Develop. 11, 393 (1972).

52. H. C. Andersen and A. J. Haley, U. S. Patent 3,079,232 (1963).

53. P. W. Palmberg and T. H. Rhodin, J. Appl. Phys. 39, 2425 (1968).

54. J. C. Tracy, and P. W. Palmberg, J. Chem. Phys. 51, 4852 (1969).

55. J. V. Florio and W. D. Robertson, Surface Sci. 18, 398 (1969).

56. T. P. Pignet, L. D. Schmidt, and N. L. Jarvis, J. Catal. 31, 145 (1973).

57. P. H. Hollway and J. B. Hudson, Surface Sci. 33, 56 (1972).

58. H. P. Bonzel and R. Ku, J. Chem. Phys. 58, 4617 (1973).

59. M. Shelef and K. Otto, J. Catal. 10, 408 (1968).

DISCUSSION

T. Kobylinski *(Gulf R & D)*

Your results confirm our work, some of which was reported two years ago at the Gordon Conference. Especially the relative rates of the NO-ammonia reaction where ruthenium without sulfur poisoning is a relatively poor catalyst. Platinum plus palladium are very good catalysts for this reaction. Also rhodium and iridium are very good. Now we did the same thing as far as the sulfur is concerned. We agree with you and our experience shows that we can load the ruthenium catalyst to about 10 wt % of sulfur and the catalyst is still active for either $NO-NH_3$ or $NO-CO-H_2$ systems. I've got a question. You show the nickel work where you used nickel and sulfided nickel and so on for these tests. Did you ever try to remove the nickel and by observing the signals could you see the disappearance of sulfur from the nickel?

Katzer

To regenerate nickel? No, we have not tried that for nickel.

J. J. Carberry *(University of Notre Dame)*

Did you analyze for sulfuric acid or SO_3 in the exit?

Katzer

We have not yet analyzed for SO_3. We've done the SO_2 analysis and we see more decrease in the SO_2 concentration across the reactor (about 40%). The SO_2, however, appears to be undergoing reduction rather than oxidation to SO_3 as far as we can tell. We do not find any H_2S but do find sulfur, and from this we speculate that the environment on the catalyst surface is mildly reducing, giving sulfur but not H_2S. We are just beginning to study the system with oxygen present ($NO-NH_3-O_2-SO_2$) and feel the situation could be quite different in this system.

Professor Ozaki *(Tokyo Institute of Technology)*

Did you use half order kinetics with respect to ammonia of the Langmuir type?

Katzer

Yes.

Ozaki

For homolytic dissociations such as those for H_2 or N_2 you can use dissociative adsorption in terms of half order kinetics. But in the case of ammonia dissociation, it must be a heterolytic process yielding NH_2 and H or something like that. My question is, what do you mean by half order kinetics?

Katzer

If you start with the mechanism which I presented involving dissociative adsorption of ammonia or dissociation of adsorbed ammonia and work through the mathematics using the assumed rate limiting steps, you obtain a rate expression of the

Langmuir-Hinshelwood type with the ammonia partial pressure raised to the one-half power in the numerator. This is what I mean by half order kinetics, because at the low concentrations of ammonia that would typically be used, the effect of an ammonia term in the denominator should be small and the rate should show one-half order dependence on the ammonia partial pressure. The observed rate does indeed show this behavior.

G. A. Somorjai *(University of California)*

Have you enhanced the rate with oxygen? Was the rate enhancement reversible or irreversible?

Katzer

It's reversible. If we take oxygen out, it goes back.

K. Otto *(Ford Motor Company)*

There's an interesting parallelism with the reaction between ammonia and nitric oxide. I think one can also consider the discussed reaction from the standpoint of the NO-NH_2 surface complex which Dr. Shelef and I have suggested for the NO-NH_3 reaction, leading mainly to nitrogen. I think it's very likely that a surface complex like this also exists here.

Katzer

The ammonia oxidation is half order in oxygen suggesting dissociative adsorption, if we will believe the Langmuir-Hinshelwod relationships, but there is no reason why we couldn't have that kind of a complex.

W. K. Hall *(University of Wisconsin)*

Why did you use SO_2 instead of H_2S? I would suspect this would give you very different results with nickel? Is it because SO_2 is the product in the exhaust gas?

Katzer

Yes. H_2S is in our plan of experiments though.

H. Wise *(Stanford Research Institute)*

In your derivative curves for Auger spectra, I think you conclude that platinum is less resistant to sulfur poisoning than palladium. Is that correct?

Katzer

It's more resistant in the sense that sulfur is not incorporated into the lattice to any significant depth, whereas in palladium it is.

Wise

Did you take into account the difference in the cross section of these metals for the Auger electron?

Katzer

No we didn't. The peak height was the same in all three cases for H_2S at the higher temperature. There was a deviation for SO_2 but with H_2S the sulfur peak height was the same with all three metals, meaning roughly the same sulfur concentration at the surface.

Wise

You just measure the peak height in the derivative spectrum?

Katzer

Yes, we measured the sulfur peak height in the derivative spectrum. We would expect the sulfur concentration on the surface to have been the same under these conditions, though, based on the work of Bénard and coworkers in Paris[*]; and our measurements are consistent with this expectation.

J. Gland *(General Motors Research Laboratoriees)*

I wondered if you observed any difference in the shape of the sulfur peak when you treated with H_2S and SO_2?

Katzer

The answer is no, although we have not compared peak shapes carefully. We tried to obtain chemical shift information here, which is maybe what you're leading to, but our techniques at that time did not provide us with unambiguous information on chemical shift.

F. Williams *(General Motors Research Laboratories)*

I'd like to make a comment on the same point. There's a note in Surface Science by Helen Farrell[**] that clearly shows that oxidized sulfur does have a completely different structure than the sulfided structure. Perhaps not the treatment with SO_2 on platinum, but surely the treatment with oxygen in a feed where you show that you recover the activity . . . if you took that same foil, for example, and put it back into the chamber you should observe, if you are forming the sulfate you want for increased activity. You should see that on the Auger spectrometer.

[*]*References 21-23, p. 165.*
[**]*H. Farrell, Surface Science 34, 465 (1973).*

D. P. McArthur *(Union Research Center)*

We also studied the poisoning effects of SO_2, water and oxygen on nickel catalysts for NO reduction. We found the relative poisoning effect was oxygen greater than water greater than SO_2, which agrees with your results.

SESSION III

SELECTIVITY, KINETICS AND POISONING

Session Chairman
G. W. KEULKS

University of Wisconsin
Milwaukee, Wisconsin

SIMULTANEOUS NO AND CO CONVERSION OVER RHODIUM

K. C. TAYLOR

General Motors Research Laboratories, Warren, Michigan

ABSTRACT

Various catalysts are examined for the simultaneous removal of CO and NO in an exhaust-like feed. Rhodium supported on alumina possesses exceptionally high activity for reducing NO but forms mostly NH_3 below 600°C. At higher temperature NH_3 is an intermediate in the reduction of NO to N_2. CO and H_2 are oxidized at comparable rates as the oxygen content of the feed is increased, so NH_3 formation is not checked by selective H_2 removal below 600°C. On the other hand, above 600°C ammonia removal is enhanced by oxygen.

INTRODUCTION

One of several methods being considered for the control of automobile exhaust emissions by catalysts is the application of a single catalyst to simultaneously convert nitric oxide, carbon monoxide, and hydrocarbons (1). This technique is frequently referred to as three-component control or as closed-loop control since its application generally requires the use of an oxygen sensor in the exhaust system with feedback to the carburetor in order to hold the air/fuel ratio of the exhaust near the stoichiometric point. Several catalysts have been reported to yield high conversions of all three species. For example, a platinum catalyst promotes these reactions at 480°C over a relatively narrow range of air/fuel ratios; moreover, the width of this band is increased on the lean side of the stoichiometric point if the catalyst temperature is lowered to 260°C (2). Catalysts which form less NH_3 on the rich side of the stoichiometric point such as Cu-Ni or Pt-Ni potentially give high conversions to nitrogen over a wider air/fuel ratio range (3). A wide range for the simultaneous control of all three pollutants is desirable since it alleviates the requirements for

References pp. 186-187.

air/fuel ratio control. Laboratory studies which compare conversions for a range of oxygen concentrations around the stoichiometric point can be used for the preliminary evaluation of catalytic activity for this application. Several workers have so examined the activity of noble metals (e.g., Pt, Rh, Ru) (2,4). Rhodium showed promise because of its low temperature NO activity (4) and its selectivity for N_2 formation near stoichiometry (5).

In order to develop an understanding of the chemical reactions which are essential for three-component control, we have examined the activity of several catalysts with a synthetic exhaust-like feed stream. Simultaneous reactions of CO and NO over $Rh-Al_2O_3$ are compared with $Pt-Al_2O_3$ and $Pd-Al_2O_3$ in order to demonstrate the activity and selectivity characteristics which provide best control of NO_x near the stoichiometric air/fuel ratio.

EXPERIMENTAL

The rhodium catalyst was prepared by impregnating preformed alumina spheres (Kaiser KC/SAF, surface area $250 \, m^2/g$) with an aqueous solution of rhodium chloride ($RhCl_3 . 4 \, H_2O$, ROC/RIC). After impregnation, the catalyst was dried overnight and then calcined in an air stream (GHSV = 500) for 4 hours at 500°C. The platinum catalyst and palladium catalyst were similarly prepared using aqueous solutions of chloroplatinic acid (10% solution $H_2PtCl_6 . 6H_2O$, Matheson, Coleman, and Bell) and palladium chloride (5% solution $PdCl_2$ Matheson, Coleman, and Bell), respectively. The noble metal content of the catalysts was determined by x-ray fluoresence to be 0.1% Rh for the $Rh-Al_2O_3$ catalyst, 0.11% Pt for the $Pt-Al_2O_3$ catalyst, and 0.09% Pd for the $Pd-Al_2O_3$ catalyst.

The catalytic reactor and techniques for continuous gas analysis have been described by Klimisch and Barnes (7). For the variable temperature studies, the catalysts were generally heated to 650°C in the feed stream under study, then the temperature of the reactor was lowered in steps to measure steady state conversion at each temperature. For other experiments, the temperature was held constant and the inlet oxygen in the feed stream was varied. The gas hourly space velocity for most experiments was 38,000.

RESULTS AND DISCUSSION

From the standpoint of pollution control, a closed-loop catalyst ideally should convert NO, CO, and hydrocarbons completely to N_2, CO_2, and H_2O at the stoichiometric air/fuel ratio. On the rich side of the stoichiometric point, there is necessarily excess reducing agent. Preferably the excess reducing agent is H_2 rather than CO or hydrocarbons, and NO should be reduced only to N_2. On the lean side of the stoichiometric point, there is necessarily excess oxidizing agent. Preferably, this is oxygen while NO is completely reduced to N_2, and hydrocarbons and CO are completely oxidized. In order to achieve this, the oxidation of CO would have to predominate over the oxidation of hydrogen with a rich feed, and NO would have to compete effectively with O_2 as oxidizing agent with a lean feed.

To try to understand the selectivity differences actually observed when NO is reduced over the noble metals, some of the individual chemical reactions have been examined. First, the two reducing agents (CO and H_2) were compared by examining the reduction of NO by H_2 and by a H_2-CO mixture. Water was omitted from the feed so the H_2 and CO levels would not be influenced by the water-gas shift reaction and so the hydrogen concentration might be monitored.

For Rh-Al_2O_3, the selectivity of the NO-H_2 reaction changed with temperature such that more NH_3 was formed at 427°C (Fig. 1) than at 538°C (Fig. 2). The oxygen concentration range in which most (>80%) of the inlet NO is reduced to N_2 is wider at the higher temperature. As was found previously for ruthenium (4,8,9), the addition of CO caused an increase in NH_3 formation (Figs. 3 and 4).

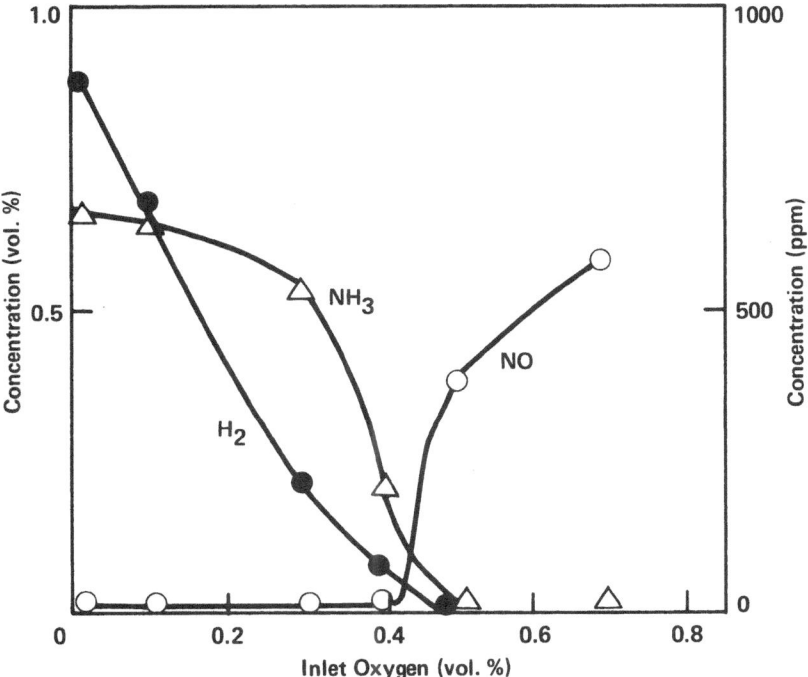

Fig. 1. NO reduction by H_2 with variable oxygen concentration at 427°C over Rh-Al_2O_3. Feed stream: 0.1% NO, 1.0% H_2, and 0-1.0% O_2 in N_2. GHSV = 38,000.

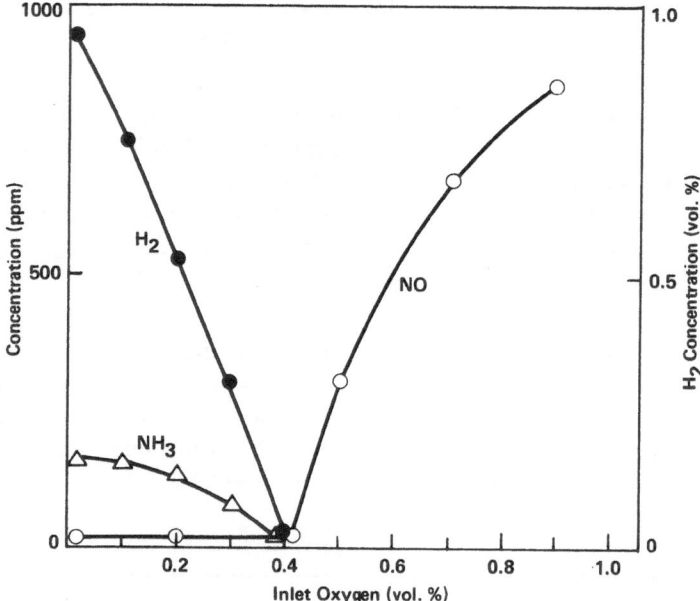

Fig. 2. NO reduction by H_2 with variable oxygen concentration at 538°C over Rh-Al$_2$O$_3$. Feed stream: 0.1% NO, 1.0% H_2, and 0-1.0% O_2 in N_2. GHSV = 38,000.

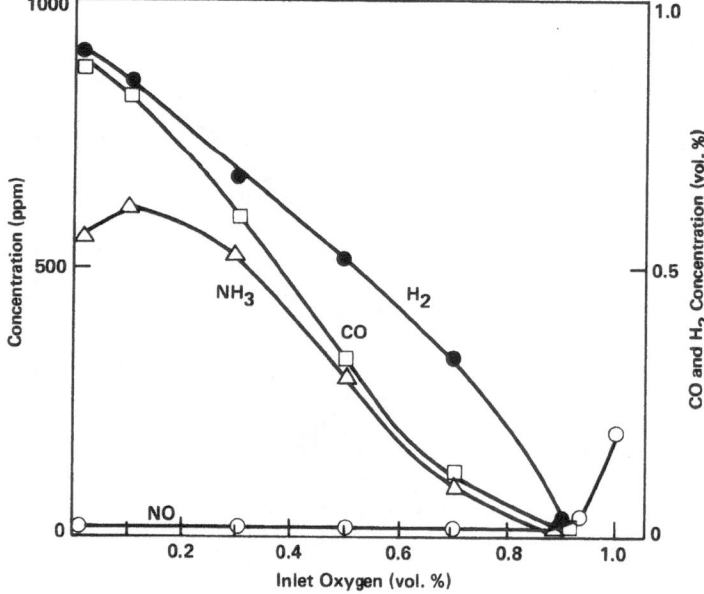

Fig. 3. NO reduction by CO and H_2 with variable O_2 at 538°C over Rh-Al$_2$O$_3$. Feed stream: 0.1% NO, 1.0% H_2, 1.0% CO, and 0-1.0% O_2 in N_2. GHSV = 38,000.

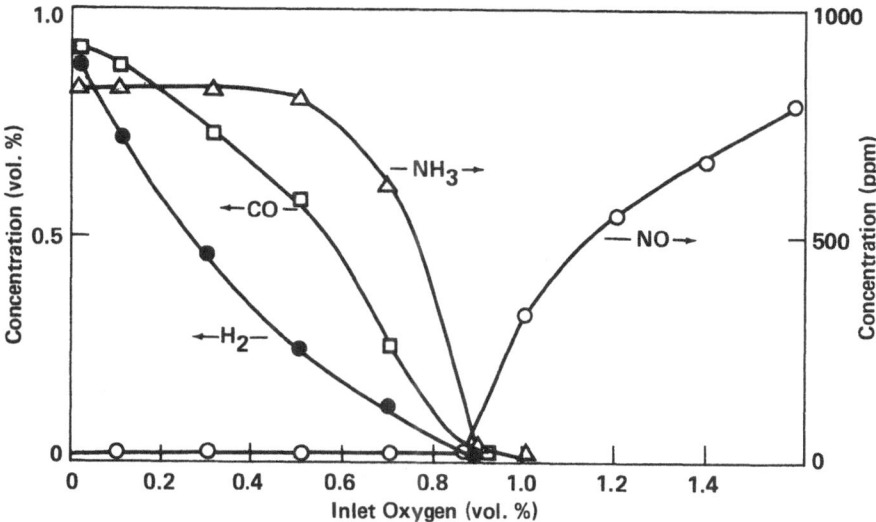

Fig. 4. NO reduction by CO and H_2 with variable O_2 at 427°C over Rh-Al$_2$O$_3$. Feed stream: 0.1% NO, 1.0% H_2 1.0% CO, and 0-1.0% O_2 in N_2. GHSV = 38,000.

A comparison of these results for Rh-Al$_2$O$_3$ with the results of similar experiments for Pd-Al$_2$O$_3$ and Pt-Al$_2$O$_3$ was made to see if the variation in the NH$_3$/N$_2$ product ratio with inlet oxygen is related to available hydrogen. Results obtained with

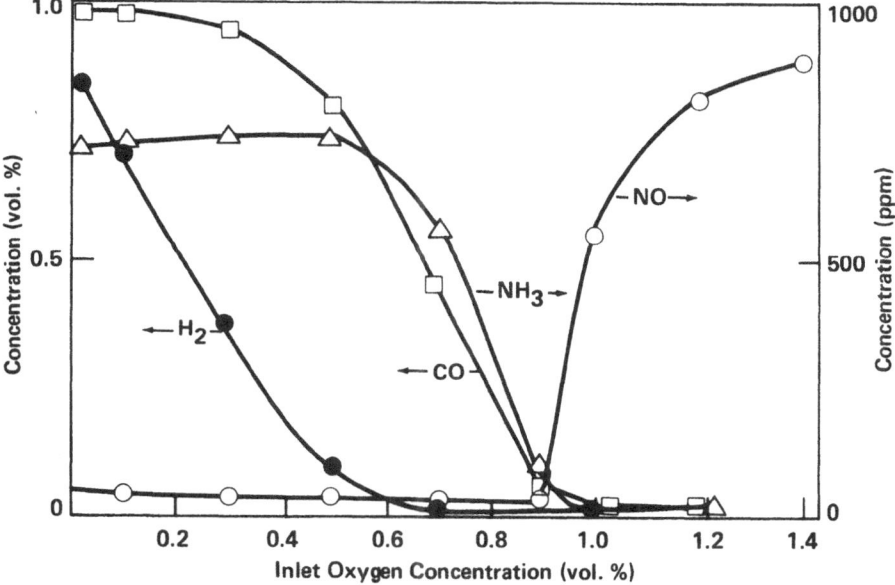

Fig. 5. NO reduction by CO and H_2 with variable O_2 at 538°C over Pd-Al$_2$O$_3$. Feed stream: 0.1% NO, 1.0% H_2, and 1.0% CO in N_2. GHSV = 38,000.

References pp. 186-187.

Pd-Al$_2$O$_3$ suggested that preferential removal of H$_2$ by reaction with oxygen can influence the NH$_3$/N$_2$ selectivity. Fig. 5 shows that as oxygen is added to a H$_2$-CO-NO feed over this catalyst H$_2$ removal preceeds CO removal. The NH$_3$/N$_2$ product ratio is constant at low oxygen levels and only declines at higher levels where the hydrogen reacts completely. With a Pt-Al$_2$O$_3$ catalyst (Fig. 6), H$_2$ and CO are removed to a more equal extent as the oxygen concentration is increased. The NH$_3$ curve is more complex for the platinum catalyst; it appears that the dominant product is NH$_3$ whenever sufficient H$_2$ remains in the feed. The results indicate that the H$_2$-O$_2$ reaction is faster than the CO-O$_2$ reaction over palladium for this system and that NH$_3$ formation appears to be related to H$_2$. Over Pt-Al$_2$O$_3$ and Rh-Al$_2$O$_3$ the gas phase CO and H$_2$ were removed together as inlet oxygen was increased.

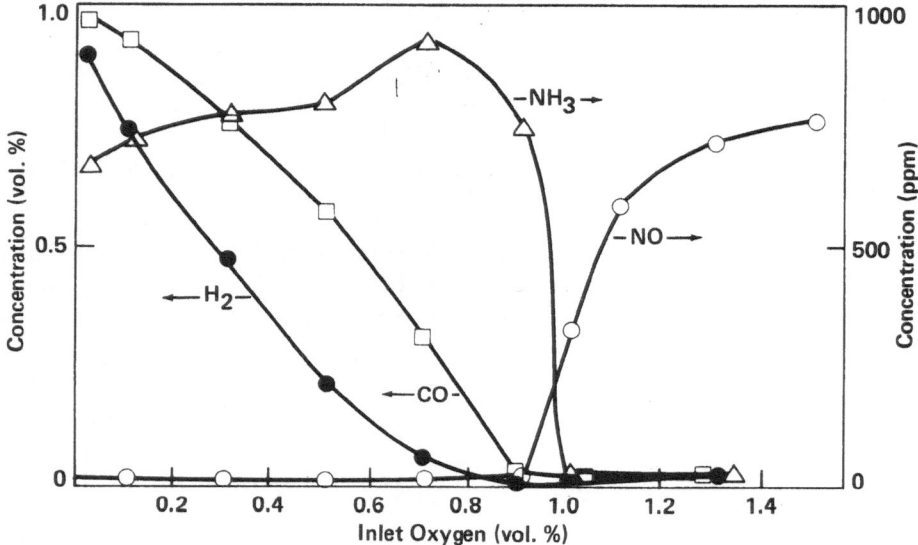

Fig. 6. NO reduction by CO and H$_2$ with variable O$_2$ at 538°C over Pt-Al$_2$O$_3$. Feed stream: 0.1% NO, 1.0% H$_2$, and 1.0% CO in N$_2$. GHSV = 38,000.

If the reaction of oxygen with hydrogen limits NH$_3$ formation, such an explanation also presupposes that O$_2$ removal is faster than NO removal. At low temperature, both Pt and Pd show selectivity for NO over O$_2$ (< 250°C) but this selectivity disappears at higher temperature. The Pt data has been reported by several authors (2). Similar data for rhodium is shown in Table 1. Here under net reducing conditions O$_2$ and NO removal proceed simultaneously. Estimation of the oxygen consumption in Exp. 1 and 2 (10% and 22% of inlet, respectively) suggests that NO actually slows down the O$_2$ reaction. The decrease in H$_2$ removal with added CO is also noted here (Exp. 1 and 5). Using such a high space velocity, it is important to point out that the reactor containing SiC in place of the catalyst did not reduce NO at 232°C.

TABLE 1

Effect of NO on CO and H_2 Oxidation at 232°C over 0.1% $Rh-Al_2O_3$

Exp. No.	Feed stream[a]				% Inlet CO Reacted	% Inlet H_2 Reacted	% Inlet NO Reacted	% Inlet NO to NH_3
1	1% CO	1% H_2	0.1% NO	0.3% O_2	7	17	55	35
2	1% CO	1% H_2	–	0.3% O_2	5.5	7.5	–	–
3	1% CO	–	–	0.3% O_2	17.5	–	–	–
4	–	1% H_2	–	0.3% O_2	–	38	–	–
5	–	1% H_2	0.1% NO	0.3% O_2	–	47	49	15

a. In each case the balance of the feed was made up to 1 atmosphere with N_2. GHSV = 304,000.

In comparing the product distribution data for these catalysts at several temperatures, we find that the N_2 selectivity profile changes with $Rh-Al_2O_3$ in comparison with $Pt-Al_2O_3$ but only above 500°C. Fig. 4 shows that $Rh-Al_2O_3$ is initially like Pt in that the selectivity of the NO reduction reaction is insensitive to added oxygen at 427°C. But at 538°C (Fig. 3), the product distribution changes in that the product NH_3 declines as H_2 is removed even though the available H_2 is comparable to that for the Pt catalyst result. We can not say for sure that it is the hydrogen which determines the level of product NH_3. However, evidence that the H_2 concentration influences the amount of product NH_3 over $Rh-Al_2O_3$ was obtained by omitting H_2 from the feed. The NH_3 formed at 538°C was 30% lower for a feed containing 0.1% NO, 1.0% CO, 10% CO_2 and 10% H_2O compared to the same feed plus 0.3% H_2.

The $Rh-Al_2O_3$ catalyst was also examined with a synthetic exhaust-like feed stream (i.e., C_3H_6, H_2O, and CO_2 were added to the feed). The activity and selectivity characteristics under reducing conditions and near the stoichiometric point are comparable in several respects to the results shown earlier for $NO-H_2-O_2$ and $NO-CO-O_2$ The temperature dependence of the reduction reaction without oxygen is presented in Fig. 7. The results for $Rh-Al_2O_3$ are distinguishable from those for a $Pt-Al_2O_3$ catalyst (Fig. 8) by a much lower temperature for reduction of NO (225°C for 50% removed versus 475°C for Pt). Both catalysts form NH_3, but as has been noted by others (5), NO is reduced primarily to N_2 over rhodium above 600°C. Over Pt, NH_3 is still the major product at higher temperatures. Rhodium, however, does not perform as well as ruthenium which has even better selectivity for N_2 formation at low temperatures (3,8,9). Some loss in activity for CO removal seems to coincide with the onset of propylene removal. Nitric oxide may be reduced by propylene here though it should be noted that the selectivity of the reduction (NH_3/N_2) is unchanged. There is not sufficient NO present to oxidize the propylene to CO_2 and H_2O. Some other reaction such as a H_2O-HC reaction must account for the complete removal of propylene. This reaction appears to be inhibiting the water-gas shift reaction below 500°C. The shift reaction is desired here as a means of reducing the concentration of CO.

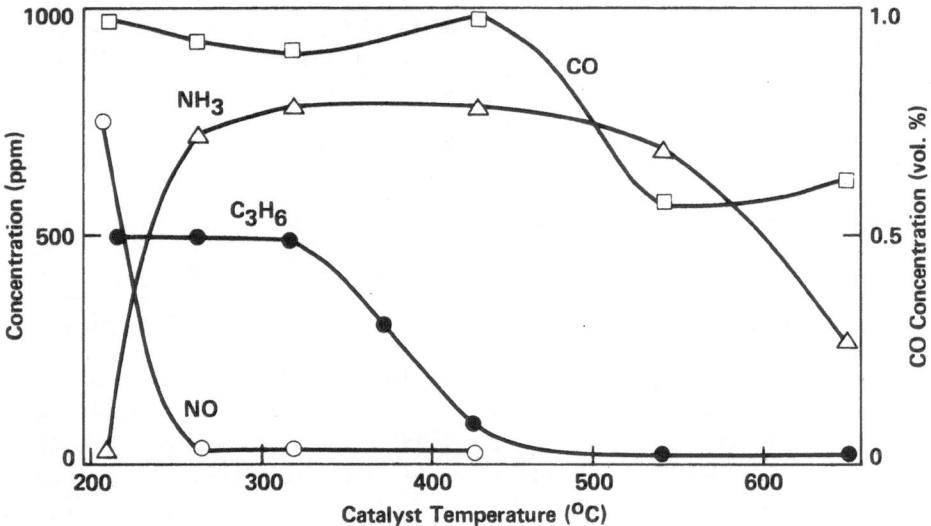

Fig. 7. Nitric oxide reduction over 0.1% Rh-Al$_2$O$_3$. Feed stream: 0.1% NO, 1.0% CO, 0.05% C$_3$H$_6$, 0.3% H$_2$, 10% CO$_2$, and 10% H$_2$O in N$_2$. GHSV = 38,000.

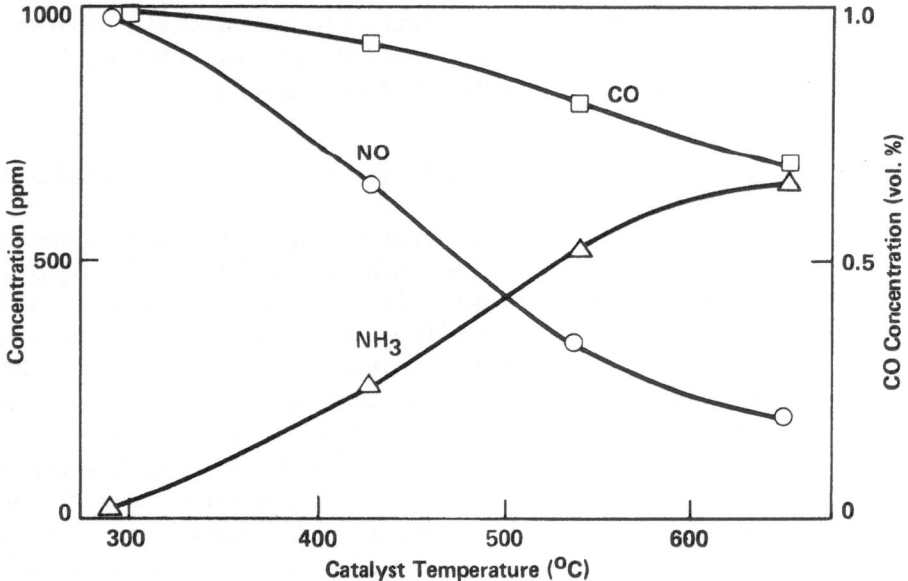

Fig. 8. Nitric oxide reduction over 0.11%% Pt-Al$_2$O$_3$. Feed stream: 0.1% NO, 1.0% CO, 0.3% H$_2$, 10% CO$_2$, and 10% H$_2$O in a N$_2$ atmosphere. GHSV = 38,000. ○NO, △NH$_3$. The catalyst was reduced at 650°C in this feed stream before taking this steady state conversion data.

To examine the ability of the Rh-Al$_2$O$_3$ catalyst to simultaneously control NO and CO in an exhaust-like feed, oxygen was added in steps until the feed became net oxidizing. At 538°C (Fig. 9), the removal of CO increased with oxygen content, presumably via CO oxidation. Separate experiments (Table 1) show that Rh-Al$_2$O$_3$ like Pt-Al$_2$O$_3$ readily promotes CO oxidation. Hydrogen concentration could not be measured here, but it was shown earlier that H$_2$ also decreases with oxygen addition. Presumably the water-gas shift reaction contributes to the overall result in as much as the CO removal by oxygen and the shift reaction appear to be superimposed. Most important and as seen for the NO-CO-H$_2$ feed, the added oxygen improved the selectivity of the reduction of NO to N$_2$ with no apparent loss in activity for NO reduction as long as the feed stream was net reducing. Comparison of the products with and without added propylene revealed that propylene does not influence the magnitude of the CO and NH$_3$ concentrations close to the stoichiometric point ($\pm$ 0.2% O$_2$).

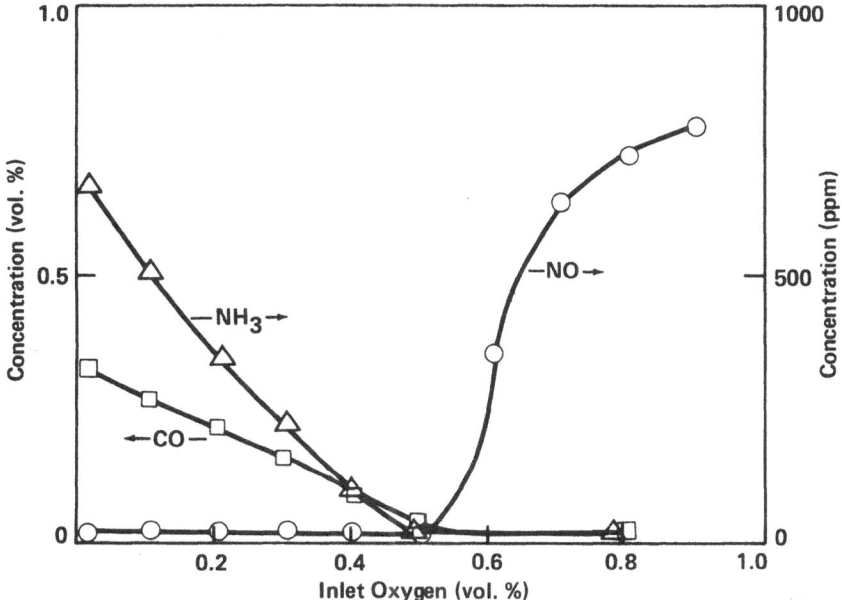

Fig. 9. Effect of oxygen concentration on nitric oxide reduction over Rh-Al$_2$O$_3$ at 538°C. Feed stream: 0.1% NO, 1.0% CO, 0.3% H$_2$, 10% CO$_2$, 10% H$_2$O, and O-1.0% O$_2$ in N$_2$. GHSV = 38,000.

At higher temperatures (e.g., 650°C) very little NH$_3$ is formed over Rh-Al$_2$O$_3$ at any oxygen concentration for the net reducing feed stream (Fig. 10). While at 425°C, the selectivity for N$_2$ formation does not improve with added oxygen except near the stoichiometric point (Fig. 11). The Pt-Al$_2$O$_3$ catalyst (Fig. 12) exhibited poor N$_2$ selectivity at 427°C and 538°C alike.

References pp. 186-187.

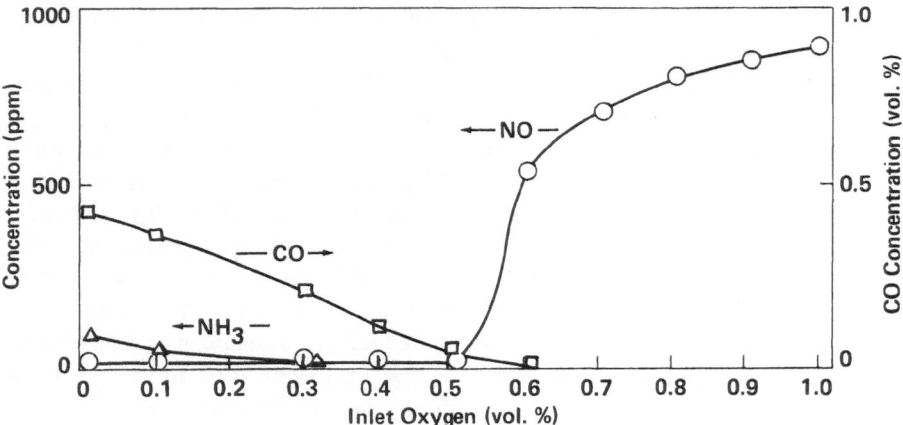

Fig. 10. Effect of oxygen concentration on nitric oxide reduction over $Rh-Al_2O_3$ at 650°C. Feed stream: 0.1% NO, 1.0% CO, 0.3% H_2, 10% CO_2, 10% H_2O, and 0-1.0% O_2 in N_2. GHSV = 38,000.

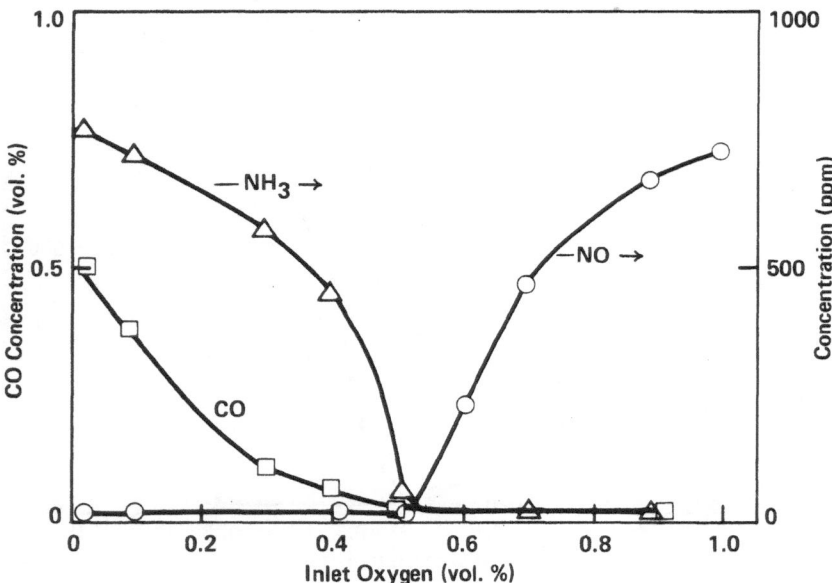

Fig. 11. Effect of oxygen concentration on nitric oxide reduction over $Rh-Al_2O_3$ at 427°C. Feed stream: 0.1% NO, 1.0% CO, 0.3% H_2, 10% CO, 10% H_2O, and 0-1% O_2. GHSV = 38,000.

We have thus seen that over $Rh-Al_2O_3$ there is less NH_3 near stoichiometry at higher temperatures for both the exhaust-like feed (Figs. 3, 9, and 10) and for the simple feed. In view of the better selectivity for N_2 formation above 500°C over $Rh-Al_2O_3$ compared to $Pt-Al_2O_3$ and $Pd-Al_2O_3$, we looked for a change in the reaction pathway as an explanation.

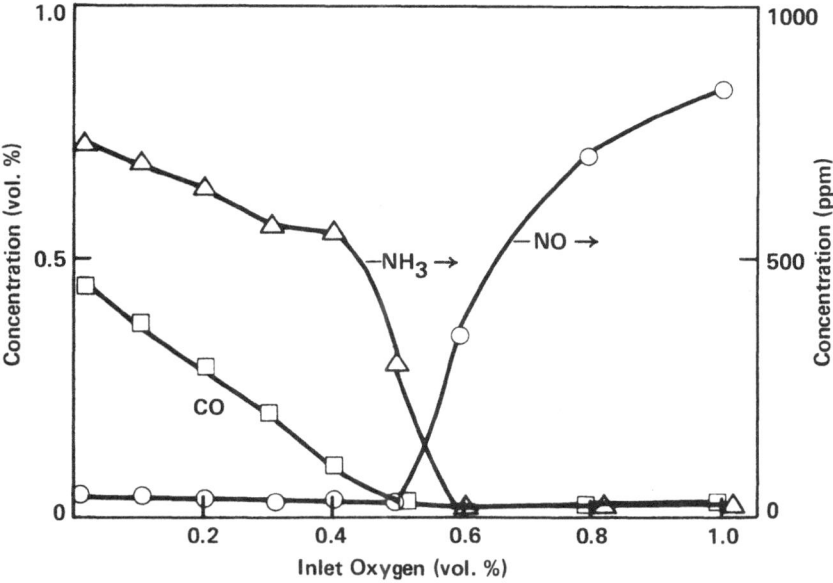

Fig. 12. Effect of oxygen concentration on nitric oxide reduction over Pt-Al_2O_3 at 538°C. Feed stream: 0.1% NO, 1.0% CO, 0.3% H_2, 10% CO_2, 10% H_2O, and 0-1.0% O_2 in a N_2 atmosphere.

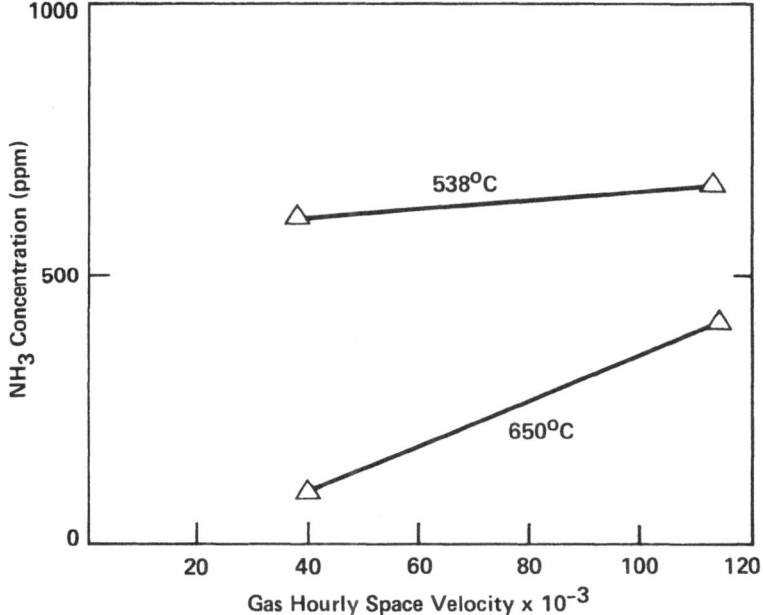

Fig. 13. Effect of space velocity on NH_3 formation and CO removal over Rh-Al_2O_3 at 538°C and 650°C. Feed stream: 0.1% NO, 1.0% CO, 0.3% H_2, 10% CO_2 and 10% H_2O in N_2.

References pp. 186-187.

First, we looked for an NH_3 intermediate with the reducing feed stream. Fig. 13 shows that the NH_3 in the product was increased with space velocity at 650°C though the increase was less at 538°C. Thus, NH_3 appears to be a gas phase intermediate for the reduction of NO to N_2 with the reducing feed stream over Rh-Al_2O_3 at the higher temperatures. Variable space velocity experiments showed that NH_3 is not a gas phase intermediate with Pt-Al_2O_3. The addition of NH_3 to the exhaust-like feed (Fig. 14) showed that NH_3 removal was only appreciable above 550°C, and again there is no NH_3 removal at lower temperatures. These NH_3 addition experiments are completely consistent with the space velocity results discussed above. It is well known that rhodium catalysts promote NH_3 decomposition (9) Fig. 15 shows the temperature dependence of the decomposition of 500 ppm NH_3 in N_2 with and without water vapor in the feed as well as in a feed containing O_2, CO, H_2, CO_2, and H_2O. While this experiment does not prove NH_3 decomposition is the path followed for removal of the NH_3 intermediate at 650°C, NH_3 decomposition activity is a property which the rhodium catalyst has in common with other catalysts for which the NH_3 intermediate path also has been proposed (3).

The tendency toward a steady improvement in the selectivity for N_2 formation with added oxygen at the higher temperature for rhodium suggests that oxygen participation in the removal of an NH_3 intermediate is one possible explanation. Thus, the reaction of 800 ppm NH_3 in the presence of 500 ppm O_2 was examined.

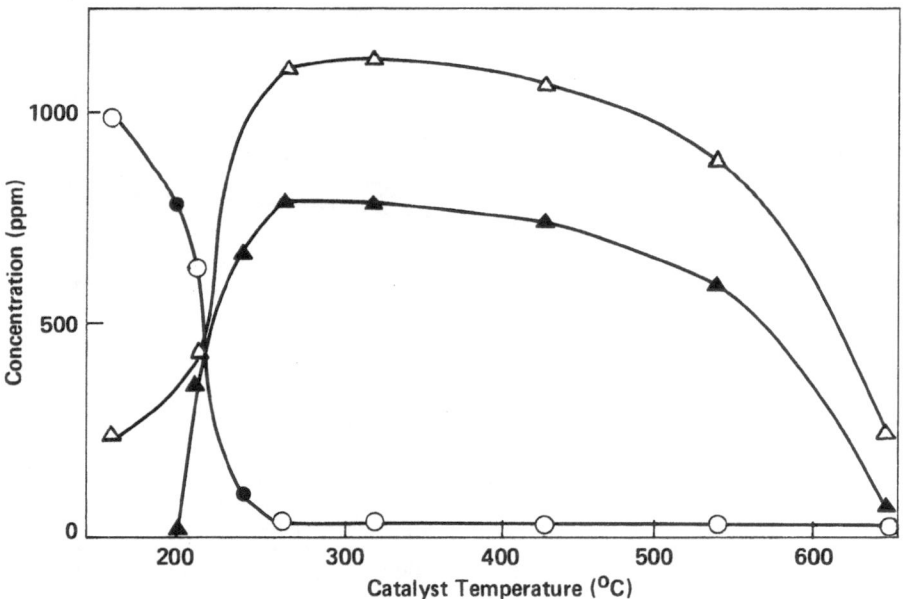

Fig. 14. NO reduction with NH_3 in the feed stream over Rh-Al_2O_3. Feed stream: 0.1% NO, 1.0% CO, 0.3% H_2, 10% H_2O, 10% CO_2, and 0 or 0.05% NH_3 in a N_2 atmosphere. GHSV = 38,000. ○●NO △▲NH_3. Open symbols no NH_3. Solid symbols 0.05% NH_3 in the feed.

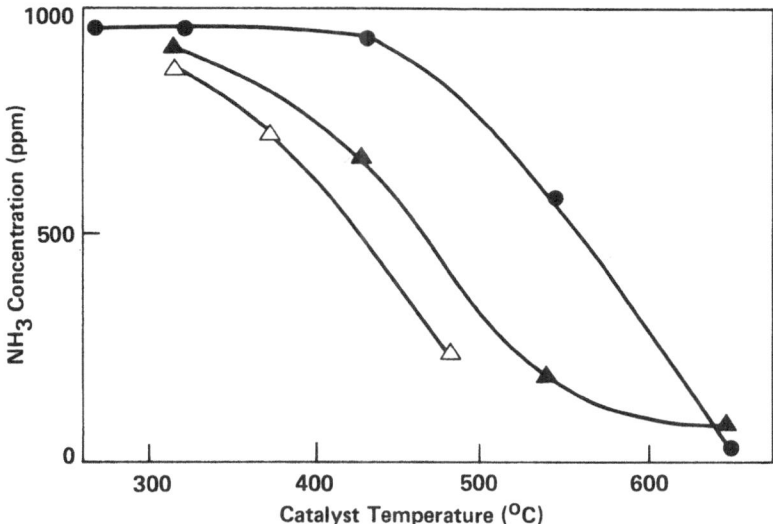

Fig. 15. NH$_3$ decomposition over Rh-Al$_2$O$_3$. Feed stream: $\triangle$ 0.1% NH$_3$ in a N$_2$ atmosphere. $\blacktriangle$ 0.1% NH$_3$ and 10% H$_2$O in N$_2$. $\bullet$ 0.08% NH$_3$, 0.3% O$_2$, 0.3% H$_2$, 10% CO$_2$, and 10% H$_2$O in N$_2$. GHSV = 38,000.

Fig. 16 shows that NH$_3$ was removed at a lower temperature and N$_2$ was the only nitrogen containing product. It might be proposed here that if O$_2$ is available over the catalyst subsequent to NO reduction, an NH$_3$-O$_2$ reaction could explain the improved

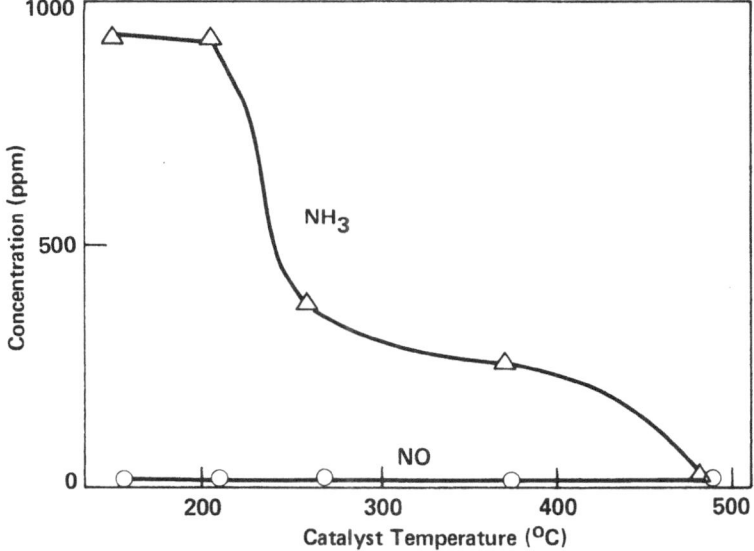

Fig. 16. NH$_3$ removal in the presence of O$_2$ over Rh-Al$_2$O$_3$. Feed stream: 800 ppm NH$_3$ and 500 ppm O$_2$ in a N$_2$ atmosphere. GHSV = 38,000. $\triangle$ NH$_3$, $\bigcirc$ NO.

References pp. 186-187.

selectivity for N_2 formation. Another possible explanation might be that the O_2 removes CO or H_2 which otherwise inhibit NH_3 decomposition. For example, when 1000 ppm NH_3 was reacted with 0.15% O_2 and 0.5% H_2 or CO (Fig. 17), again no NO was formed while NH_3 was completely converted to N_2 above 500°C. The higher level of oxygen is then compensated for by the reducing agent so selectivity for N_2 formation is maintained.

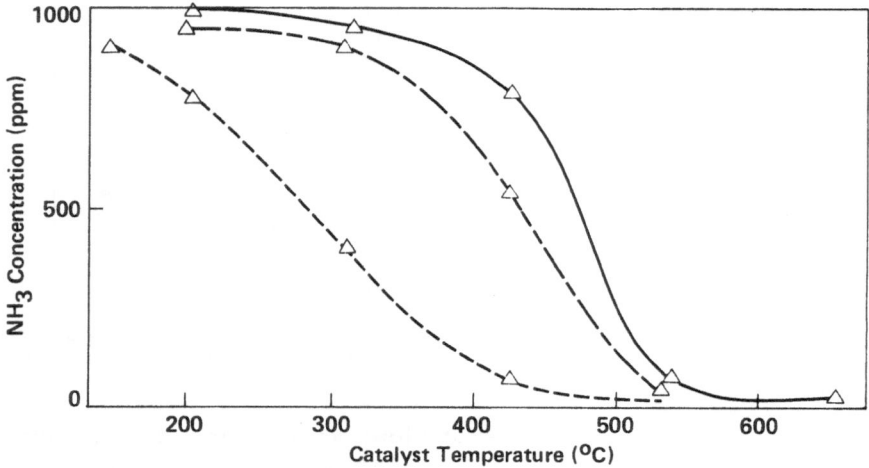

Fig. 17. NH_3 removal in the presence of O_2 and H_2 or CO over Rh-Al$_2$O$_3$. Feed stream:_____ 1000 ppm NH_3, and 0.5% H_2 in a N_2 atmosphere.____.____ 1000 ppm NH_3, 0.15% O_2, and 0.5% CO in N_2,_ _ _ _ _1000 ppm NH_3, 0.15% O_2, and 0.5% H_2 in N_2. GHSV = 38,000.

CONCLUSIONS

The basic chemical reactions involved in the simultaneous conversion of CO and NO have been investigated. Rh shows high temperature selectivity for N_2 rather than NH_3 formation. This selectivity appears to be related to the ability of Rh to decompose NH_3. In addition, its improved selectivity in the presence of oxygen appears to result from an oxygen assisted decomposition of ammonia rather than a selective removal of H_2, the precursor of ammonia.

REFERENCES

1. *J. O. Logan and C. G. Gerhold, Statement to Presidential Task Force on Air Pollution, Washington, D. C., March 6, 1970.*
2. *J. H. Jones, J. T. Kummer, K. Otto, M. Shelef, E. E. Weaver, Environ. Sci. Technol. 5, 790 (1971).*
3. *R. L. Klimisch and K. C. Taylor, Environ. Sci. Technol. 7, 127 (1973).*
4. *T. P. Kobylinski and B. W. Taylor, J. Catal. 33, 376 (1974).*

5. D. J. Ashmead, J. S. Campbell, P. Davies, K. Farmery, presented to Soc. Auto. Eng., Detroit, Mich., February, 1974. Paper No. 740249.
6. R. L. Klimisch and G. J. Barnes, Environ. Sci. Technol. 6, 543 (1972).
7. M. Shelef and H. S. Gandhi, Ind. Eng. Chem. Prod. Res. Develop. 11, 393 (1972).
8. K. C. Taylor and R. L. Klimisch, J. Catal. 30, 478 (1973).
9. A. Amano and H. Taylor, J. Amer. Chem. Soc. 76, 4201 (1954).

DISCUSSION

K. Hellman *(Environmental Protection Agency)*

Would you care to speculate on the effects of hydrocarbons in the feed gas on the reactions that you observed?

Taylor

I had one slide which showed NO reduction in an exhaust-like feed containing propylene. We saw that the propylene was removed completely at about 400° or 500°C. This feed was net reducing and there was not enough NO to oxidize the propylene. We have shown in separate experiments that the rhodium catalyst promotes the propylene — water reaction as well as the reaction of water with other hydrocarbons. From our laboratory experiments we conclude then that propylene and other hydrocarbons are removed efficiently at low temperatures.

C. H. Lee *(Gould Inc.)*

We have studied the NO reduction mechanism with our nickel-copper catalyst. With this metal catalyst we have seen similar results as have been presented today. We see that oxygen enhances the ammonia decomposition. With the addition of O_2, we see that the catalyst forms a lot less ammonia. I think the similarity between rhodium and nickel-copper is not totally unexpected, since rhodium is, among the noble metals, one of the most "base" metals with respect to catalytic behavior. We have developed evidence which shows that this oxygen effect could result from (1) a surface state effect or (2) direct reaction between oxygen and ammonia both leading to the formation of nitrogen and water. I wonder is you have checked whether or not the rhodium surface can be oxidized and the oxidized surface then enhance the ammonia decomposition.

Taylor

We have talked before about the reaction of oxygen with ruthenium catalysts which leads to an enhancement in their catalytic activity. The 0.1% rhodium catalyst used here seems to be very active and we do not see anything like the dual activity states which we saw with ruthenium. We have not done experiments to determine whether the rhodium surface is oxidized or reduced. Some time ago we did work on copper-nickel and platinum-nickel catalysts (3) and we found also that is we varied the oxygen concentration over these catalysts we saw about the same result as you are referring to.* That is, the ammonia concentration comes down quite nicely as we approach stoichiometry. I think it is sort of interesting to speculate that we may have the same sort of mechanism going on with the rhodium and the copper-nickel catalysts. All of these catalysts are ammonia decomposition catalysts. While it is not really fair to draw conclusions from catalyst comparisons of this sort the ammonia reaction does deem to be a common property of these catalysts which show good conversion of both NO and CO around stoichiometry.

G. J. K. Acres *(Johnson Matthey and Co., Ltd.)*

I think there may be another explanation regarding ammonia formation under your test conditions.

For example, you found that if you take the water out of your reaction mixture, then even in the absence of oxygen you form very little ammonia. In the presence of water and at oxygen concentrations close to stoichiometric, again, little ammonia is formed.

We agree that reforming and water-gas shift reactions take place which produce hydrogen. This in turn reacts with adsorbed nitrogen atoms to produce ammonia. However, in the presence of oxygen, the hydrogen is preferentially removed and little or no ammonia is formed.

Taylor

I think we can use these results to argue the other way though. The water may indeed participate in ammonia formation via an isocyanate mechanism as well as by its role in the shift reaction or a reforming reaction. We have shown that water interfers with NH_3 decomposition and that NH_3 reacts with oxygen under net reducing conditions to form N_2. We initially considered the argument you propose but found no evidence that selective H_2 removal by O_2 is responsible for lower ammonia formation near to stoichiometry.

W. Mannion *(Englehard Minerals and Chemicals Corp.)*

I wonder if you've had a chance to age these thermally, possibly, with automotive exhaust to see if the window stays the same width.

Taylor

This work's been done, but I don't have the results at my disposal today.

*R. L. Klimisch and K. C. Taylor, Env. Sci. and Tech., 7, 127 (1973).

D. P. McArthur *(Union Research Center)*

Kathy, if you keep the CO/O_2 ratio constant but increase your CO level what happens to your CO activity?

Taylor

We have not done that experiment.

G. H. Meguerian *(Amoco Oil Company)*

If you keep the CO-oxygen ratio constant and increase the CO level, the selectivity remains constant.

THE CATALYTIC REDUCTION OF NITRIC OXIDE OVER SUPPORTED RHODIUM AND COPPER NICKEL CATALYSTS

T. OHARA

Japan Catalytic Chemical Ind. Co., Ltd., Osaka, Japan

ABSTRACT

Catalytic reduction of nitric oxide with CO and H_2 reductants over Cu and/or Ni-Al_2O_3 catalysts and a Rh-Al_2O_3 catalyst was investigated. The comparative reactivity of CO and H_2 over the base metal catalysts and noble metal catalyst showed the same trends.

Among the various parameters, the presence of water and other reaction conditions (such as oxidizing or reducing atmosphere) caused substantial changes in the catalysts and the reactivity of the reducing agents.

Furthermore, nitric oxide (NO) reduction efficiencies over the base metal catalysts and noble metal catalyst depended upon the state of the catalytic elements, which depended upon the catalyst preconditioning.

INTRODUCTION

During the past several years, purification of auto exhaust gas by means of catalytic devices has been extensively studied. Catalytic elemination of carbon monoxide and hydrocarbons is approaching the final application stage, even though there remain some minor technical problems. On the contrary, a reliable catalyst for nitrogen oxide removal with superior durability has not been reported to date, and seems to be blocked by some technical problems which are in urgent need of a solution.

In the history of NO_x catalyst research, there have been two reaction schemes proposed; direct decomposition and reduction. Because of its slow reaction rate, direct decomposition of NO is hardly considered to be applicable. Reduction however seems to be more promising because it can be accomplished by various reductants including CO, H_2, and hydrocarbons which are present in exhaust gas.

In practical systems, other components such as CO_2 and H_2O would make the following reactions possible;

$$NO + CO \rightarrow 1/2\ N_2 + CO_2 \tag{1}$$

$$C_nH_m + (2n + m/2)NO \rightarrow (n + m/4)N_2 + nCO_2 + m/2\ H_2O \tag{2}$$

$$NO + H_2 \rightarrow 1/2\ N_2 + H_2O \tag{3}$$

$$NO + 5/2\ H_2 \rightarrow NH_3 + H_2O \tag{4}$$

$$CO + H_2O \rightarrow CO_2 + H_2 \tag{5}$$

While the products of equation (1), (2) and (3) are environmentally acceptable, the formation of NH_3 by equation (4) creates a major problem. In a dual converter system for complete purification of auto exhaust, any NH_3 produced in the reduction stage is reoxidized to NO in the second stage, thereby decreasing the overall efficiency of NO elimination.

It is often noted that the most effective reduction catalysts must possess low NH_3 formation activity. Among proposed NO_x reduction catalysts, monel metal catalyst (1), one of the promising alloy catalysts, showed poor activity at low temperatures. Platinum and palladium catalysts which are considered to be successful oxidation catalysts unfortunately show high NH_3 formation (2).

Ruthenium, as a promising element with low NH_3 formation (2), is not practical to use because of the subliming property of its oxide, RuO_4 (3), and the toxicity of this oxide. Base metal catalysts have generally lower activity than noble metal catalysts. Many base metal catalysts tend to undergo continuous chemical change under practical usage, such as spinel formation with alumina substrate.

In the present study, the author chose Cu and/or $Ni-Al_2O_3$ catalysts as typical base metal catalysts and a $Rh-Al_2O_3$ catalyst as a representative of noble metal catalysts. Catalytic reduction of NO by reductants such as H_2 and CO over these catalysts was investigated as a function of H_2O, which presumably accelerated NH_3 formation by the water-gas shift reaction, and as a function of added O_2 gas which presumably suppresses NH_3 formation (4).

The study was extended to oxidative and reductive heat treatment effects on catalytic activity as they relate to the possible conditions in exhaust gas purification. The characteristic behavior of Cu and/or $Ni-Al_2O_3$ and $Rh-Al_2O_3$ catalysts under these conditions will be a major topic of this paper.

EXPERIMENTAL

Catalyst Preparation — *Cu and/or Ni-Al$_2$O$_3$ Catalysts* — Active alumina pellets (American Cyanamid Company's extrudate, Aeroban®, diameter 3 mm, length 5 mm, bulk density 0.64 g/cm^3 surface area 90 m^2/g) were impregnated with an aqueous solution of cupric nitrate and/or nickel nitrate. The concentration of each solution was adjusted to yield 7.8 wt % deposition as metal on the finished catalysts. The impregnated pellets were dried at 120°C for 15 hours and then treated under the following conditions:

Method (1). Calcined at 50°C for 3 hours in an air stream.
　　　　　　(denoted as Cu/Ni − O, Cu − O and Ni − O)
Method (2). After method (1), additionally treated at 550°C for 3 hours in 10% H$_2$ − 90% N$_2$ gas stream.
　　　　　　(denoted as Cu/Ni-R, Cu-R and Ni-R)
Method (3). After method (2), additionally treated at 800°C for 2 hours in 10% H$_2$ − 90% N$_2$ gas stream.
　　　　　　(denoted as Cu/Ni-RS, Cu-RS and Ni-RS)
Method (4). After method (2), additionally treated at 980°C for 24 hours in an air stream.
　　　　　　(denoted as Cu/Ni-OS, Cu-OS and Ni-OS)

Rh-Al$_2$O$_3$ Catalyst — The same active alumina pellets were impregnated with an aqueous solution containing rhodium trichloride, then dried. Final treatments were carried out in the same manner as for the Cu/Ni-Al$_2$O$_3$ catalyst except for method (2) and are denoted as Rh − O, Rh-RS and Rh-OS, respectively. Instead of the previously mentioned method (2), impregnated pellets were reduced at 550°C for 3 hours in 10% H$_2$ − 90% N$_2$ gas immediately after drying, and are denoted as Rh-R. Two kinds of Rh-Al$_2$O$_3$ catalysts were prepared according to the previous procedure. They were different in Rh loading, that is 0.008 and 0.08 wt %, and are denoted as Rh-R or Rh-OS and Rh-IIR or RH-IIOS.

In Table 1, there are the comparative data of BET surface area (5) and bulk density for the catalysts used in this study.

Apparatus and Procedure — The experiments were carried out by using the continuous flow system in Fig. 1. The reactor which consisted of 25 mm ID stainless steel pipe was situated in an electrically heated furnace. 10 cm^3 of catalyst was supported on a piece of stainless steel screen. 200 cm^3 of porcelain chips were tightly placed in a concentric preheating zone which surrounded the reaction tube to ensure complete mixing of reactants before they entered the catalyst bed.

A thermocouple was located at the entrance of the catalyst bed. This temperature reading was used for regulating reaction temperatures from 300 to 600°C with an electronic temperature controller.

References p. 212.

TABLE 1

Bulk Density and Surface Area of Catalysts

Catalyst	Bulk Density (g/ml)	Surface Area (m²/g)
Cu/Ni-R	0.72	82.8
Cu/Ni-O	0.73	83.5
Cu/Ni-RS	0.71	80.4
Cu/Ni-OS	0.87	18.1
Cu-R	0.71	84.7
Cu-O	0.73	85.2
Cu-RS	0.70	81.5
Cu-OS	0.83	11.8
Ni-R	0.70	77.0
Ni-O	0.72	79.2
Ni-RS	0.71	77.5
Ni-OS	0.72	65.3
Rh-R	0.64	87.8
Rh-O	0.66	82.9
Rh-OS	0.67	72.3
Rh-II R	0.63	92.6
Rh-II OS	0.65	74.3
Al₂O₃	0.64	90.3

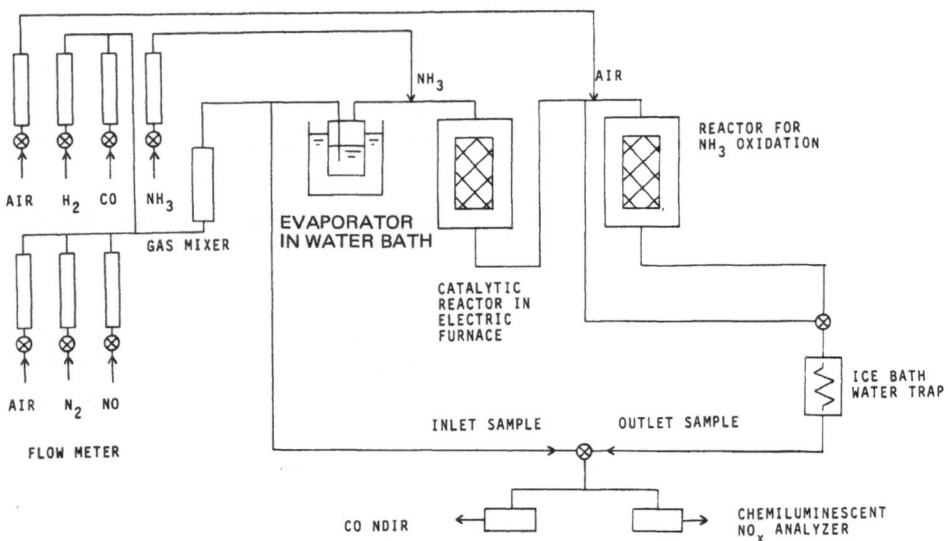

Fig. 1. Experimental setup for catalyst studies.

The inlet gas concentration was calculated by measuring the flow rate of cylinder gases through calibrated rotameters. The cylinder gases used were:

$$100 \% \text{ CO}, 30\% \text{ H}_2 - \text{N}_2 \text{ balance}, 5\% \text{ NO} - \text{N}_2 \text{ balance},$$
$$5\% \text{ NH}_3 - \text{N}_2 \text{ balance}, 100\% \text{ N}_2, 100\% \text{ air}.$$

Steam was added by bubbling the previously mixed reaction gas (except NH_3) through hot water at 50°C. The gas was again thoroughly mixed in the preheating zone, then introduced to the catalyst bed from the bottom at a flow rate corresponding to 45,000 GHSV.

Analysis – The detector used for NO_x was a YANAGIMOTO (Japan) ECL-7S chemiluminescent NO_x analyzer. In order to determine NH_3 concentration, the gas mixture was introduced over a Pt-Pd-Al_2O_3 catalyst (Pt: 0.14%, Pd: 0.05%) with 6 vol. % of oxygen to convert the NH_3 to NO_x, then analyzed by the former NO_x analyzer. While conversion efficiency of NH_3 to NO_x over this catalyst was 99% without NO, it decreased to 92% in the presence of NO. In the following data, the net conversion value of NO was adjusted according to these NH_3 conversion efficiencies.

Although N_2O may be formed over Rh-Al_2O_3 catalyst at 300°C (6), the author has not analyzed for this compound.

Each NO_x concentration was measured at the reasonable steady state of the reaction. As Bauerle pointed out (7), it took a fairly long time (sometimes nearly 1 hour) to obtain steady state over the Cu/Ni-Al_2O_3 catalyst. It is worth noting that measurement precision for base metal catalysts is possibly in error by ±3%.

For the Rh-Al_2O_3 catalyst, steady state was easily obtained and showed good reproducibility.

RESULTS AND DISCUSSION

NO Reduction with CO and H_2 over Cu/Ni-Al_2O_3 and Rh-Al_2O_3 — NO reduction over the Cu/Ni-Al_2O_3 and Rh-Al_2O_3 catalysts was examined as a function of CO and H_2 concentration.

In Fig. 2, the results for the NO-CO reaction are shown and in Fig. 3, those for the NO-H_2 reaction. The open circles with dashed lines represent gross NO conversion values and solid circles with solid lines represent net NO conversion. The difference between them corresponds to NH_3 formation.

Using CO as the reductant, NO conversion over the Cu/Ni-Al_2O_3 catalyst increased with CO concentration rise. On the contrary, NO conversion to N_2 over the Rh-Al_2O_3 catalyst dropped considerably at 300°C. This implies that chemisorption of CO on the RH-Al_2O_3 catalyst is extremely strong as compared to CO chemisorption on the Cu/Ni-Al_2O_3 catalyst.

References p. 212.

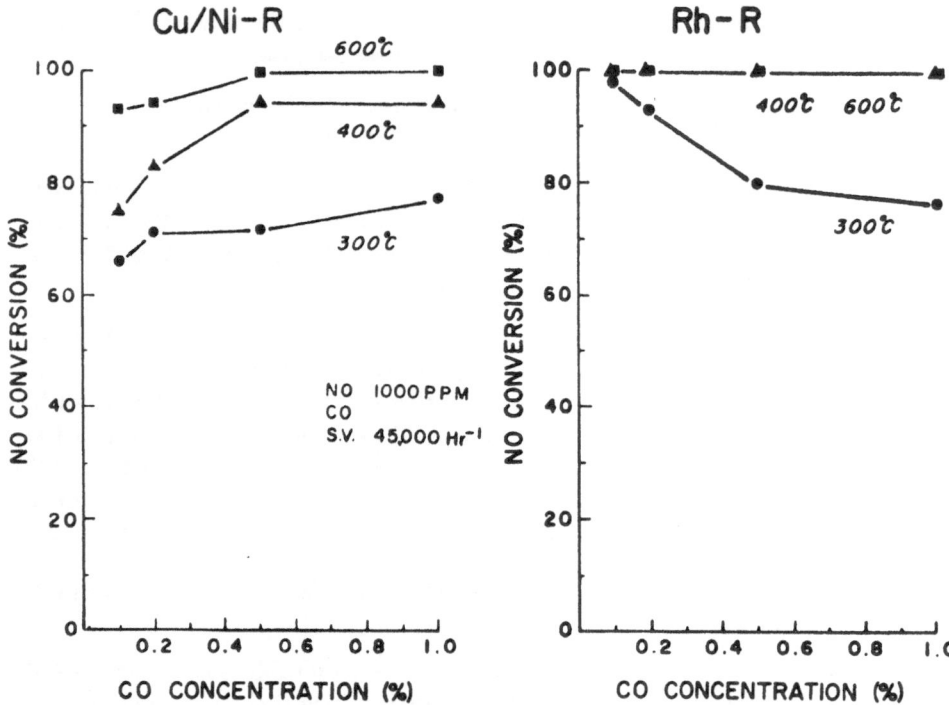

Fig. 2. Conversion of NO to N_2 over Cu/Ni-R and Rh-R as a function of CO concentration. Feed stream: 1000 ppm NO, variable CO, and balance N_2. Space velocity: 45,000 hr-1.

At higher than 400°C, the Rh-Al$_2$O$_3$ catalyst showed almost 99% NO conversion efficiency at any CO concentration. NO chemisorption may easily take place to accelerate the reaction by increasing the temperature. This could be speculated to occur by assuming that the chemisorption of CO is lessened by the temperature increase, or that the rate of the CO-NO reaction on the catalyst surface is accelerated.

In the case of H$_2$ as reductant, NO conversion over the Cu/Ni-Al$_2$O$_3$ catalyst increased with an increase in H$_2$ concentration which is very similar to the case of CO as reductant. The Rh-Al$_2$O$_3$ catalyst also promoted NO conversion without any interference from further addition of H$_2$ which contrasts with the interference caused by CO at 300°C. One can hardly see a difference in reactivity between CO and H$_2$. However, NO conversion efficiencies over both catalysts at or near the stoichiometric point reveal the superiority of CO over H$_2$.

When H$_2$ is used, NH$_3$ was formed over both catalysts, but the amount of NH$_3$ formed over the Cu/Ni-Al$_2$O$_3$ catalyst was slightly less than over the Rh-Al$_2$O$_3$ catalyst.

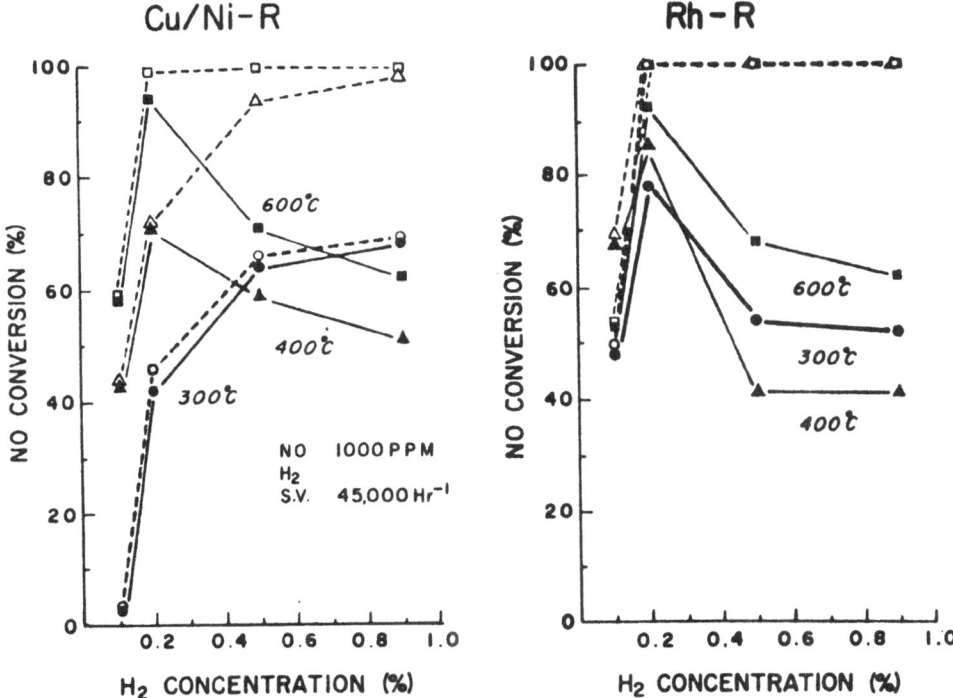

Fig. 3. Conversion of NO over Cu/Ni-R and Rh-R as a function of H_2 concentration. Feed stream: 1000 ppm NO, variable H_2, and balance N_2. Space velocity: 45,000 hr^{-1}. --- Gross NO conversion,—Net NO conversion.

NH_3 formation over both the Cu/Ni-Al$_2$O$_3$ and Rh-Al$_2$O$_3$ catalysts reached a maximum at 400°C. The fact that NH_3 formation over Cu/Ni-Al$_2$O$_3$ at 300°C is very small is perhaps because NO chemisorption on the catalyst surface becomes dominant due to relatively weak H_2 chemisorption. Consequently, the increase in the number of adjacent chemisorbed NO molecules favors preferential reduction of NO to N_2.

On the Rh-Al$_2$O$_3$ catalyst, an appreciable amount of NH_3 formation was observed even at 300°C. It is considered that the higher activity of the Rh-Al$_2$O$_3$ catalyst, as compared to the Cu/Ni-Al$_2$O$_3$ catalyst, produced H_2 chemisorption as well as NO chemisorption at unexpectedly high rates; therefore, NH_3 formation and N_2 formation occur simultaneously.

The decrease of NH_3 formation over both Cu/Ni-Al$_2$O$_3$ and Rh-Al$_2$O$_3$ at 600°C can be attributed to acceleration of the rate of NH_3 decomposition in a reductive atmosphere at this same temperature (Table 2).

O_2 Effects in NO Reduction Reaction – In Fig. 4 and Fig. 5, there are plots of the effect of additional oxygen gas using reaction conditions of the preceding section.

References p. 212.

TABLE 2

NH$_3$ Conversion over Cu-Ni-Al$_2$O$_3$ and Rh-Al$_2$O$_3$

(NH$_3$, 1000 ppm; O$_2$, 0-0.2 %)

Catalyst	Temp. °C	300			400			600		
	O$_2$ Conc. %	0	0.1	0.2	0	0.1	0.2	0	0.1	0.2
Cu/Ni-R	A	13	70	70	13	95	99	95	99	100
	B	0	1	1	0	1	4	0	5	24
Cu-R	A	8	32	52	21	95	100	86	100	100
	B	0	0	0	0	1	5	0	5	21
Ni-R	A	6	17	16	6	35	47	98	99	99
	B	0	1	1	0	2	4	0	12	46
Rh-R	A	11	99	99	17	99	100	94	100	100
	B	0	3	5	0	3	28	0	10	75

A, NH$_3$ Conversion; B, Conversion to NO, %

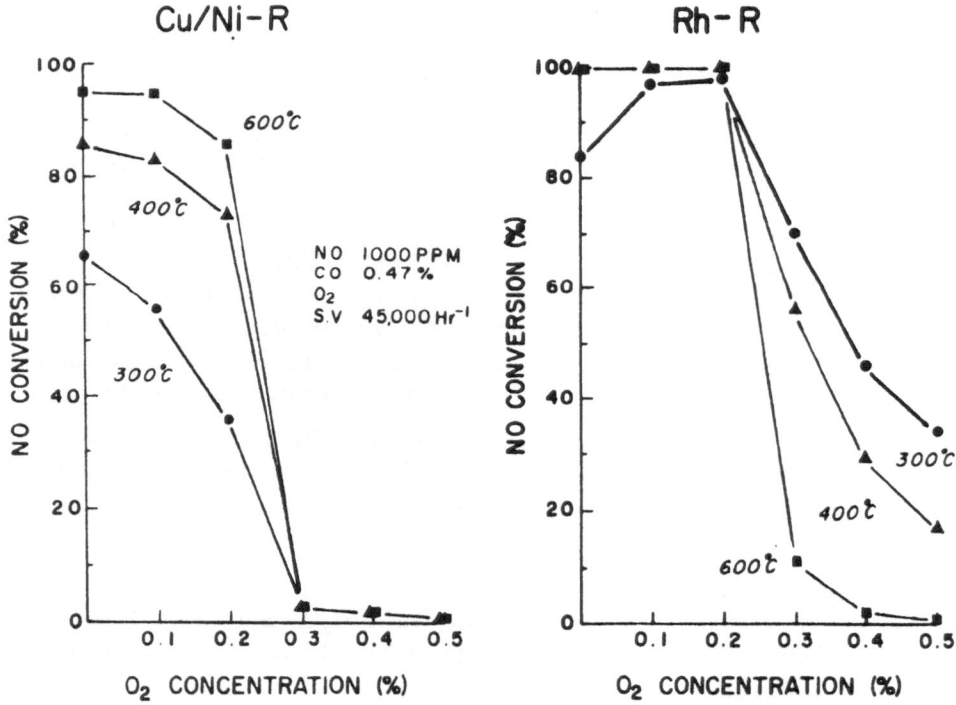

Fig. 4. The effect of oxygen on the reduction of NO by CO over Cu/Ni-R and Rh-R. Feed stream: 1000 ppm NO, 0.47% CO, variable O$_2$, and the balance N$_2$. Space velocity: 45,000 hr^{-1}.

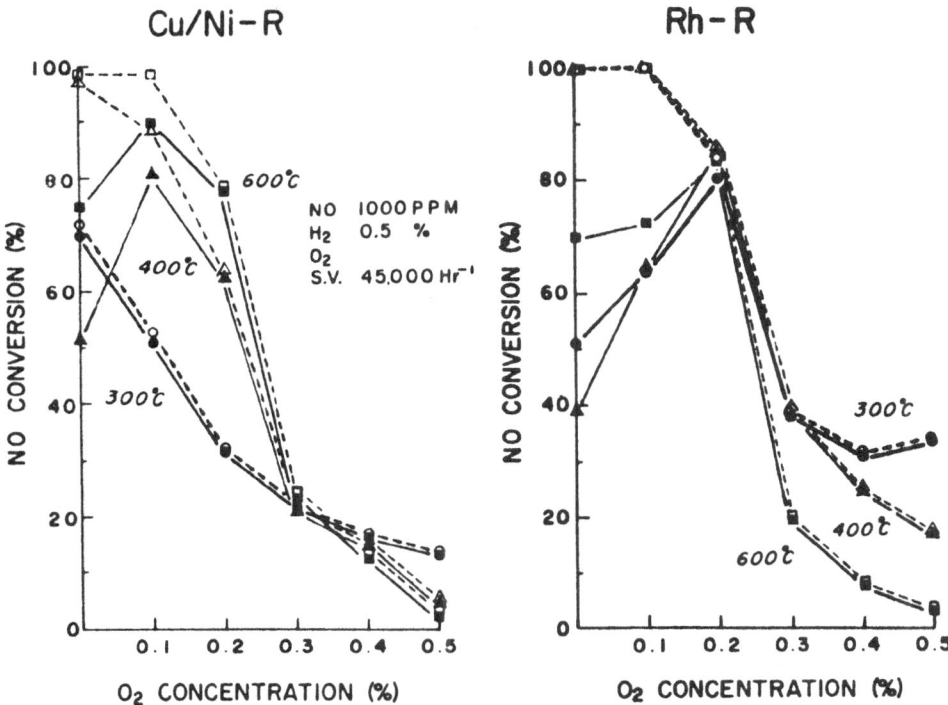

Fig. 5. The effect of oxygen on the reduction of NO by H_2 over Cu/Ni-R and Rh-R. Feed stream: 1000 ppm NO, 0.5% H_2, variable O_2, and balance N_2. Space velocity: 45,000 hr^{-1}. - - - Gross NO conversion, — Net conversion.

When CO was used as reductant, almost the same trend was observed for NO conversion over both Cu/Ni-Al_2O_3 and Rh-Al_2O_3 in the reductive atmosphere. However, the addition of oxygen gas suppressed the inhibiting effect of CO on NO reduction over the Rh-Al_2O_3 catalyst; therefore, a high NO conversion rate was obtained. The higher activity of Rh-Al_2O_3 compared to Cu/Ni-Al_2O_3 was similar to the activity results in the preceding section.

In an oxidative atmosphere, NO could not be converted over the Cu/Ni-Al_2O_3 catalyst at any O_2 concentration and almost all the CO was consumed by O_2 combustion. The Rh-Al_2O_3 catalyst promoted the reaction to a significant extent even though NO conversion efficiency decreased with increase in O_2 concentration at the lowest temperatures. As the supposed chemisorption of CO on Rh-Al_2O_3 catalyst tends to become dominant at lower temperature ranges, NO could be converted effectively. This led to the conclusion that Rh-Al_2O_3 can be active in a wider range including slightly oxidative atmospheres, as compared to Cu/Ni-Al_2O_3 catalysts.

When H_2 was used as the reductant, Rh-Al_2O_3 showed superior activity to Cu/Ni-Al_2O_3 in the reductive atmosphere just as in the case of CO as reductant.

References p. 212.

Although CO provided a little advantage in reactivity over H_2, generally CO is preferred to H_2 as explained in the preceding section.

In oxidative atmospheres, selective reduction of NO with H_2 was observed over the $Cu/Ni-Al_2O_3$ catalyst as well as over the $Rh-Al_2O_3$ catalyst. This tendency was not observed in NO reduction with CO over the $Cu/Ni-Al_2O_3$ catalyst. Furthermore, their selectivities were improved when the temperature was lowered. This has not been completely elucidated yet.

Experimental observations on NH_3 formation in the reductive atmosphere agreed with the results obtained in the preceding section regarding the following points:

1) Extremely low NH_3 formation was observed over the $Cu/Ni-Al_2O_3$ catalyst at 300°C
2) The NH_3 formation rate reached a maximum at 400°C on both $Cu/Ni-Al_2O_3$ and $Rh-Al_2O_3$.

Therefore, the explanation postulated in the preceding section can be applied here even under the condition where O_2 is present.

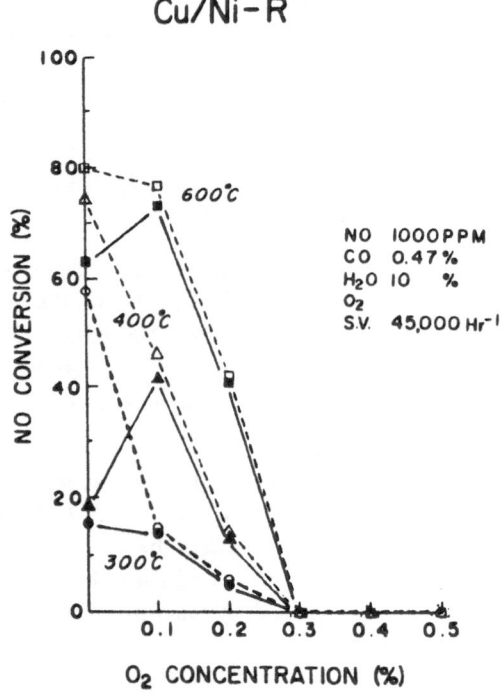

Fig. 6. The effect of H_2O on NO reduction over Cu/Ni-R. Feed stream: 1000 ppm NO, 0.47% CO, 10% H_2O, variable O_2, and balance N_2. Space velocity: 45,000 hr^{-1}. - - - Gross NO conversion,—Net NO conversion.

As seen in Fig. 5, it is notable that little NH_3 formation was observed over both catalysts in oxidizing atmospheres and also at 0.2% O_2 concentration, which is very close to the stoichiometric oxygen concentration. This implies that oxygen gas is suppressing NH_3 formation successfully.

H_2O Effect on NO Reduction Reaction with Addition of O_2 — Experimental results of H_2O addition under the reaction conditions of the preceding section are shown in Figs. 6 and 7 with CO as the reductant and in Fig. 8 with H_2 as the reductant.

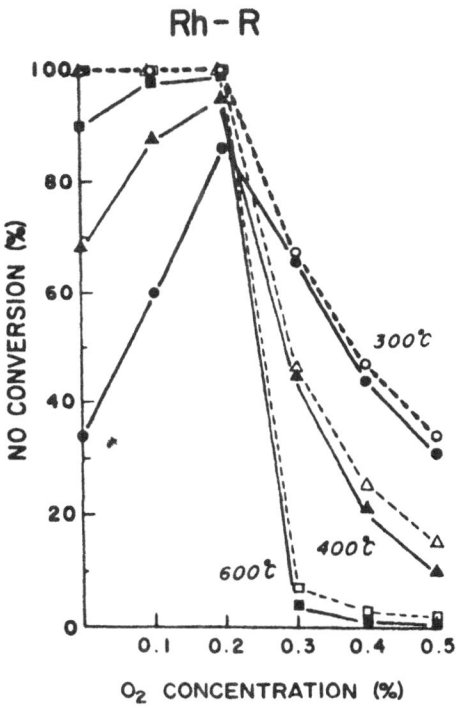

Fig. 7. The effect of H_2O on NO reduction over Rh-R. Feed stream: 1000 ppm NO, 0.47% CO, 10% H_2O, variable O_2, and balance N_2. Space velocity: 45,000 hr^{-1}. – – –Gross NO conversion,—— Net NO conversion.

Using CO as reductant over the Cu/Ni-Al$_2$O$_3$ catalyst, NO conversion activity was greatly affected by the addition of H_2O, presumably due to an H_2O poisoning effect. The comparison of the data with or without H_2O shows a similar tendency in both oxidative and reductive atmospheres. In Fig. 5, a negligible amount of NH_3 formation was detected at 300° over Cu/Ni-Al$_2$O$_3$, while in Fig. 6, NH_3 formation was observed even at 300°C without O_2 addition. This can be understood by assuming that atomic hydrogen produced by the water-gas shift reaction has a much higher reactivity than molecular hydrogen.

References p. 212.

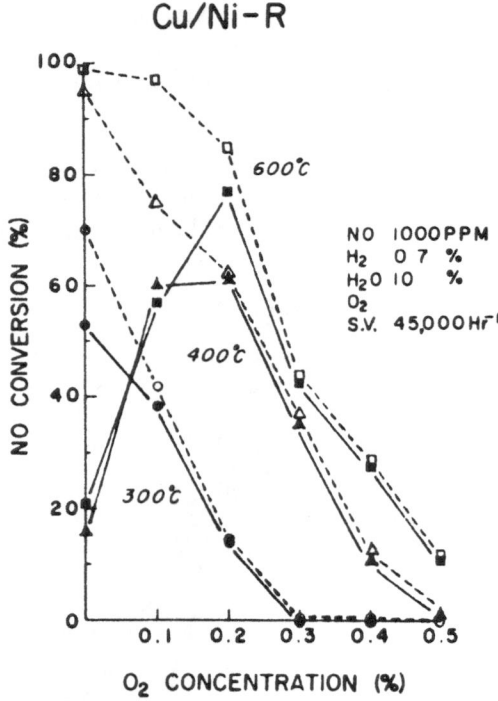

Fig. 8. The effect of H_2O on NO reduction over Cu/Ni-R and Rh-R. Feed stream: 1000 ppm NO. 0.7% H_2, 10% H_2O, variables O_2, and balance N_2. Space velocity: 45,000 hr^{-1}. – – –Gross NO conversion,—Net NO conversion.

In the case of the Rh-Al_2O_3 catalyst (Fig. 7), the large deactivation effect of H_2O as seen for Cu/Ni-Al_2O_3 was not observed. This could be explained by a weakening of the extremely strong CO chemisorption by H_2O which consequently makes NO chemisorption more effective. This explanation is supported by NO conversion efficiency at 300°C with no O_2.

Using the combination of CO and H_2O over the Rh-Al_2O_3 catalyst, an appreciable amount of NH_3 was formed at 300°C, presumably due to the contribution of the water-gas shift reaction. This is the very distinctive behavior which differs from that of molecular hydrogen on Rh-Al_2O_3 and was described above.

As discussed in the preceding section, oxygen gas obviously suppressed NH_3 formation even though H_2O was present in the feed gas. Particularly in the case of H_2 reductant, NH_3 formation over the Cu/Ni-Al_2O_3 catalyst was dependent upon O_2 concentration, for example a sharp increase in NH_3 formation was noted when O_2 was not present (Fig. 8).

Ammonia Conversion over Catalysts — It is easily understood that molecular hydrogen or atomic hydrogen produced by the water-gas shift reaction has a great influence on the NO reduction reaction. It will be worthwhile to investigate the behavior of NH_3 on the surface of the NO_x catalysts.

In Table 2, NH_3 conversion efficiencies over four catalysts, $Cu/Ni-Al_2O_3$, $Cu-Al_2O_3$, $Ni-Al_2O_3$ and $Rh-Al_2O_3$ catalysts are illustrated as a function of added O_2 concentration; namely, 0, 0.1, and 0.2 vol. % at 300, 400 and 600°C.

In this system, the following reactions were postulated:

$$2 NH_3 \rightarrow N_2 + 3 H_2 \tag{1}$$

$$2 NH_3 + 3/2 O_2 \rightarrow N_2 + 3 H_2O \tag{2}$$

$$2 NH_3 + 5/2 O_2 \rightarrow 2 NO + 3 H_2O \tag{3}$$

As seen in Table 2, NH_3 conversion was increased greatly over all four catalysts at any temperature by adding oxygen. Ammonia decomposition was found to be extremely difficult without O_2. At lower temperatures, NH_3 decomposition is accelerated by oxygen and seems to proceed via equation (2).

Among base metal catalysts, the $Cu/Ni-Al_2O_3$ catalyst showed a higher NH_3 conversion rate at 300°C than the $Cu-Al_2O_3$ catalyst, which agrees well with the synergism proposed by Klimisch (4). It is notable that Cu in the $Cu/Ni-Al_2O_3$ catalyst became the predominant element for the NH_3 conversion reaction at 400°C and its activity was almost equal to that of the $Rh-Al_2O_3$ catalyst at that point.

In spite of poor activity at lower temperatures (300 and 400°C), nickel became an effective element for NH_3 conversion above 600°C. For NO removal over the $Cu/Ni-Al_2O_3$ and $Rh-Al_2O_3$ catalysts, the lower NH_3 formation observed at 600°C was presumably due to higher NH_3 decomposition efficiency such as 94% at this temperature (Table 2).

On the other hand, $Rh-Al_2O_3$ exhibited superior NH_3 conversion activity, particularly at 300°C in the presence of O_2, while the base metal catalysts did not.

However, there is the possibility of the simultaneous occurrence of reactions (1), (2) and (3) due to the presence of O_2. In fact, NH_3 conversion to NO by reaction (3) over $Rh-Al_2O_3$ and $Ni-Al_2O_3$ with 0.1% O_2 at 600°C was found to be about 10 to 12%. With 0.2% O_2 present, $Rh-Al_2O_3$ converted NH_3 to NO by 28% at 400°C and 75% at 600°C, while the $Ni-Al_2O_3$ catalyst showed only 46% conversion at 600°C. The $Cu/Ni-Al_2O_3$ catalyst converted only 20 to 24% under the same conditions.

Selectivity for the conversion of NH_3 to NO was therefore found to be higher over $Rh-Al_2O_3$ than over Cu or $Cu/Ni-Al_2O_3$. Alternatively, the $Cu/Ni-Al_2O_3$ catalyst had superior selectivity for the NH_3 to N_2 reaction compared to the $Rh-Al_2O_3$ catalyst.

References p. 212.

Effect of Heat Treatment on Catalyst Activity — In Table 1, there is a summary of BET surface area and bulk density of catalysts before and after oxidative and reductive heat treatments at 550°C and 800°C.

It is clear that Cu/Ni-OS and Cu-OS showed the most distinctive change in physical properties, with a fairly moderate change for Ni-OS, and the smallest change for Rh-OS. X-ray diffraction analysis confirmed the formation of $CuAl_2O_4$, the conversion of a portion of active alumina to the α form in Cu/Ni-OS and Cu-OS, and the formation of $NiAl_2O_4$ and α-alumina in Ni-OS.

As for Rh-Al_2O_3, Rh $-$ O oxidized at 550°C was compared with Rh-R reduced at the same temperature, with respect to their NO conversion efficiencies with CO in the presence of O_2 and H_2O (Fig. 9). In contrast to Rh-R, Rh $-$ O showed a large drop in gross NO conversion at 300°C in the reductive atmosphere. Some possible explanations may be strong CO chemisorption on Rh $-$ O catalyst, slow water-gas shift reaction, consumption of atomic hydrogen by Rh $-$ O catalyst, or weak NO adsorption on Rh $-$ O catalyst, etc. The author will not try to specify the major reason at this point.

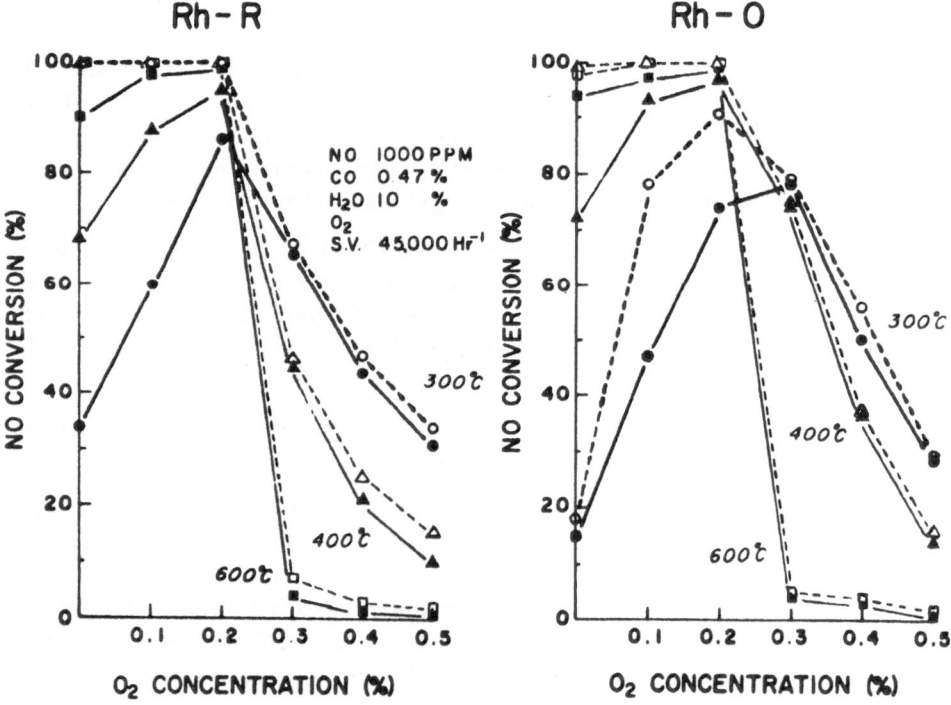

Fig. 9. The effect of catalyst oxidation or reduction at 550°C on NO conversion efficiency over Rh-R and Rh-O. Feed stream: 1000 ppm NO, 0.47% CO, 10% H_2O, variable O_2, and balance N_2. Space velocity: 45,000 hr^{-1}. ---Gross NO conversion, —Net NO conversion.

In the oxidative atmosphere, Rh − O showed slightly higher NO conversion efficiency than Rh-R, but both are considered to be of the same magnitude.

When the Rh-R catalyst was treated at 800°C in a reductive atmosphere, it provided almost the same performance as the original Rh-R catalyst. In addition, no significant change in physical properties of the catalyst was detected.

Heat treatment of the Rh-R catalyst at 980°C in an oxidative atmosphere (Rh-OS) offered interesting results, as seen in Fig. 10. When the reaction was started in oxidative atmopshere (shown in the right side of Fig. 10), NO conversion was extremely low at 300°C and 400°C with about 85% conversion at 600°C with no O_2. In contrast, the reaction started in a reductive atmosphere (shown at the left side of Fig. 10) provided the same extremely low efficiency at 300°, but effectiveness was improved at 400°C and approached the activity of the Rh-R catalyst at 600°C.

The Rh-IIOS (heavily loaded catalyst) showed better performance above 400°C as seen in Fig. 11.

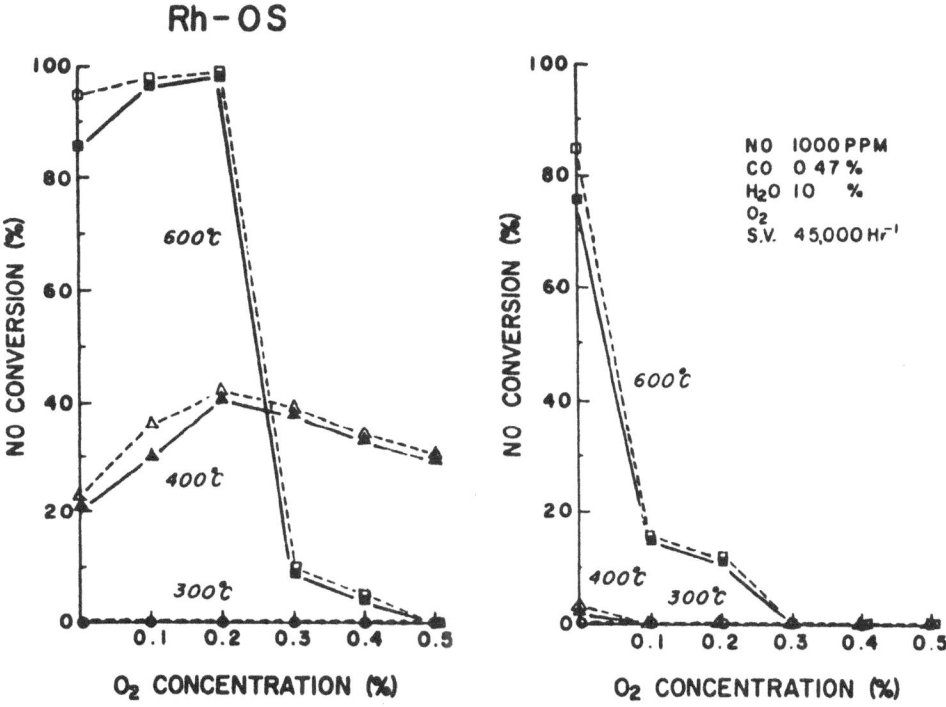

Fig. 10. The effect of an oxidizing heat treatment (980°C) on NO conversion activity over Rh-OS. Feed stream: 1000 ppm NO, 0.47% CO, 10% H_2O, variable O_2, and balance N_2. Space velocity: 45,000 hr[-1]. Left figure: Reaction started in a reductive atmosphere. Right figure: Reaction started in oxidative atmosphere. - - - Gross NO conversion,—Net NO conversion.

References p. 212.

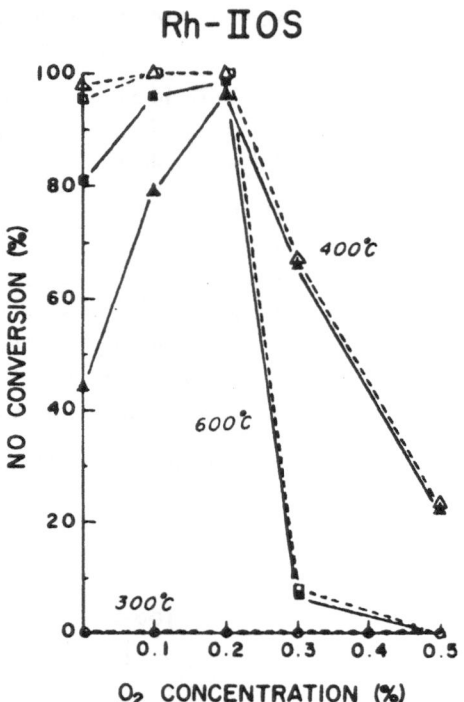

Fig. 11. The effect of an oxidizing heat treatment (980°C) on NO conversion over Rh-IIOS. Feed stream: 1000 ppm NO, 0.47% CO, 10% H_2O, variable O_2, and balance N_2. Space velocity: 45,000 hr^{-1}. – – – Gross NO conversion,——Net NO conversion.

X-ray diffraction analysis revealed that a phase transformation of active alumina to the α form occurred. Therefore, deactivation of the Rh-OS catalyst could be explained by extinction of active sites because of light loading (0.008 wt %), while the Rh-IIOS catalyst maintains a fairly large number of active sites because of the ten-fold heavier loading (0.08 wt %). It is concluded that metal loading is important for thermal stability of NO_x catalysts.

Base metal catalysts were also treated at high temperature in the same manner as Rh-Al$_2$O$_3$. As described earlier in the "Analysis" section, base metal catalysts, especially if treated at 980°C, require a very long period to reach steady state (as compared with the Rh-Al$_2$O$_3$ catalyst). Because the data for base metal catalysts have less precision, the author tried to grasp the overall comparison between base metal catalysts and noble metal catalysts, rather than a precise discussion in fine detail.

As illustrated in Fig. 12, the Cu – O catalyst showed slightly poorer activity than the reduced Cu-R catalyst. According to X-ray diffraction analysis, cupric oxide was found in the Cu – O catalyst and copper metal was found in the Cu-R catalyst. The

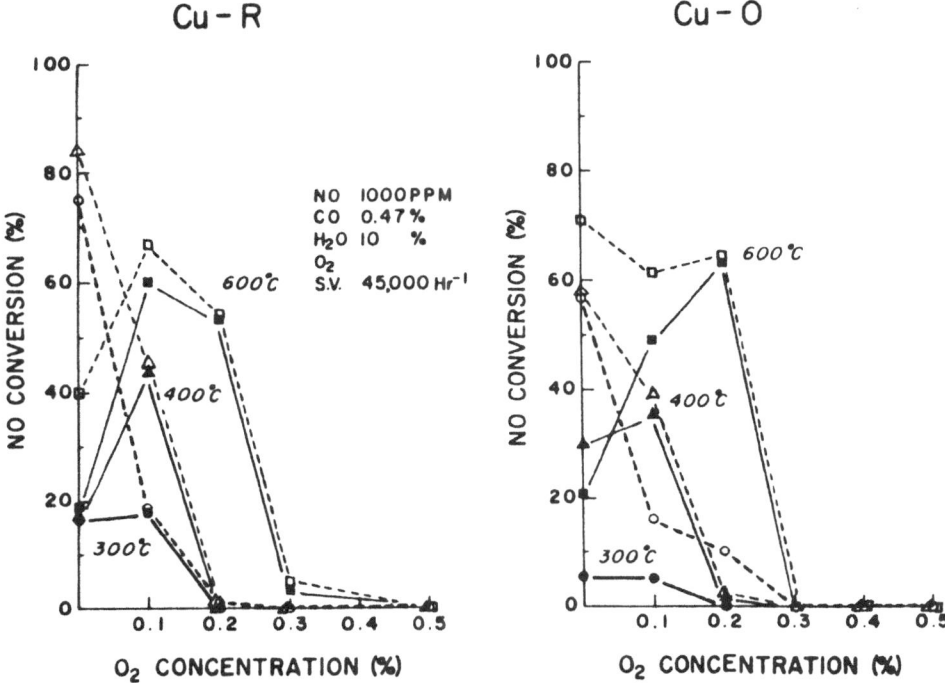

Fig. 12. The effect of catalyst oxidation or reduction at 550°C on NO conversion over Cu-R and Cu-O. Feed stream: 1000 ppm NO, 0.47% CO, 10% H₂O, variable O₂, and balance N₂. Space velocity: 45,000 hr⁻¹. $- - -$ Gross NO conversion, $-$ Net NO conversion.

conversion of Cu metal to its oxide form may contribute to lower NO conversion efficiency.

As illustrated in Fig. 13, the Cu/Ni $-$ O catalyst also showed slightly inferior activity to the Cu/Ni-R catalyst. However, the Ni $-$ O and Ni-R catalysts showed very poor activity, with less than 10% conversion, even in reductive atmosphere at 600°C. Unfortunately, little information about the Ni-Al₂O₃ catalyst was obtained.

In spite of the formation of CuAl₂O₄ as well as α-Al₂O₃ and changes in physical properties, the Cu/Ni-OS and the Cu-OS catalysts which were treated at 980°C showed fairly good NO conversion efficiencies except for Cu/Ni-OS at 600°C. Even at a low temperature such as 300°C, catalysts heat treated in this way exhibited satisfactory results compared with the original reduced catalysts. This behavior is significantly different from the behavior of the Rh-OS catalyst at 300°C. Catalytic activity survived in the base metal catalyst because there is still Cu and CuO which has not reacted with active alumina to form CuAl₂O₄ in this heat treatment. For Cu containing catalysts the copper loading is important to stabilize catalytic activity.

References p. 212.

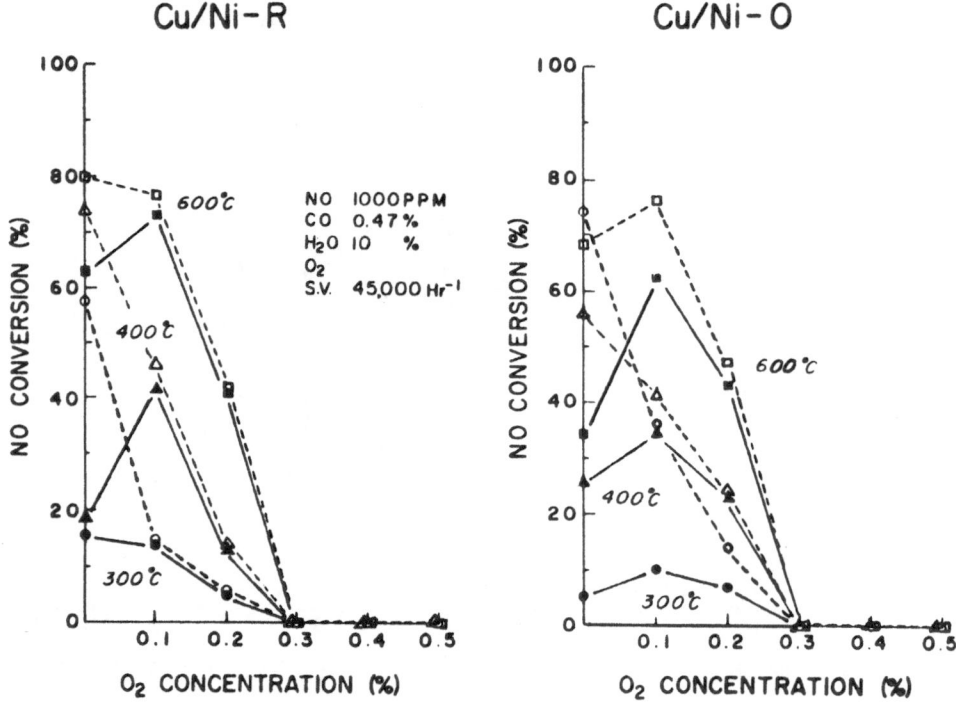

Fig. 13. The effect of catalyst oxidation or reduction at 550°C on NO conversion over Cu/Ni-R and Cu/Ni-O. Feed stream: 1000 ppm NO, 0.47% CO, 10% H_2O, variable O_2, and balance N_2. Space velocity: 45,000 hr^{-1}. – – – Gross NO conversion, — Net NO conversion.

The unexpected difference in conversion depending upon whether the starting gas is oxidizing or reducing was also observed for the base metal catalysts which had been treated at high temperature. The data is shown in Figs. 14, 15, 16 and 17. This effect was seen previously for the Rh-OS catalyst. The author must point out that only oxidative high temperature treated catalysts, including both base metal and noble metal catalysts, showed this unexplained phenomenon. Some atomic state of the catalytic element may cause this phenomenon, but adequate elucidation has not been established yet. However, one must pay careful attention to this point so that one may properly evaluate the activity of NO_x catalysts.

While Ni-OS did not show any activity for NO conversion, similar to Ni − O and Ni-R, the Ni-RS catalyst showed some activity (Fig. 18). This implies that Ni was activated in some way by the more severe reductive heat treatment. Considering the fact that NH_3 conversion over the Ni-Al_2O_3 catalyst was stimulated by high temperature and the Ni-OS catalyst showed a mild change in physical properties as compared with the Cu-OS catalyst, it is concluded that Ni in the Cu/Ni-Al_2O_3 catalyst plays an important role at high temperatures.

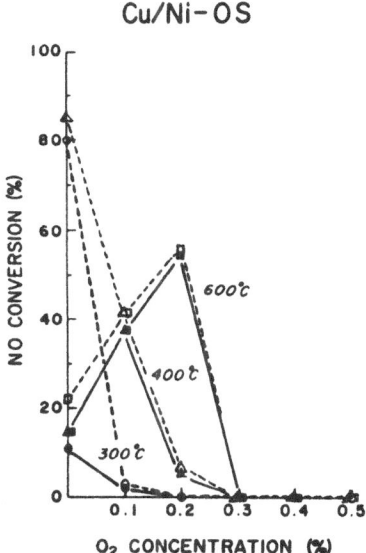

Fig. 14. The effect of initial reducing atmosphere on NO conversion over Cu/Ni-OS. Feed stream: 1000 ppm NO, 0.47% CO, 10% H_2O, variable O_2, and balance N_2. Space velocity: 45,000 hr^{-1}. ---Gross NO conversion,—Net NO conversion.

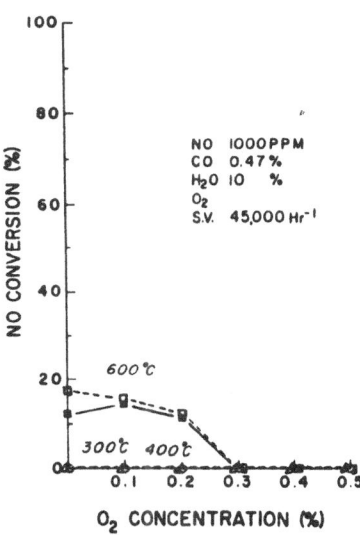

Fig. 15. The effect of initial oxidizing atmosphere on the NO conversion over catalyst Cu/Ni-OS. Feed stream: 1000 ppm NO, 0.47% CO, 10% H_2O, variable O_2, and balance N_2. Space velocity: 45,000 hr^{-1}. ---Gross NO conversion,—Net NO conversion.

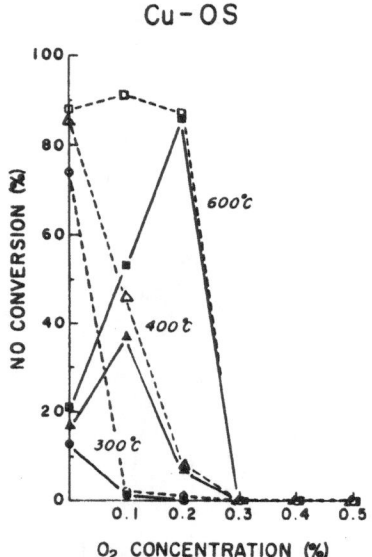

Fig. 16. The effect of initial reducing atmosphere on the NO conversion over Cu-OS. Feed stream: 1000 ppm NO, 0.47% CO, 10% H_2O, variable O_2, and balance N_2. Space velocity: 45,000 hr.$^{-1}$. – – –Gross NO conversion, —Net NO conversion.

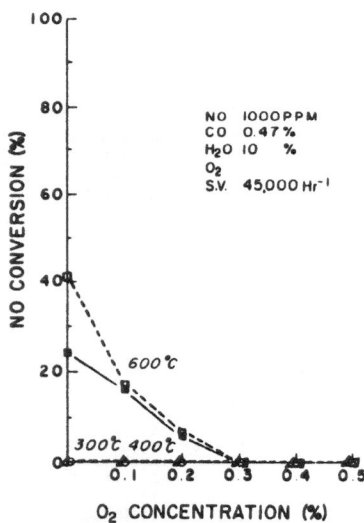

Fig. 17. The effect of initial oxidizing atmosphere on the NO conversion over Cu-OS. Feed stream: 1000 ppm NO, 0.47% CO, 10% H_2O, variable O_2, and balance N_2. Space velocity: 45,000 hr^{-1}. – – –Gross NO conversion, —Net NO conversion.

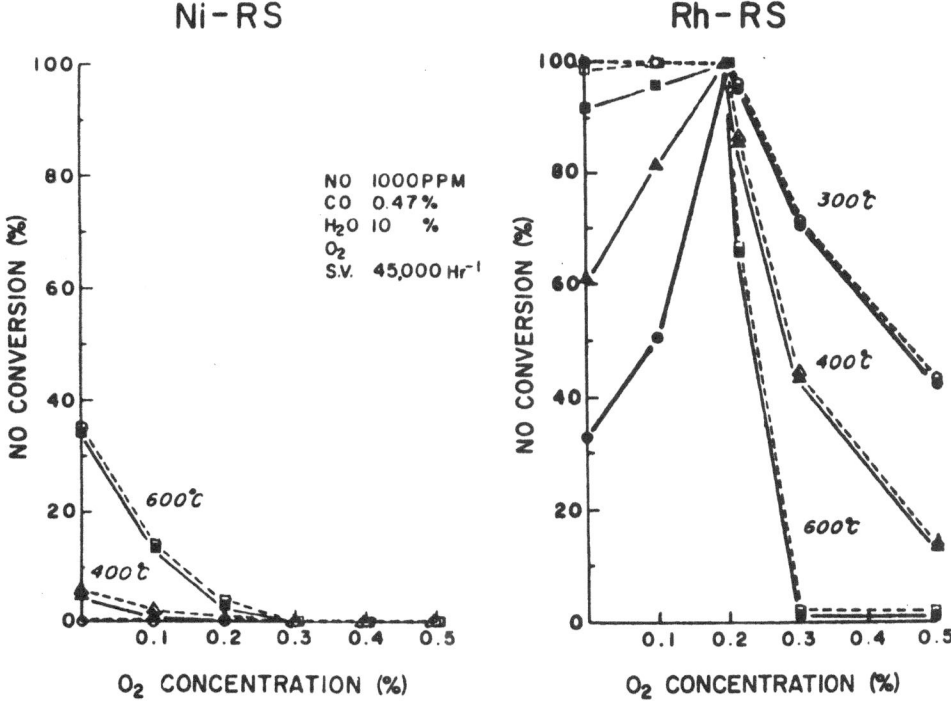

Fig. 18. The effect of a severe reductive heat treatment (800°C) on NO conversion over Ni-RS and Rh-RS. Feed stream: 1000 ppm NO, 0.47% CO, 10% H_2O, variable O_2, and balance N_2. Space velocity: 45,000 hr^{-1}. — — —Gross NO conversion,—Net NO conversion.

SUMMARY

In the study of the catalytic reduction of NO with CO and H_2 over Cu/Ni-Al$_2$O$_3$ and Rh-Al$_2$O$_3$ in the presence of H_2O and O_2, the following observations were made:

1. Carbon monoxide showed higher reducing reactivity than H_2 over both base metal and noble metal catalysts. Atomic hydrogen produced by the water-gas shift reaction exhibited stronger NH_3 formation ability than molecular hydrogen. This is particularly evident at low temperatures.

2. NH_3 formation was effectively suppressed by oxygen.

3. Rh-Al$_2$O$_3$ has greater NO reduction efficiency than Cu/Ni-Al$_2$O$_3$. At 300°C, NO conversion efficiency over Rh-Al$_2$O$_3$ was inhibited at high CO concentration, presumably due to strong CO chemisorption. Furthermore, the Rh-Al$_2$O$_3$ catalyst could promote selective NO reduction with CO and H_2 even in a slightly oxidizing atmosphere.

References p. 212.

4. H_2O suppressed the NO conversion ability of $Cu/Ni-Al_2O_3$ very distinctly, while no significant effect of water was observed over the $Rh-Al_2O_3$ catalyst.

5. In general, the catalytic decomposition of NH_3 in the presence of O_2 was accomplished successfully at high temperatures. $Rh-Al_2O_3$ showed high activity for NH_3 conversion, but the unfavorable selectivity to NO increased at high temperatures. $Cu/Ni-Al_2O_3$ showed the same level of NH_3 conversion as the $Rh-Al_2O_3$ catalyst above 400°C with O_2 present. However, $Cu/Ni-Al_2O_3$ showed better selectivity for the NH_3 to N_2 reaction compared to the $Rh-Al_2O_3$ catalyst.

6. In NO reduction, $Cu/Ni-Al_2O_3$ formed less NH_3 than the $Rh-Al_2O_3$ catalyst.

7. In general, catalysts treated at high temperature in an oxidative atmosphere tended to regain activity to some extent by further treatment in a reductive atmosphere without O_2.

8. The Rh-OS catalyst lost its original catalytic activity at 300°C and was not revived by treatment in a reductive atmosphere. The $Cu/Ni-Al_2O_3$ catalyst, however, showed significant revival of activity at 300°C by treatment in a reductive atmosphere. This might be attributed to a different atomic state of the catalyst and different catalyst loading.

9. After oxidative high temperature treatment, the $Cu/Ni-Al_2O_3$ catalyst formed $CuAl_2O_4$ with obvious changes in physical properties. Therefore, one may increase copper content in the finished catalyst in order to obtain an NO_x catalyst with high thermal stability.

REFERENCES

1. L. S. Bernstein et al., SAE 710014 (1971).
2. M. Shelef, H. S. Gandhi, Ind. Eng. Chem. Prod. Res. Develop. 11, 393 (1972).
3. E. W. Bell, M. Tagami, J. Phys. Chem., 67, 2434 (1963), also T. P. Kobylinski, B. W. Taylor, J. E. Young, SAE 740250 (1974).
4. R. L. Klimisch. K. C. Taylor, Spring Meeting of the California Catalysis Society, California, April 28-29 (1972), also Env. Sci. & Tech. 7, 127 (1973).
5. S. Brunauer, P. H. Emmett, E. Teller, J. Amer. Chem. Soc., 60, 309 (1938).
6. G. L. Bauerle, G. R. Service, K. Nobe, Ind. Eng. Chem. Prod. Res. Develop., 11, 54 (1972).
7. G. L. Bauerle, L. L. Sorensen, K. Nobe, ibid., 13, 61 (1974).

DISCUSSION

M Shelef *(Ford Motor Co.)*

In your oxidizing treatment you've shown that the rhodium lost its activity and did not regain it on reduction. You are probably sintering the metal area which you do not measure since you measure the BET area. But you say that in copper-nickel containing catalysts you have lost activity but you also have regained it. How do you explain this fact?

Ohara

The metallic content of the rhodium catalyst is very low, about 0.008 wt %, but for the copper-nickel catalyst, the copper content is very high, i.e. 7-8 wt % for copper in order to achieve effective reaction. I assume all the copper doesn't form copper aluminate, so the amount of copper available is still high.

H. Wise *(Stanford Research Institute)*

From your experience can you decide whether there is some advantage in introducing nickel into the copper-nickel system over copper?

Ohara

I already explained that nickel has activity at higher temperature ranges and so I think the copper-nickel bi-component system is the best catalyst.

E. L. Holt *(Exxon Research and Engineering Co.)*

We have done extensive work in catalysts with copper-nickel supported on alumina and we have found that nickel aluminate is the preferred species that's formed when you high-temperature treat this type of catalyst. Did you find nickel aluminate?

Ohara

Yes.

Holt

I further suggest that the explanation for the various catalyst activities as a function of pre-treatment is explained not only by preferential nickel-aluminate formation versus copper-aluminate formation, but I believe the surface of your copper-nickel catalyst on the support is also changing as a function of atmosphere. For example, in oxidative atmospheres we have found that you will preferentially tend to form a nickel-oxide layer. The copper will tend to migrate away from the surface as a result, and this may explain why you got different activities as a result of both your pre-treatment and the atmosphere that you exposed the catalyst to during your activity tests.

G. H. Meguerian *(Amoco Oil Co.)*

We have also found, in the case of nickel-copper, that, if you have large copper clusters you lose activity very fast. You have to have a very finely dispersed state in order to keep activity.

NITRIC OXIDE AND PEROVSKITE-TYPE CATALYSTS: SOLID STATE AND CATALYTIC CHEMISTRY

R. J. H. VOORHOEVE, J. P. REMEIKA and L. E. TRIMBLE

Bell Laboratories, Murray Hill, New Jersey

ABSTRACT

The solid state properties of perovskite-type compounds of transition metals which are of importance in NO catalytic chemistry are: (1) the presence of mixed valence ions (e.g., Mn^{3+}/Mn^{4+}), (2) the binding of oxygen in the lattice, and (3) the binding of N-containing fragments to the surface. On the surface, both molecular and dissociative chemisorption of NO contribute to the reduction of NO. The importance of the mixed valence and of the binding of lattice oxygen is linked to the simultaneous presence of cation vacancies and oxygen vacancies. Systematic substitution of cations in the dodecahedral and octahedral positions of the perovskite structure is used to demonstrate the importance of the solid-state chemistry for the catalytic activity and selectivity.

INTRODUCTION

The reduction of NO with CO and H_2 is of importance in the catalytic treatment of exhaust gases of conventional spark-fired internal combustion engines. On the other hand, the catalytic decomposition of NO yielding N_2, N_2O and O_2 is preferred for the treatment of stack gases and of the exhaust of diesel engines and stratified-charge internal combustion engines. On oxide catalysts, the two reactions are directly related to each other and to the exchange of oxygen between the gaseous reactants and the oxide lattice (1). The adsorption of NO, the oxidation of a reduced oxide catalyst with NO, the reduction of the oxide catalyst with H_2 or CO and the interaction of NO or its fragments with CO and H_2 are the elements of a seemingly complex chemical process. The use of catalysts with a simple basic structure which can be modified in their electronic properties without changing the geometry or even the chemical identity of the active site may elucidate the interaction between the

References pp. 230-231.

catalytic chemistry at the surface and the solid state chemistry of the catalyst. The spinel oxides have been initially used for such a study (2). However, the members of the extensive class of ABO_3 perovskite-type oxides (3) appear to fulfill the requirements especially well since their 3d and 4d transition metal ions are found in only one crystallographic position (the octahedrally coordinated B site). The valence of the B ion may be varied by changing the nature of the catalytically inactive A ion. For example, in the perovskite-type cobaltites and cobaltates the valence of Co can vary from +2 to +4 in the compounds $(La^{3+}, Th^{4+})(Co^{3+}, Co^{2+})O_3^{2-}$, $La^{3+}Co^{3+}O_3^{2-}$, and $(La^{3+}, Sr^{2+})(Co^{3+}, Co^{4+})O_3^{2-}$. Substitution of part of the Co ions with other transition metal ions can yield ordered perovskite catalysts in which the Co-Co distance is double that in $LaCoO_3$ as in Ba_2CoWO_6 (4). Similar permutations are possible in the perovskite-type chromites, titanates, manganites and ruthenates.

The present paper emphasizes the catalytic and solid state chemistry of the manganites. Their catalytic chemistry has been studied for a number of years for both CO oxidation (5-7) and NO reduction (8,9). Starting with $LaMnO_3$ as the parent compound, the Mn valence has been varied by introducing Pb^{2+}, Sr^{2+}, or alkali ions on the La site, or by introducing cation vacancies. In addition, partial substitutions on the Mn position have been used to alter the properties of the catalytic site in a more direct way.

EXPERIMENTAL

The catalytic rate measurements were performed on about 0.3 cm^3 of the oxide catalysts in the form of loose powders on a porous quartz disc. A feed mixture of 0.13% NO, 0.4% H_2 and 1.3% CO in He was passed through the catalyst bed in a downward flow in the reduction experiments. For NO catalytic decomposition, a mixture of 0.13% NO in He was used. Temperature control of the reactor and the analysis of the feed and exit gases by gas chromatography have been described in detail (4,10).

The preparation of the powders of $LaMnO_{3.01}$, $LaMnO_{3.15}$, $La_{0.8}M_{0.2}MnO_3$ (M = Na, K, Rb) and $La_{0.8}K_{0.2}Mn_{0.94}Ru_{0.06}O_3$ has been described elsewhere (9,11). $La_{0.8}K_{0.2}Mn_{0.9}Rh_{0.1}O_3$ was prepared in the same way as its Ru analog, with a final firing in air at 1100°C. $La_{0.6}Sr_{0.4}MnO_3$ was prepared by firing the appropriate mixture of La_2O_3, $SrCO_3$ and Mn_2O_3 in air at 1375°C, with intermediate regrinding and refiring until a single phase product was obtained. The electrical resistivity of the sample was measured on a pressed pellet by a four-probe method and compared well with previous measurements (12). The behavior is metallic at low temperature (T ⟨ 370 K) and semiconducting at T ⟩ 390 K. At 400°C, $\rho = 0.01$ Ωcm.

The preparation of single crystals of $La_{0.7}Pb_{0.3}MnO_3$ from a $PbO-B_2O_3$ flux was described before (8). $La_{0.7}Pb_{0.3}Mn_{0.97}Ni_{0.03}O_3$ crystals were obtained in the same way, by adding 5 at. % Ni as NiO to the Mn_2O_3 used. The single crystals of these two compounds were freed from the flux by leaching with acetic acid. Crushed and sieved crystals (size 37-250 μ) were etched in 5% HNO_3 prior to use (8). Single crystals of (La, Bi, K) MnO_3 were prepared by melting 6.11 g La_2O_3, 0.86 g K_2CO_3 and 3.95 g Mn_2O_3 with an excess of Bi_2O_3 (30-50 g), heating to 1250°C for 4 h in a covered Pt crucible in air and cooling at 5°C/h. The crystals were freed from the flux by leaching in dilute hydrochloric acid and were crushed to a fine powder (< 20 μ). They were not etched prior to use.

The surface areas of the catalysts were determined by N_2 adsorption in a Shell-Perkin Elmer Sorptionmeter. X-ray diffraction patterns were obtained with a Philips Debye-Scherrer camera. In Table 1, the catalysts are listed with some of their properties.

TABLE 1

Perovskite-Type and Related Catalysts for NO Reduction

Surface Area

No.	Formula[a]	m²/g (fresh catalyst)	X-ray Structure[b]	Ref.[c]
1	$LaMnO_{3.01}$	0.40	O	9
2	$LaMnO_{3.15}$	0.40	R	9
3	$La_{0.8}K_{0.2}Ru_{0.06}Mn_{0.94}O_3$	1.1	R	11
4	$La_{0.8}K_{0.2}Rh_{0.1}Mn_{0.9}O_3$	0.70	R	–
5	$La_{0.7}Pb_{0.3}Ni_{0.1}Mn_{0.9}O_3$[d]	5.3	R	–
6	$La_{0.7}Pb_{0.3}MnO_3$[d]	1.0	R	8
7	$La_{0.6}Sr_{0.4}MnO_3$	0.6	C	–
8	$La_{0.8}Na_{0.2}MnO_3$	0.90	R	9
9	$La_{0.8}K_{0.2}MnO_3$	2.5	R	9
10	$La_{0.8}Rb_{0.2}MnO_3$	0.88	R	9
11	$La_{0.85}Bi_{0.08}K_{0.07}MnO_3$[d]	0.22	R	–
12	$SrRuO_3$[d]	0.6	O	11
13	$SrRuO_3$	1.0	O	11
14	$Pb_2Ru_2O_{7-x}$[d]	0.4	C	11

a) From chemical analysis, except for No. 1 and 2 where the oxygen content is derived from the lattice parameters, and No. 4, 12, 13 and 14, where the nominal composition is given.
b) All are perovskites except No. 14, which has the cubic pyrochlore structure. O means orthorhombic distortion, R rhombohedral, C is cubic, not distorted.
c) Reference to earlier publication on the same catalyst.
d) Single crystals. All other catalysts were powder preparations.

References pp. 230-231.

Structure and Composition of Perovskite Catalysts — The crystal structure of the perovskites is very close to cubic, with a close packing of oxygen ions and large A ions, with B ions in octahedral coordination by oxygen. The A ions have 12 O^{2-} as nearest neighbors, and the O^{2-} are six-coordinated by four A ions and two B ions (Fig. 1). Very stable perovskites are formed with the rare earths La, Nd and Pr or the alkaline earths Ca, Sr and Ba as the A ions. The B ions in stable perovskites are for example Ti^{4+}, Cr^{3+}, Mn^{3+}, Fe^{3+}, Co^{3+}. However, other valence states of these ions can occur to a certain degree in the B-sites. Also, Ni^{3+} and V^{3+} form fairly stable perovskites $LaNiO_3$ and $LaVO_3$. Ni^{2+} can be stabilized in compounds such as Ba_2NiWO_6 (13). Of importance for the present paper is also the existence of fairly stable perovskites of some of the noble metals, exemplified by $SrRuO_3$ (14), and $LaRhO_3$ (15).

The surface structure of perovskites has not been studied extensively. The (100) cube faces constitute the surfaces of the flux-grown single crystals used in the present study, and crushed single crystals also expose cube faces. Calculations of the surface structure of (100) $BaTiO_3$ surfaces indicated that the square TiO_2 array was the outer surface rather than the Ba-containing plane and that the Ti ions and oxygen ions are displaced towards the lattice by ~ 0.23 Å and ~ 0.021 Å, respectively (16). LEED studies of $BaTiO_3$ have shown that the (100) surface is stable at high temperature (17).

In view of the data on the surface structures of perovskites, it is assumed that the square array of B-O-B ions is the catalytically active surface for these compounds.

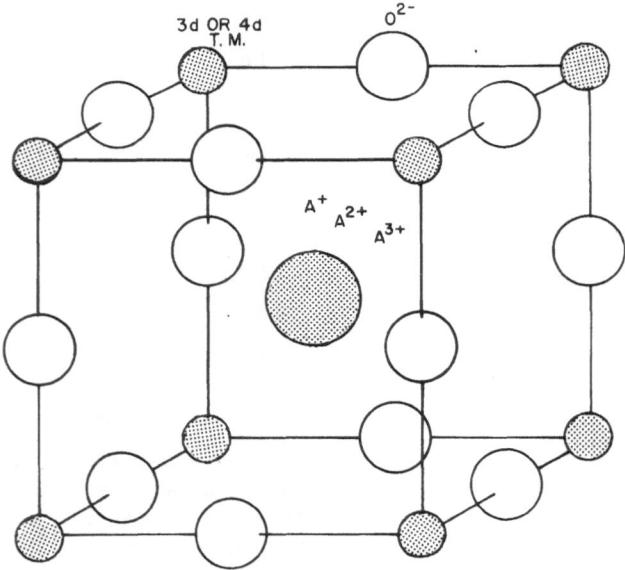

Fig. 1. The perovskite structure.

Participation of lattice oxygen in catalytic reactions then takes the form of reducing this surface by creating an anion vacancy. In the case of the manganites, the oxygen ion in the surface is coordinated by two Mn ions, either Mn^{3+} or Mn^{4+} and by two A-ions, e.g., two La^{3+} ions, or a La^{3+} ion and a K^+ ion, etc. $LaMnO_3$ is particularly interesting because it can accommodate a sizeable fraction of cation vacancies on the La-position. For example, $LaMnO_{3.12}$ was found to have 5% of the La^{3+} positions vacant (18). In such a case, part of the surface oxygen ions may be coordinated by a vacancy ▫. Evidently, the binding energy for the oxygen will depend on the Mn^{4+}/Mn^{3+} ratio in the solid and on the chemical nature of the occupants, if any, of the A sites.

The structure of the perovskites is in general not quite cubic, but shows small distortions. In addition, the substitutions of various ions on the A and B sites cause small variations in lattice parameters (9). In the light of reports that the lattice parameter of oxide catalysts might have an important bearing on the catalytic activity (1) or selectivity (19) for the reactions of NO, it is of interest to compare the compounds $BaRuO_3$, $SrRuO_3$ and $Pb_2Ru_2O_{7-x}$. All three have structures in which Ru ions are octahedrally coordinated by O^{2-}. However, only $SrRuO_3$ has a perovskite structure (14), whereas $BaRuO_3$ has a nine-layer hexagonal structure (20) and $Pb_2Ru_2O_{7-x}$ has a cubic pyrochlore structure (21). The Ru-Ru distances in the three compounds are quite different (Table 2). Nevertheless, it was found (11) that their catalytic activity and selectivity are quite similar in the reduction of NO with CO and

TABLE 2

Properties of Ruthenate Catalysts in NO Reduction

Catalyst[a]	Structural Date			Catalytic Data[b]		
	Structure	Ru-Ru $\overset{\circ}{A}$	Ref.	NO Reaction Rate at 225°C	NH_3 yield, % 300-400°C	Ref.
$SrRuO_3$ powder	Perovskite	3.92	(14)	2.5	30	(11)
$SrRuO_3$ crystals	Perovskite	3.92	(14)	7.8	32	(11)
$PbRu_2O_{7-x}$ crystals	Pyrochlore	3.62	(21)	5.5	30	(11)
$BaRuO_3$ powder	9-layer hexagonal	2.55	(20)	–	25	(19)

a) *Powders were prepared by solid state reaction, whereas crystals were obtained from a flux and were crushed before use.*
b) *In 0.13% NO, 0.4% H_2 and 1.4% CO in He at GHSV = 50,000 h^{-1} except for $BaRuO_3$ which was measured in 0.1% NO, 1.43% H_2 and 1.5% CO in N_2 at GHSV = 20,000 h^{-1} on a monolith support. Rate in ml NO at NTP per m^2 catalyst surface per h. NH_3 yield in % of NO converted.*

H_2 and in fact, vary more between different preparations of the same compound (Table 2). It may be concluded that the much smaller variations of the geometry within the series of perovskite catalysts to be discussed below are of no importance for their catalytic properties.

Lattice Oxygen and Oxygen Vacancies — The comparison of $LaMnO_{3.01}$ which has few, if any, Mn^{4+} ions or La vacancies, with $LaMnO_{3.15}$ in which up to 9% of the La^{3+} positions are vacant shows the importance of lattice oxygen in a most striking manner (9). The NO conversion on the two catalysts used under identical test conditions is compared in Fig. 2. It is evident that $LaMnO_{3.15}$ is a much more active catalyst than the nearly stoichiometric $LaMnO_{3.01}$. The differences between the two catalysts comprise the Mn^{4+} content, the cation vacancy concentration and the electrical conductivity (9). To establish which of these is the important factor, a series of catalysts was compared with about equal Mn^{4+}/Mn^{3+} ratio and similar electrical conductivities, but in which the charge-compensating defects (i.e., the substitution on the A-site) were different. These were $LaMnO_{3.15}$ (30% Mn^{4+}), $La_{0.8} M_{0.2} MnO_3$ (M = Na, K or Rb; 40% Mn^{4+}) and $La_{0.6} Sr_{0.4} MnO_3$ (40% Mn^{4+}). In Fig. 3, the effective conversion in the NO reduction with the standard $NO-CO-H_2-He$ feed gas is expressed as the fraction of the inlet NO converted into $N_2 + N_2O$. The reduced space velocities (i.e., volume of gas treated per m^2 of catalyst surface in the reactor and per h) were

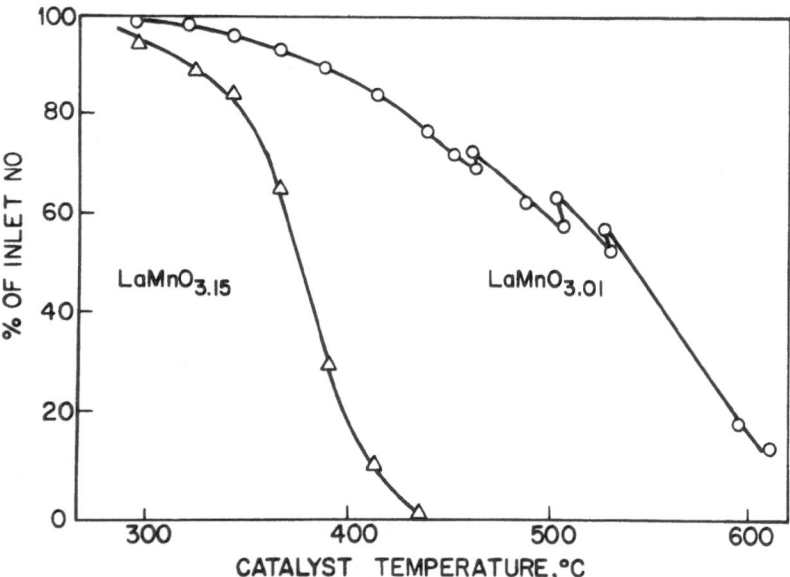

Fig. 2. Catalytic reduction of NO. NO in reactor effluent from $LaMnO_{3.15}$ (rhombohedral) and $LaMnO_{3.01}$ (orthorhombic). RSV = 23,500 ml $h^{-1}m^{-2}$ based on surface area of fresh catalysts. $NO-CO-H_2-He$ feed mixture. Measured at ascending temperature on fresh catalysts (9).

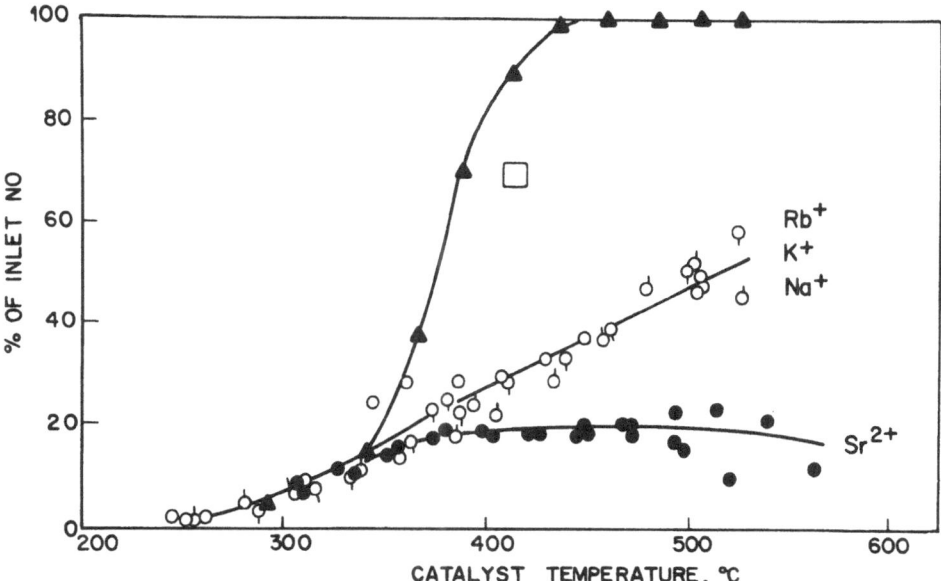

Fig. 3. Catalytic reduction of NO. Yield of $N_2 + N_2O$ over $(La,A')MnO_3$ with 30-40% Mn^{4+}. A' is La-vacancy (□) or Rb, K, Na, Sr. RSV = 23,000 ml $m^{-2}h^{-1}$ for □, 13,000-14,000 for the rest. Standard NO-CO-H_2-He feed.

RSV = 13,000-14,000 ml $h^{-1}m^{-2}$ for the catalysts with Na^+, K^+, Rb^+ and Sr^{2+} substitutions and RSV = 23,500 ml $h^{-1}m^{-2}$ for the catalyst with La vacancies (□). As to their effective NO conversion the catalysts clearly rank in the order $(La,\square) \rangle$ $(La,Na) \approx (La,K) \approx (La,Rb) \rangle (La,Sr)$. Catalysts with Pb^{2+} substituted for La^{3+} were very similar to (La,Sr) in this reaction. The data in Fig. 3 might suggest that it is the effective charge on the compensating defect which determines the catalytic properties. However, this is incorrect, as shown by the results with a catalyst containing both Bi and K as A-site substitutions. The resulting compound is not very stable in the NO-CO-H_2-He mixture beyond 400°C, but its activity in the NO conversion outranks that of $LaMnO_{3.15}$ considerably below 400°C, even when used at twice the reduced space velocity (Fig. 4).

The results in Fig. 2-4 clearly indicate the importance of the binding energy of oxygen in the reduction of NO. The former is reflected in the reaction enthalpy for the formation of an oxygen vacancy (22) V_o :

$$\text{Mn}\!\!-\!\!\!-\!\!\text{O}\!\!-\!\!\!-\!\!\text{Mn} \quad\quad \text{Mn}\!\!-\!\!\!-\!\!\!\overset{}{\underset{o}{V}}\!\!-\!\!\!-\!\!\text{Mn} \quad + \text{O}_g \qquad\qquad (1)$$
$$\overset{\diagup\ \ \diagdown}{\text{A}\ \ \text{A}'} \quad\to\quad \overset{\diagup\ \ \diagdown}{\text{A}\ \ \text{A}'}$$

At constant Mn^{4+}/Mn^{3+} ratio, the oxygen binding in a series of manganites will rank in the order of the contributions of A and A' to the binding energy. Since in the

References pp. 230-231.

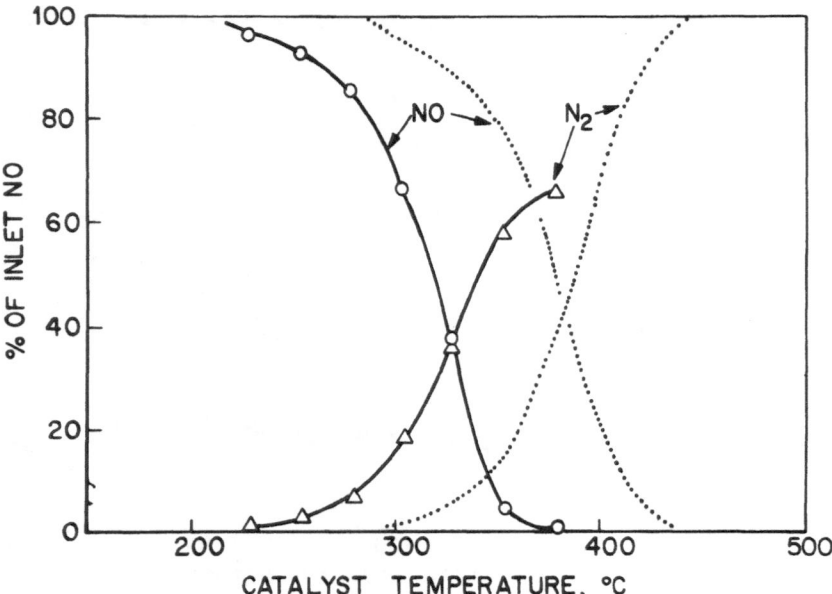

Fig. 4. Catalytic reduction of NO. Yield of N_2 over $(Bi,K,La)MnO_3$. Standard NO-CO-H_2-He feed at RSV = 51,600 ml $m^{-2}h^{-1}$. Fresh catalyst. Dashed lines: N_2 yield for $(La,\square)MnO_{3.15}$ corresponding to Figs. 2 and 3.

perovskite the A-site is coordinated by 12 O^{2-}, the contribution of one A ion to the binding energy of oxygen (or the reaction enthalpy of the reverse of Eq. (1) is

$$\triangle(A\text{-}O) = \frac{1}{12}(\triangle H_f - m\triangle H_s \frac{1}{n}D_o)/m \qquad (2)$$

where $\triangle H_f$, $\triangle H_s$ and D_o are, respectively, the enthalpy of formation on one mole of the oxide $A_m O_n$, the enthalpy of sublimation of the metal A and the dissociation energy of O_2. For a surface O^{2-} with neighbors A and A ', the total contribution of A-O bonds to the reaction enthalpy is $\triangle(A\text{-}O) + \triangle(A'\text{-}O)$. These values are listed in Table 3 for the AA ' combinations used in the experiments in Figs. 3-4. The catalysts are listed in order of decreasing $N_2 + N_2O$ yield. This ranking is seen to yield a ranking in order of increasing reaction enthalpy for the oxygen release of Eq. (1). The binding energy of oxygen in the oxide lattices had been correlated with activity for oxidation-reduction reactions before (23-26), but these correlations had been ambiguous because the comparison was within a series of oxides with different active centers. The present correlation is significant because the same active center is involved in all manganites and the oxygen binding energy in the correlation is for a specific oxygen ion, rather than the average for all oxygen in the lattice. This specific binding energy has been suspected before to control the catalysis of the oxidation of benzaldehyde (27). For the present reaction of NO reduction, the release of oxygen

TABLE 3

Binding of Lattice Oxygen in $(A,A')MnO_3$ Perovskites

Rank order[a] of Catalysts $(A,A')MnO_3$	Contributions to binding energy (Eq. 1) in kcal/mol		
	$\triangle(A\text{-}O)$	$\triangle(A'\text{-}O)$	$\triangle(A\text{-}O) + \triangle(A'\text{-}O)$
$(Bi,K)MnO_3$[b]	-17.2	-7.8	-25.0
$(La,\square)MnO_3$	-33.7	0	-33.7
$(La,Rb)MnO_3$	-33.7	-7.4	-41.1
$(La,K)MnO_3$	-33.7	-7.8	-41.5
$(La,Na)MnO_3$	-33.7	-8.7	-42.4
$(La,Pb)MnO_3$	-33.7	-13.1	-46.8
$(La,Sr)MnO_3$	-33.7	-19.9	-53.6

[a] *In order of decreasing $N_2 + N_2O$ yields.*

[b] *Catalyst No. 11. Since Bi and La are both trivalent, both (La,K) and (Bi,K) combinations are present. For the catalytic activity, the combination with the lowest value of $\triangle(A\text{-}O) + \triangle(A'\text{-}O)$ is determining and that is (Bi,K).*

from the lattice is concluded to be an important factor in determining the rate of conversion of NO to N_2 and N_2O.

Reduction of the Catalyst by NO-CO-H_2 and Oxidation by NO — It is known that the partial reduction of some perovskite-type catalysts greatly affects the catalytic activity (4,28). In the NO reduction, this is demonstrated by the behavior of $La_{0.8}K_{0.2}Mn_{0.94}Ru_{0.06}O_3$ (Cat. No. 3, Table 1). The degree of reduction of the catalyst was increased by treatment at $T \geq 600°C$ in the NO-CO-H_2-He mixture and as a result, the activity in NO reduction increased. In Fig. 5, the NO conversions are given for two runs following the reduction. In the two runs, the degrees of reduction differ, more than compensating for the differences in space velocity. N_2 is the major product, with small amounts of N_2O being produced also. To study NO decomposition on this same catalyst, 0.13% NO in He was reacted at an RSV = 4,700 $ml/m^2 \cdot h$. The temperature was scanned from 100°C to 500°C in two consecutive runs. The products of the NO conversion were NO, N_2O and N_2 and minor amounts of an unknown component, possibly NO_2. No O_2 was produced at all and the activity for the NO conversion decreased as the catalyst surface was oxidized by the NO conversion. In Fig. 6, the second scan from 100°C to 500°C is shown. In the first scan, the activity was appreciably higher, with the NO being fully converted at $T \geq 160°C$. The second run is shown because it is believed to correspond more closely to the oxidation state of the catalyst as it was used in the experiments of Fig. 5. The oxygen taken up by the reduced catalyst during oxidation with NO amounted to 40-50 monolayers, indicating the participation of lattice oxygen. Comparing Figs. 5 and 6, it is at once apparent that the NO and N_2 lines in both show a general similarity but

References pp. 230-231.

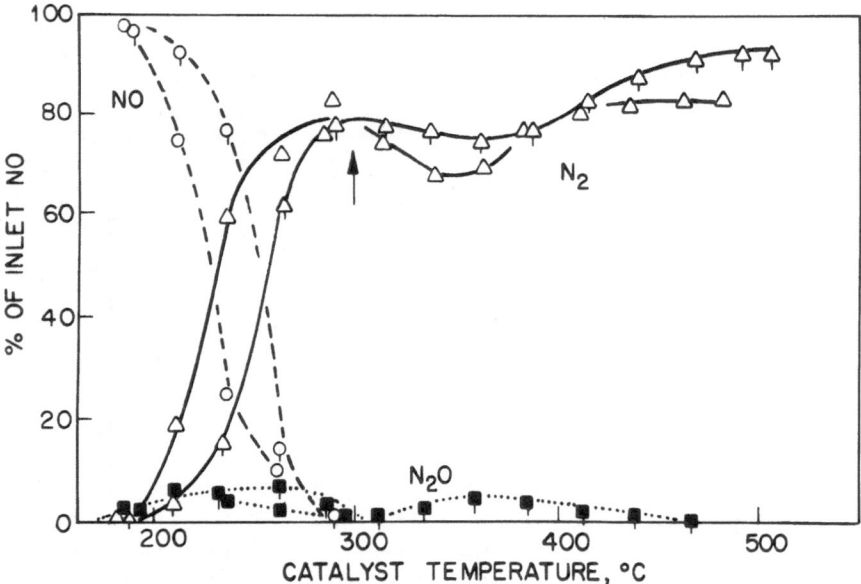

Fig. 5. Catalytic reduction of NO. Composition of reactor effluent over reduced $La_{0.8}K_{0.2}$ $Mn_{0.94}Ru_{0.06}O_3$. Standard NO-CO-H_2-He feed. O■△Run 6, RSV = 42,600 ml m^{-2}h^{-1}, ○■△Run7, RSV = 16,000 ml m^{-2}h^{-1}. The oxidation state of the catalyst for both runs is likely to be different.

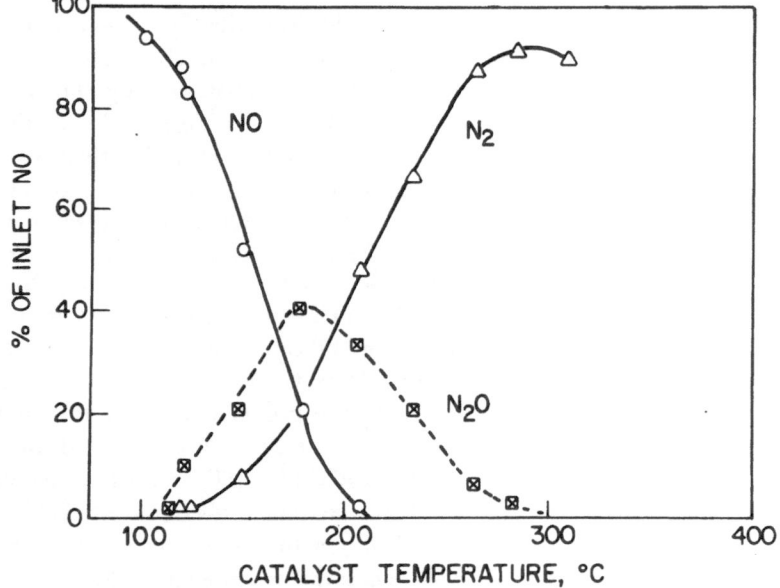

Fig. 6. "Catalytic decomposition" of 1% NO in He over reduced $La_{0.8}K_{0.2}Mn_{0.94}Ru_{0.06}O_3$. RSV = 4,700 ml m^{-2}h^{-1}. No O_2 was formed. This is the second run. On the first run, the curves were all shifted to 60° lower temperatures and the N_2O yield was 50% at 125°C.

that in the presence of NO only, more N_2 is formed at lower temperatures. The results indicate strongly the participation of lattice oxygen vacancies in the production of N_2 and N_2O and the participation of molecularly adsorbed NO in the formation of N_2O.

In the second run of the experiment on NO decomposition (Fig. 6), the NO conversion rate dropped dramatically shortly after the data in Fig. 6 were taken. This is due to poisoning of the surface by oxygen and the activity has been restored by reduction. In the temperature range studied ($T \leq 550°C$), the surface does not release oxygen to a sufficient degree, so that the number of available oxygen vacancies is small. This is similar to the situation on many binary oxide catalysts (1,29).

NO Molecular Adsorption and Conversion on Manganites Containing Ni, Ru or Rh — It was established in earlier work (9) on the adsorption of NO on $La_{0.8}K_{0.2}MnO_3$ that NO adsorbs in a molecular form, and also in a dissociative form, with its oxygen becoming part of the oxide lattice. Similarly, NO dissociatively adsorbed on MnO was found to oxidize Mn^{2+} to Mn^{3+}, and molecular NO adsorption was found on Mn^{3+} (30).

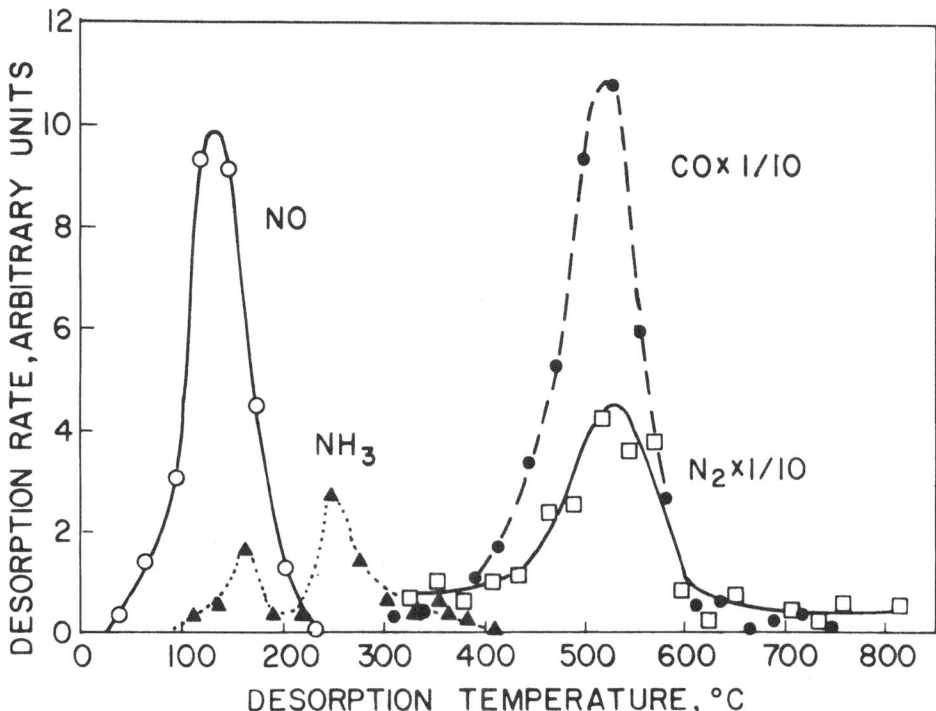

Fig. 7. Mass spectrometric desorbed gas analysis on $La_{0.8}K0.2Mn_{0.94}Ru_{0.06}O_3$. N_2 and CO peaks scaled down by 10X. CO_2 and H_2O peaks omitted.

To pursue the NO adsorption on the manganites further, a sample of $La_{0.8} K_{0.2} Mn_{0.94} Ru_{0.06} O_3$ (No. 3) was used as a catalyst in the $NO-CO-H_2$-He mixture at 620°C for 1 h, and was then cooled down in the same mixture to 150°C, at which point the catalyst was soaked in 1% NO in He for 1 h and cooled to room temperature. Still in 1% NO in He, the sample was transferred to a silica tube attached to a desorbed gas analysis train with a mass spectrometer. The tube was then evacuated to 10^{-4} torr and the sample heated at 2°C/min to obtain the desorption spectrum of nitrogen compounds (Fig. 7). Desorption of molecular NO shows a peak at 150-200°C. Since reversible dissociative adsorption of NO is very unlikely, this peak shows the presence of molecular NO on the surface. Desorbed gas analysis on identically treated $La_{0.7} Pb_{0.3} MnO_3$ and $La_{0.8} K_{0.2} MnO_3$ similarly shows a NO peak in the same temperature range. On all these catalysts, NO is therefore assumed to adsorb in molecular form on low valency metal ions, predominantly Mn^{3+}, as the nitrosyl group. The evolution of N_2 at higher temperatures (Fig. 7) shows the presence of a substantial coverage of the surface with N-containing fragments which are tightly bound. The simultaneous desorption of N_2 and CO at 450-600°C in the ratio N:CO = 1:1 suggests the presence of a surface compound decomposing at that temperature. This may be an isocyanate compound or other N-C-O compound. Note that on this Ru-containing catalyst, the NH_3 desorption is over beyond 400°C, while in similar experiments on $La_{0.8} K_{0.2} MnO_3$ the NH_3 desorption persists until 600°C. This is in line with the NH_3 yields on both catalysts in the $NO-CO-H_2$ reaction (9,11).

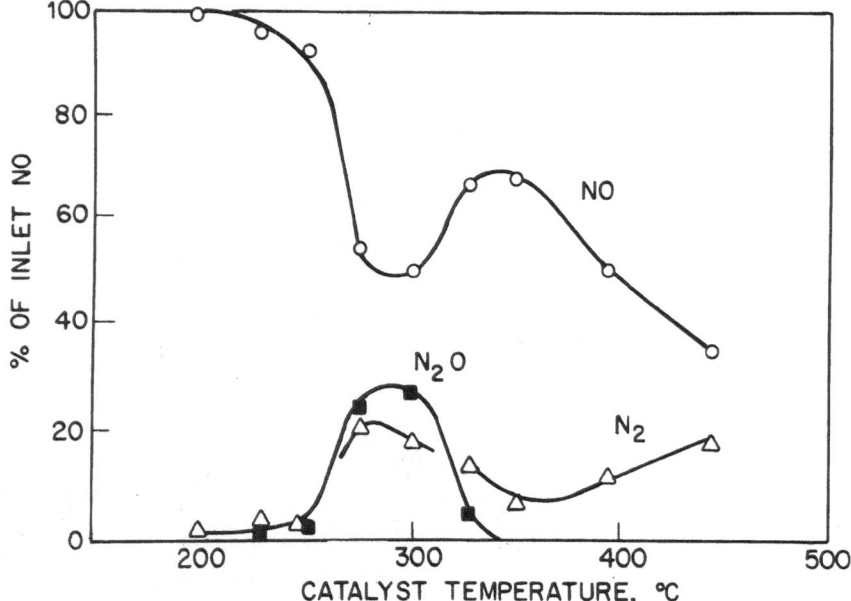

Fig. 8. Catalytic reduction of NO over $La_{0.7}Pb_{0.3}Mn_{0.97}Ni_{0.03}O_3$. Composition of reactor effluent. Standard $NO-CO-H_2$-He feed. RSV = 2,200 ml m^{-2}h^{-1}.

The decrease of the coverage of the catalyst surface with NO between 200 and 400°C is reflected in the reaction rate and the product composition of the catalytic reduction. This was found for $La_{0.7}Pb_{0.3}MnO_3$, for $La_{0.8}K_{0.2}MnO_3$ and for several other Mn-containing perovskites. In Fig. 8, the example is $La_{0.7}Pb_{0.3}Mn_{0.97}Ni_{0.03}O_3$ (No. 5). A maximum in the NO reaction rate is shown by the minimum in the NO content of the reactor effluent at 290°C. The conversion in that range yields N_2O and N_2, in agreement with a mechanism:

$$N_{ads} + NO_{ads} \rightarrow N_2O \qquad (3)$$

and

$$N_2O_{ads} \rightarrow N_2 + O_{ads} \qquad (4)$$

or

$$N_2O_{ads} + CO_{ads} \rightarrow N_2 + CO_2 \qquad (5)$$

At higher temperature, T ⟩ 350°C, another mechanism of NO conversion appears to be operating. Without the presence of Ni in the catalyst, the N_2O peak is less

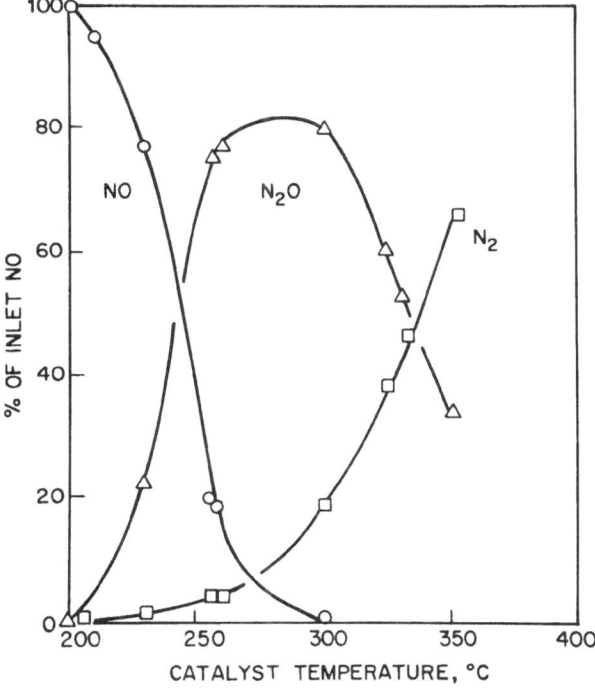

Fig. 9. Catalytic reduction of NO over $La_{0.7}Pb_{0.3}MnO_3$. Composition of reactor effluent. Feed 50% NO and 50% CO, RSV = 700 ml m^{-2}h^{-1}.

References pp. 230-231.

pronounced, but in $La_{0.7}Pb_{0.3}MnO_3$ it still occurs at the same temperature. The role of Ni, which in the perovskite structure is likely to be present as Ni^{3+} (31.32), is not clear. The N_2O yield in the range of 250-350°C is strongly enhanced by higher NO and CO concentrations. On $La_{0.7}Pb_{0.3}MnO_3$, the yield may reach 80% for a feed with 50% NO and 50% CO (Fig. 9). It was shown before (Fig. 6) that the N_2O yield is very high also in the absence of CO. This might be indicative of a reaction such as Eq. (5), proceeding when CO is competing effectively with molecular adsorption of NO.

On $La_{0.7}Pb_{0.3}MnO_3$ and $La_{0.8}K_{0.2}MnO_3$, the NO conversion rate reaches a relative maximum at 250-300°C (Not shown here. In Figs. 3 and 9 these extremes at 250-300°C are not in evidence since the NO conversion is too low or too high, respectively). The presence of Ru in the catalysts, as in $La_{0.8}K_{0.2}Mn_{0.94}Ru_{0.06}O_3$ (No. 3), does not change the temperature at which the NO conversion reaches a relative maximum (290°C) but does affect the product distribution (Fig. 10). More N_2 than N_2O is now formed. This may be due in part to an increased rate of N_2O decomposition or reduction. On Ru-containing manganites, N_2O has been shown to be an intermediate in the production of N_2 (11). It is of interest that the relative maximum in N_2 yield at 290°C does not depend on the degree of reduction of the catalyst. The more reduced and more active catalyst in Fig. 5 shows a relative maximum at the same temperature.

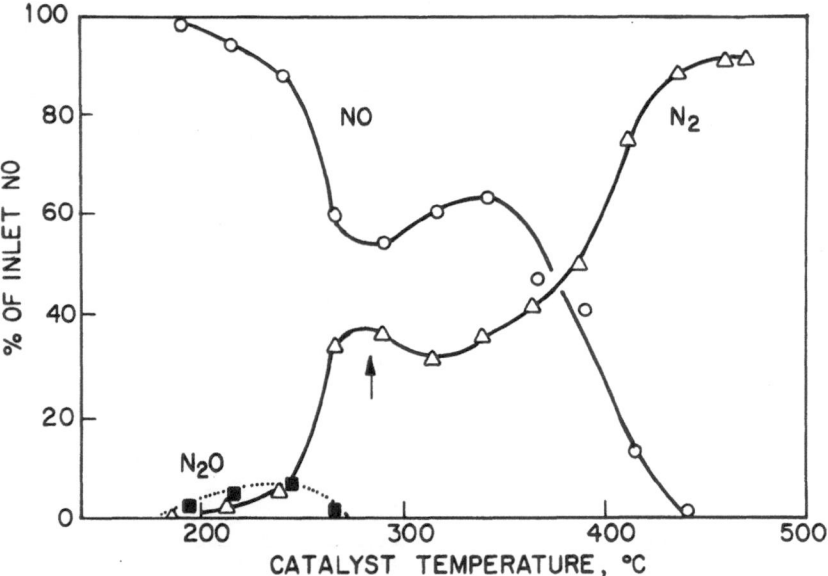

Fig. 10. Catalytic reduction of NO over $La_{0.8}K_{0.2}Mn_{0.94}Ru_{0.06}O_3$. Composition of reactor effluent. Fresh catalyst, oxidation state similar to that in Figs. 6 and 7. Standard NO-CO-H_2-He feed, RSV = 42,600 ml $m^{-2}h^{-1}$.

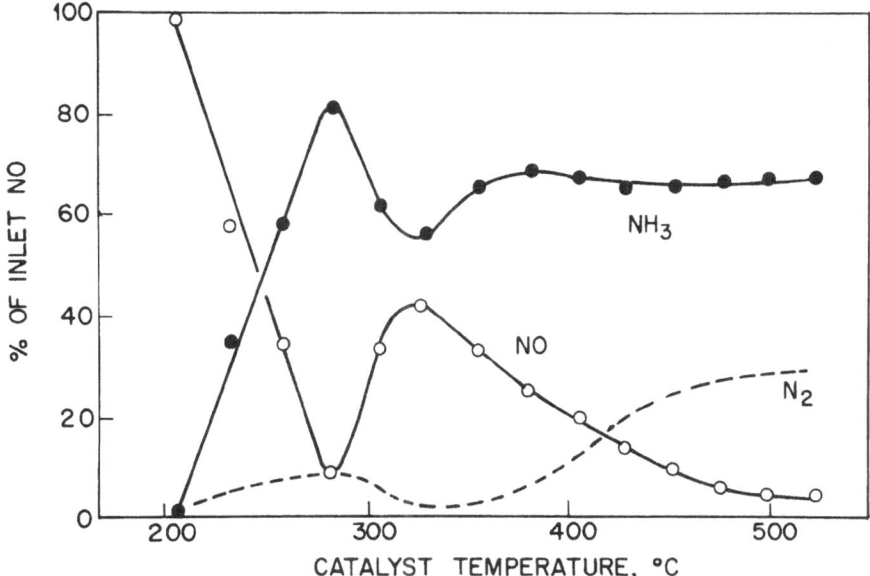

Fig. 11. Catalytic reduction of NO over $La_{0.8}K_{0.2}Mn_{0.9}Rh_{0.1}O_3$. Composition of reactor effluent. Fresh catalyst. Standard NO-CO-H_2-He feed, RSV = 15,600 ml $m^{-2}h^{-1}$.

The presence of Rh, as in $La_{0.8}K_{0.2}Mn_{0.9}Rh_{0.1}O_3$, used in the same test gases and procedures, leads to a greatly enhanced yield of NH_3 (Fig. 11). But again, the NO conversion has a relative maximum at 290°C, showing the effect of molecular NO adsorption on the conversion rate. Rh^{3+} may serve as a hydrogenation catalyst, dissociating H_2 which then reacts with NO_{ads} to form NH_3. It is well known that Rh is a much more active hydrogenation catalyst than Ru (33), and Rh metal used as a catalyst in NO reduction yields appreciable amounts of NH_3 (34). As the catalyst is used in subsequent experiments in the NO-CO-H_2-He mixture, the relative extreme at 290°C shifts to lower temperature, i.e., 230°C.

These experiments demonstrate that the molecular NO adsorption, presumably on Mn^{3+}, consistently enters into the mechanism of the catalytic reduction of NO on Mn-O-TM compounds (TM is a 3d or 4d transition metal), but that the selectivity of the reactions is strongly influenced by the nature of TM.

CONCLUSIONS

In the perovskite systems $ARuO_3$ and $AMnO_3$, where A is La, Sr, Ba, Na, K, Rb or Pb and mixtures thereof, the catalytic activity for NO reduction is determined by molecular adsorption of NO on a transition metal ion of low valence, combined with dissociative adsorption of NO on a lattice oxygen vacancy. The effective rate of conversion of NO into N_2 and N_2O is related to the heat of formation of an oxygen vacancy. In the series (A,A') MnO_3, the effective conversion rate decreases in the rank order of A,A' combinations (Bi,K) > (La,) > (La,Rb) $\approx$ (La,K) $\approx$ (La,Na) > (La,Pb) $\gtrless$ (La,Sr), which coincides with the rank order of increasingly tight binding of lattice oxygen.

The dissociative adsorption of NO on the surface of a reduced oxide is a fast process. The number of active oxygen vacancies is limited by the release of oxygen.

In the catalytic reduction of NO over (La,K) (Mn,TM)O_3 and (La,Pb) (Mn,TM)O_3, where TM is Ni, Ru or Rh, the low-temperature reaction rate is determined by molecular adsorption of NO, probably on Mn^{3+}. However, the product distribution between N_2O, N_2 and NH_3 is strongly influenced by the nature of the transition metal ion.

The variations in the geometry of the perovskite lattice produced by the various substitutions do not affect the catalytic processes in a perceptible way.

ACKNOWLEDGMENTS

It is a pleasure to thank P. K. Gallagher for the desorbed gas analysis experiments on the NO-treated catalyst, F. J. DiSalvo and J. V. Waszczak for the measurements of electrical conductivity, Mrs. A. S. Cooper for x-ray diffraction data, T. Y. Kometani for chemical analysis of the compounds, and F. Schrey for determinations of the surface areas.

REFERENCES

1. E. R. S. Winter, J. Catal. 22, 158 (1971).
2. J. W. Linnett and M. M. Rahman, Trans. Faraday Soc. 67, 191 (1971).
3. J. B. Goodenough and J. M. Longo, Landolt-Bornstein, New Series, Group III, Vol. 4a, p. 126, Springer-Verlag, Berlin (1970).
4. R. J. H. Voorhoeve, L. E. Trimble and C. P. Khattak, Mater. Res. Bull. 9, 655 (1974).
5. G. Parravano, J. Am. Chem. Soc. 75, 1497 (1953).
6. R. J. H. Voorhoeve, J. P. Remeika, P. E. Freeland and B. T. Matthias, Science 177, 353 (1972).
7. D. W. Johnson, Jr. and P. K. Gallagher, Thermochimica Acta 7, 303 (1973).
8. R. J. H. Voorhoéve, J. P. Remeika and D. W. Johnson, Jr., Science 180, 62 (1973).
9. R. J. H. Voorhoeve, J. P. Remeika, L. E. Trimble, A. S. Cooper, F. J. DiSalvo and P. K. Gallagher, to be published.
10. R. J. H. Voorhoeve and L. E. Trimble, to be published, J. Catalysis.
11. R. J. H. Voorhoeve, J. P. Remeika and L. E. Trimble, Mater. Res. Bull. 9, 1393 (1974).
12. G. H. Jonker and J. H. van Santen, Physica 16, 337 and 599 (1950).
13. D. E. Cox, G. Shirane and B. C. Frazer, J. Appl. Phys. 38, 1459 (1967).
14. J. J. Randall and R. Ward, J. Am. Chem. Soc. 81, 2629 (1959).
15. A. Wold, B. Post and E. Banks, J. Am. Chem. Soc. 79, 6365 (1957).
16. L. I. Ahmad, Surface Sci. 12, 437 (1968).
17. D. Aberdam, G. Bouchet and P. Ducros, Surface Sci. 27, 559 (1971).
18. B. C. Tofield and W. R. Scott, J. Solid State Chem 10, 183 (1974).

19. *M. Shelef and H. S. Gandhi, Platinum Met. Rev. 18, 2 (1974).*

20. *P. C. Donohue, L. Katz and R. Ward, Inorg. Chem. 4, 306 (1965).*

21. *J. M. Longo, P. M. Raccah and J. B. Goodenough, Mater. Res. Bull. 4, 191 (1969).*

22. *P. Kofstad, Non-stoichiometry, Diffusion and Electrical Conductivity in Binary Metal Oxides, Wiley, New York (1972).*

23. *O. V. Krylov, Catalysis by Non-Metals, Academic Press, New York (1970).*

24. *G. K. Boreskov, Advan. Catal. 15, 332 (1964).*

25. *Y. Morooka and A. Ozaki, J. Catal. 5, 116 (1966).*

26. *Y. Morooka, Y. Morikawa and A. Ozaki, J. Catal. 7, 23 (1965).*

27. *W. M. H. Sachtler, G. J. H. Dorgelo, J. Fahrenfort and R. J. H. Voorhoeve, Rec. Trav. Chim. 89, 460 (1970).*

28. *S. C. Sorenson, J. A. Wronkiewicz, L. B. Sis and G. P. Wirtz, Amer. Cer. Soc. Bull. 53, 446 (1974).*

29. *A. Amirnazmi, J. E. Benson, and M. Boudart, J. Catal. 30, 55 (1973).*

30. *H. C. Yao and M. Shelef, J. Catal. 31, 377 (1973).*

31. *A. Wold and R. J. Arnott, J. Phys. Chem. Solids 9, 176 (1959).*

32. *J. B. Goodenough, A. Wold, R. J. Arnott and N. Menyuk, Phys. Rev. 124, 373 (1961).*

33. *G. C. Bond, Catalysis by Metals, Academic Press, New York (1962).*

34. *T. P. Kobylinski and B. W. Taylor, J. Catal. 33, 376 (1974).*

DISCUSSION

V. Haensel *(UOP, Inc.)*

Rudie, two questions. First, in the $LaMnO_{3.15}$ catalyst which you've shown, how long do you hold it at any individual temperature to measure the NO change?

Voorhoeve

Typically about 20 minutes, in some cases we take a lunch break and leave it there for 2 hours.

Haensel

And what happens?

Voorhoeve

In the $LaMnO_3$, I showed you the variations. You see less of a variation on the $LaMnO_{3.15}$. It is not due to the fact that it is more stable but due to the fact that as you reduce it, it goes through a phase transformation, and the phase transformation gives you a higher surface area. By serendipity or by sheer luck, the product of the excess oxygen times the surface area, (which product is a measure of the active oxygen) is constant. That's why you don't really see a change. In general, these catalysts I was talking about have been selected to show the points I wanted to make, without regard to their stability, and hence, without regard to their practical usefulness. For instance, the bismuth potassium manganite is a very nice catalyst for some reactions, but not for automotive exhaust applications because beyond 450°C it starts to deteriorate very quickly due to bismuth mobility and formation of a bismuth oxide layer on the surface.

Haensel

What happens with water vapor?

Voorhoeve

Again, to keep the whole system simple, we don't have water vapor added to all these different things. In the lanthanum manganite, $LaMnO_{3.15}$, vacancy ... the one you were primarily interested in ... water vapor has very little effect. The first run you do ... you don't find any ammonia whatsoever. It's only when you start changing the composition a little bit and form binary oxides that you produce ammonia. We find 10 to 20% of ammonia. Now if you take the lanthanum manganite, $LaMnO_3$ without vacancies, even without H_2O added, 50% of the conversion is to ammonia. Now if you put water on that one the N_2 formation is killed and you get only ammonia formed. So it's very dependent on whether or not you have vacancies to control the selectivity of the catalysts.

P. Emmett (Portland State)

I have a question regarding your curves. I judge they were reactions of hydrogen and NO in most cases, because you were getting some ammonia too.

Voorhoeve

We get some ammonia, but NO ... or hydrogen and CO are very competitive on these systems. At lower temperatures, on the order of 300°C, we generally have the CO reaction, mostly. At higher temperatures you get about equal reaction.

Emmett

Was your reaction between hydrogen, CO, and NO? Now, I'm puzzled about the one case in which you mentioned the catalyst was killed by picking up oxygen. Does it pick up oxygen in the presence of hydrogen?

Voorhoeve

That was the reaction of only NO with a catalyst, it was NO decomposition. I had nothing else in there.

M. Shelef (Ford Motor Co.)

We have observed a peculiar phenomenon which has a bearing on the perovskite with vacancies. I think Klaus Otto has noted that $BaRuO_3$ will soak up NO, probably as nitrosyl, much as beta-alumina can soak up NO^+ (nitrosyl) instead of K^+ or Na^+, and so on. Because it also soaks up H^+, we think it's like a sieve which has a very loose structure. I wonder if your vacancies cannot be filled to a certain extent, by soaking up NO^+?

Voorhoeve

I think stereochemically, it's not very likely that you get NO soaked up beyond the first layer. I think, indeed, you can have some hydrogen protons in there.

M. Shelef

It is consistent and very puzzling that you get much more inside then you expect only from the surface.

Voorhoeve

My thermal desorption experiment really isn't well calibrated so ... I haven't mentioned whether it was a monolayer or 2 or 10. (Added in writing: If more than a monolayer, the formation of surface compounds is more likely than incorporation of NO in the perovskite structure, which is close-packed, quite unlike β-Al_2O_3.)

THE AMMONIA ROUTE TO NO$_x$ CONVERSION IN AUTO EXHAUST CATALYSIS

H. WISE.

Stanford Research Institute, Menlo Park, California

ABSTRACT

Because of the problem of ammonia formation during the conversion of NO$_x$ by reaction with reducing gases in a dual bed catalytic converter the need has arisen for an NH$_3$ oxidation catalyst of high N$_2$-selectivity. Examination of a number of candidate systems has demonstrated that several metal-oxide catalysts are available that exhibit high N$_2$-selectivity, activity, and stability in ammonia oxidation. However selective oxidation of NH$_3$ to N$_2$ runs counter to the complete oxidation of HC and CO. The properties of some oxidation catalysts for ammonia, and for HC and CO will be discussed with special application to the auto-exhaust problem.

INTRODUCTION

An approach to the control of NO$_x$ emissions in auto exhaust has been the development of an NO$_x$ catalyst that exhibits high N$_2$ selectively. Unfortunately the demands of high N$_2$ selectivity and extended durability under reducing and oxidizing conditions have not led to a satisfactory solution in terms of the dual-bed system (i.e., an NO$_x$ reduction catalyst in tandem with a HC and CO oxidation catalyst). Ruthenium-based catalysts were found to exhibit high activity and good N$_2$ selectivity during NO$_x$ reduction (1,2,3). However their lack of durability under redox conditions severely limits their utility in auto exhaust applications. As a result of these developments it may be worthwhile to consider in some detail the NH$_3$ route to NO$_x$ conversion as a possible alternate approach to auto-exhaust control. Such a concept appears attractive if a highly selective catalytic system is available to oxidize ammonia to nitrogen, preferably in the same section of the dual-bed reactor system as used for the oxidation of HC and CO.

References p. 243.

AMMONIA OXIDATION CATALYSTS

In the past the catalytic oxidation of ammonia has emphasized the production of nitric oxide rather than the formation of nitrogen. Still the desire to find substitutes for the noble-metal catalysts used in NO-production and related studies on the mechanism of ammonia oxidation have demonstrated that certain metal-oxide catalysts favor the partial oxidation of ammonia to N_2 rather than complete oxidation to NO. As a matter of fact recent studies on the oxidation of olefins and of ammonia have suggested (4,5,6) a parallelism between the degree of oxidation of olefins and ammonia as measured by the type of products formed. For example in the case of butene and ammonia one may write the following progressive oxidation sequence:

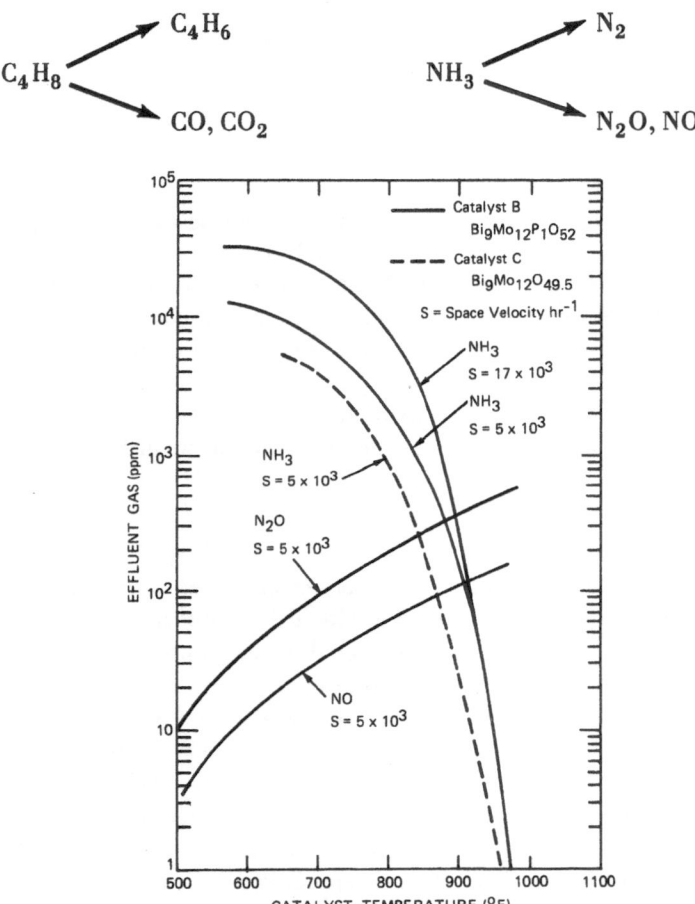

Fig. 1. Residual ammonia and NO_x levels at different operating conditions. (Initial ammonia concentration = 3.4 vol %).

In this reaction sequence step 1 involves hydrogen abstraction from the reactant molecules, while steps 2 and 3 represent oxygen addition. Just as in the case of NH$_3$ oxidation the products formed in the presence of such noble-metal catalysts as Pt and Pd are predominantly the oxides of nitrogen, the complete oxidation of C$_4$H$_8$ (and hydrocarbons in general) is favored by these same catalysts, a process successfully employed in the HC/CO catalytic converter of the 1975 automobile.

The question now arises whether those catalysts that are able by oxidative dehydrogenation to produce butadiene from butene will form N$_2$ from NH$_3$. For this purpose we may wish to consider a group of transition metal oxides which have become of considerable commercial importance to the petrochemical industry. Foremost in this group of catalysts are those used in allylic oxidation of olefins, such as the system based on the mixed oxides of Bi and Mo. Detailed investigation (7-10) of the kinetics of formation of conjugated dienes from mono-olefins demonstrated that this catalytic system favors partial dehydrogenation rather than complete oxidation of the olefins. Similarly in studies of ammonia oxidation with Bi/Mo oxide catalysts we observed high specificity for nitrogen formation (11) (Fig. 1). The catalysts (Bi/Mo = 3/4) on a silica support were similar to those developed for the partial oxidation of propylene (12). It is apparent that more than 90 vol% conversion is attainable at temperatures of the order of 400°C with N$_2$O levels less than 200 ppm, and NO less than 50 ppm. Similar results were reported by Russian workers (13) on a catalyst having a Bi/Mo = 1/2.

A number of different molybdenum-containing mixed oxides were examined in our laboratory as catalysts for ammonia oxidation (14). Among these, three compounds were found to exhibit superior specificity and activity as shown by the data in Fig. 2. A measure of performance is given in Table 1 in which the temperature

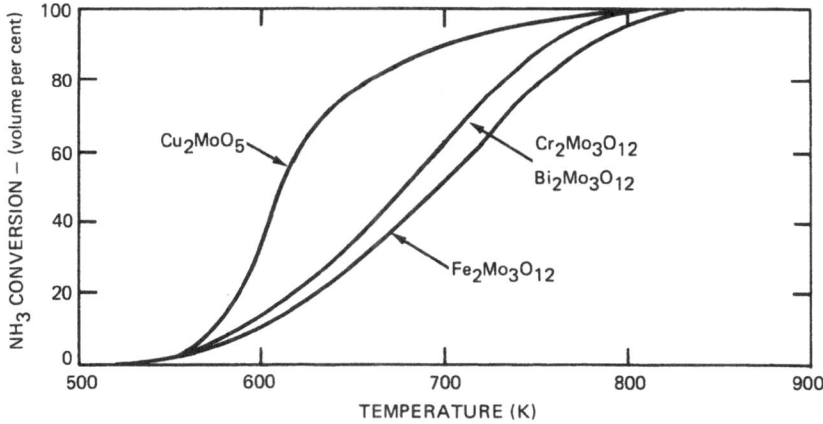

Fig. 2. Activity of catalysts for ammonia oxidation [Reactant gas composition (vol %): NH$_3$ = 7.1; O$_2$ = 28.4; He 64.4; space velocity = 6 x 10^5 hr^{-1}]

References p. 243.

requirements for different degrees of NH_3 conversion and various space velocities are tabulated for Cu_2MoO_5 ($2CuO \cdot MoO_3$) and $Fe_2Mo_3O_{12}$ ($Fe_2O_3 \cdot 3MoO_3$). Stability tests on these two catalysts by short-term cycling (10 minutes) to temperatures up to 1000K demonstrated less sintering and higher sustained activity for $Fe_2Mo_3O_{12}$ than Cu_2MoO_5.

TABLE 1

Temperature Requirements for Ammonia Oxidation at
Different Space Velocities (14)

Catalyst	Space Velocity (hr^{-1})	Temperature*		
		T_{90}	T_{80} (K)	T_{50}
Cu_2MoO_5	10^5	745	715	685
	10^4	680	654	629
	10^3	625	600	580
$Fe_2Mo_3O_{12}$	10^5	754	735	700
	10^4	655	642	620
	10^3	585	600	558

Subscripts indicate the NH_3 conversion to nitrogen in volume percent.

Another base-metal system of interest for the catalytic oxidation of ammonia is composed of the mixed oxides of Cu and Cr. It may be recalled that this system received some earlier attention as a potential candidate for HC and CO oxidation in auto exhaust applications (15). In examining the properties of this catalyst for ammonia oxidation we compared two oxides, one the spinel $CuCr_2O_4$, the other the mixed oxide $Cu_2O \cdot Cr_2O_3$. These solids differ not only in terms of their structural characteristics, but also in the valence state of the Cu-cation. The results of our study (16), as summarized in Table 2, point to some interesting differences between these

TABLE 2

N_2 Specificity in Ammonia Oxidation Catalyzed
by Cu/Cr Mixed Oxides at 673K (16)

NH_3/O_2*	Specificity (%)	
	$Cu_2 \cdot Cr_2O_3$	$CuCr_2O_4$
0.25	93	67
0.50	95	77
0.75	95	82
1.00	96	86
1.25	97	89

Total pressure = 760 torr; O_2 = 40 torr; He = difference.

two catalysts, especially in terms of their N$_2$-selectivity. In the case of the Cu^{1+} containing catalyst the N$_2$-specificity is high (in excess of 90 vol%) with N$_2$O as the minor byproduct. With the Cu^{2+} containing spinel the N$_2$-specificity is considerably lower with N$_2$O and NO appearing as products with N$_2$ in admixture. Although Cu$_2$O·Cr$_2$O$_3$ exhibits superior properties in terms of selectivity, its stability under redox conditions is impaired (Table 3), most likely due to the reaction in the presence of oxygen:

$$Cu_2O·Cr_2O_3 + 1/2\,O_2 \rightleftharpoons 2\,CuO + Cr_2O_3 \ .$$

In an earlier study by Johnstone and coworkers (17) the kinetics of ammonia oxidation were examined in the presence of a mixed-oxide catalyst composed of Mn and Bi (atom ratio 1/2.4). The authors report N$_2$O and N$_2$ as the products of this

TABLE 3

Oxygen Loss from Copper Catalysts (16)
by Reaction with CO*

Temperature K	Initial Catalyst Mass Loss (wt %/min x 10^2)			
	Cu$_2$O · Cr$_2$O$_3$		CuCr$_2$O$_4$	
	Untreated	O$_2$-treated	Untreated	O$_2$-treated
723	0	0.9	0.7	1.0
808	0.37	2.2	13.6	17.0

*10.6 vol % CO in He.

reaction, with a product distribution a function of the initial NH$_3$/O$_2$ ratio and temperature (Table 4). These results demonstrate high conversion efficiency at modest temperatures (180-220°C) and increasing N$_2$-selectivity as the O$_2$/NH$_3$ ratio and the temperature are raised.

TABLE 4

Oxidation of Ammonia Catalyzed by
Manganese Oxide-Bismuth (17)

NH$_3$ O$_2$ vol %		Temperature °C	Conversion vol %	N$_2$ Specificity (%)
10	90	180	99.38	46.7
		200	99.80	83.8
		220	99.95	67.5
20	80	180	99.28	45.6
		200	99.59	88.4
		220	99.80	90.4
33.3	66.7	180	99.17	88.7
		200	99.48	80.0
		220	99.68	91.9

References p. 243.

Finally the ammonia oxidation studies with NiO as a catalyst (4) merit attention. In this work it was reported that the product distribution (N_2O vs. N_2) is governed by the electronic defect structure of the solid catalyst. Addition of monovalent metal oxides (such as Li_2O) to nickel oxide favors the formation of N_2O, while the addition of trivalent oxides (such as Cr_2O_3) yields N_2 as the predominant product.

MECHANISTIC CONSIDERATIONS

In attempting to elucidate those properties of the catalysts that provide high activity and selectivity in ammonia oxidation, we may examine the information available for allylic oxidation of olefins in view of the apparent analogy in the mechanism of these two processes. A measure of the similarity in kinetic parameters is to be found in Table 5 which lists the orders of reaction and the activation energies reported for propylene oxidation to acrolein and ammonia oxidation to N_2 for a group of metal-oxide catalysts. For the two reacting systems (NH_3 and C_3H_6) the reaction order and the activation energies are similar for each of the catalysts examined. One may conclude similarities in reaction mechanism exist, at least as far as the rate-determining step is concerned. Also it is of interest to note that for all the molybdate catalysts the reaction is of first order in reducing agent and of half-order in oxygen, while the cuprous-oxide catalyst exhibits a zero-order dependency on reducing agent, and first-order on oxygen.

TABLE 5

Heterogeneous Oxidation Kinetics

Catalyst	Reaction Order				Activation Energy (Kcal/mole)	
	NH_3	O_2	C_3H_6	O_2	NH_3	C_3H_6
Cu_2MoO_5	1	1/2	—	—	35.0	—
$Fe_2Mo_3O_{12}$	1	1/2	—	—	25.4	—
$Bi_9Mo_{12}P_1O_{52}$	1	1/2	1(a)	1/2(a)	23.6	18-23
Cu_2O	0(b)	1(b)	0(c)	1(c)	10.0	11-15(d)

(a) C. R. Adams, Third Congr. on Catalysis, North-Holland Publ. Co., Amsterdam, 1965, Vol. 1, p. 240.
(b) L. L. Holbrook and H. Wise, J. Catalysis 27, 322 (1972).
(c) B. J. Wood, R. S. Yolles, and H. Wise, J. Catalysis 15, 355 (1969).
(d) R. J. Sampson and D. Shooter in Oxidation and Combustion Reviews, C. F. H. Tipper, ed. Elsevier Publ. Co., Amsterdam, 1965.

In view of their importance to the petrochemical industry, the molybdate catalysts, especially the Bi/Mo system, have been the subject of considerable research, as reviewed by Schuit (18). Specifically the availability of lattice oxygen and its ease of replacement by gaseous oxygen has been singled out as the unique property responsible for specificity in allylic oxidation. According to this redox mechanism, generally referred to in terms of the Mars-van Krevelen model (19), reduction of the metal oxide catalyst by the olefin is followed by reoxidation with gas-phase oxygen. Considerable evidence for this mechanism has been obtained from experiments at high temperatures (above 700 K) involving isotopic molecular oxygen (O_2^{18}) with analysis of the distribution of the labelled oxygen among the oxidation products (20,21). In an extension of this work to a silica-supported Bi/Mo catalyst it was concluded that under the conditions of temperature and pressure employed in the industrial manufacture of acrolein the gas-phase oxygen contributes predominantly to the oxidation of the propylene rather than the lattice oxygen (22).

What evidence do we have for a similar reaction mechanism in the case of ammonia oxidation? One set of experiments involves the product distribution derived from exposure of the catalyst to pulses of ammonia in the absence of gas-phase oxygen (Table 6). The remarkable stability of Cu_2MoO_5, $Fe_2Mo_3O_{12}$, and $Bi_2Mo_3O_{12}$ is apparent as these catalysts become depleted of lattice oxygen (Fig. 3). Of further interest is the degree of ammonia conversion attained in pulse experiments in which oxygen was added to the ammonia ($NH_3/O_2 = 1/4$). As shown in Table 7 a marked enhancement in ammonia conversion takes place on oxygen addition, except for the case of the two copper-containing catalysts for which the degree of ammonia oxidation is already very high. Similar observations of the increase in NH_3 conversion on Bi/Mo-oxide catalysts of different Bi/Mo ratio have been reported by Giordano and Zema (4). These authors demonstrated a large increase in the rate of NH_3-oxidation in the presence of oxygen (21 vol%). At 625 K the increase was more than five-fold for $Bi_2O_3 \cdot 2\ MoO_3$, and nearly three-fold for $Bi_2O_3 \cdot 3MoO_3$. Thus lattice-oxygen depletion becomes less significant in the presence of sufficient gas-phase oxygen.

TABLE 6

Ammonia Conversion* by Lattice Oxygen at 675 K (14)

Conversion (vol.%) Catalyst	Pulse			
	1	10	20	30
Cu_2MoO_5	90	90	90	80
$CuMoO_4$	70	61	68	67
$Fe_2Mo_3O_{12}$	24	25	25	25
$Bi_2Mo_3O_{12}$	12	10	12	11
$Cr_2Mo_3O_{12}$	13	2	nil	nil

*Pulse: 0.1 cc NH$_3$; 0.9 cc He (NTP); catalyst mass: 0.10 g.

References p. 243.

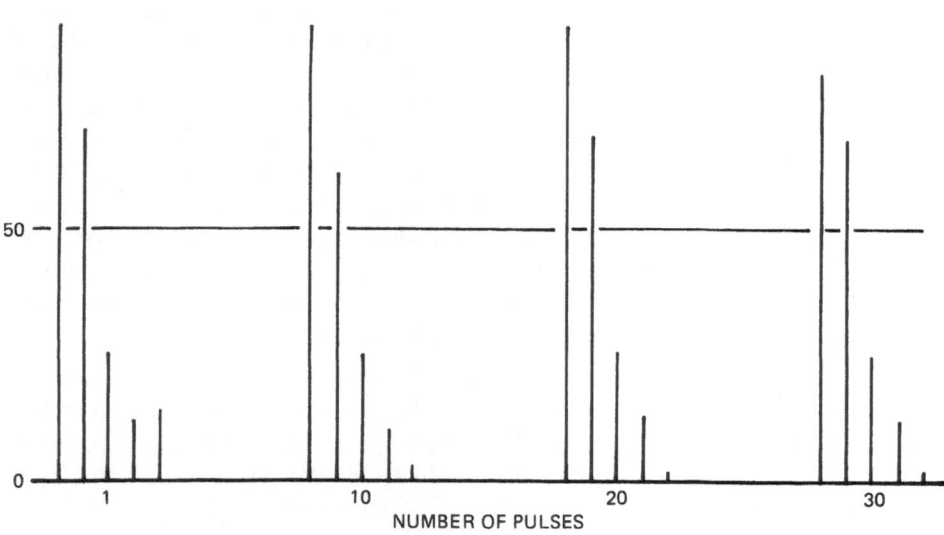

Fig. 3. NH$_3$ oxidation by lattice oxygen Cu$_2$MoO$_5$/CuMoO$_4$/Fe$_2$Mo$_3$O$_{12}$/Bi$_2$Mo$_3$O$_{12}$/Cr$_2$Mo$_3$O$_{12}$

TABLE 7

Ammonia Conversion* in Presence of
Oxygen at 675 K (14)

Catalyst	Ammonia Conversion (vol %)	
	O$_2$/NH$_3$ = 4	O$_2$/NH$_3$ = 0
Cu$_2$MoO$_5$	88	90
CuMoO$_4$	77	70
Fe$_2$Mo$_3$O$_{12}$	35	24
Bi$_2$Mo$_3$O$_{12}$	47	12

*Pulse: 0.07 cc NH$_3$, 0.28 cc O$_2$, 0.65 cc He;
Catalyst mass = 0.10 g.

Again in analogy to the allylic oxidation mechanism of olefins, the conversion of ammonia to N$_2$ probably involves hydrogen abstraction as a primary step. Furthermore the kinetic dependency of the rate on (O$_2$)$^{1/2}$ indicates dissociative chemisorption. Thus a reaction stoichiometry of the following form may be written:

$$O_2 \rightleftharpoons 2\,O_a \tag{1}$$

$$NH_3 + O_a \rightarrow NH_{2a} + OH_a \tag{2}$$

Further hydrogen abstraction will lead to the nitrogen adatom as a precursor to N$_2$-formation. At the same time it is to be expected that oxygen addition at any of the intermediate steps may result in the formation of oxygen − containing nitrogen compounds such as N$_2$O and NO. At this time insufficient information is available to develop a complete reaction mechanism, except for the rate-determining step given by reaction (2).

It may be concluded that a number of catalysts are available for ammonia oxidation. Specifically a number of molybdenum − containing mixed oxides exhibit high activity and N$_2$ specificity for this reaction. It should be realized, however, that these same catalysts cause only partial oxidation of hydrocarbons. As a result the utilization of the ammonia route to NO$_x$ reduction in auto-exhaust control may require different catalysts for NH$_3$ and for HC/CO oxidation.

ACKNOWLEDGMENT

Partial support of this research by Climax Molybdenum Company of Michigan is gratefully acknowledged.

REFERENCES

1. M. Shelef and H. S. Gandhi, Ing. Eng. Chem. Prod. Res. Deve. 11, 2 (1972); Plat. Metals Rev. 18, 2 (1974).
2. R. L. Klimisch and K. C. Taylor, Env. Sci. Techn. 7, 127 (1973).
3. T. P. Kobylinski and B. W. Taylor, J. Catal. 33, 376 (1974).
4. N. Giordano, La Chimica e L'Industria 51, 1189 (1969).
5. B. J. Wood, H. Wise, and R. S. Yolles, J. Catal. 15, 355 (1969).
6. L. L. Holbrook and H. Wise, J. Catal. 27, 322 (1972).
7. C. R. Adams in Third Internat. Cong. on Catalysis, eds. W.M.H. Sachtler, G.C.A. Schuit, and P. Zwietering, North Holland Publ. Co., Amsterdam, 1965; Vol. 1, p. 240.
8. W.M.H. Sachtler and N. H. de Boer, ibid, Vol. 1, p. 252.
9. L. Ya. Margolis, J. Catalysis 21, 93 (1971).
10. H. A. Voge and C. R. Adams, Adv. Catalysis 17, 151 (1967).
11. J. Hunter, K. Sancier and H. Wise (to be published).
12. F. Veatch, J. L. Callahan, E. C. Milberger and R. W. Foreman, Second Internat. Cong. Catalysis, 1961, Vol. 2, p. 2647.
13. T. G. Alkhazov et al., Kinetics and Catalysis (Engl. transl.) 11, 99, (1970).
14. E. Farley, R. T. Rewick, and H. Wise (to be published).
15. R. S. Yolles and H. Wise, Crit. Rev. Env. Control 2, 125 (1971).
16. S. H. Inami and H. Wise (to be published).
17. H. F. Johnstone, E. T. Houvouras and W. R. Schowalter, Ind. Eng. Chem. 46, 702 (1954).
18. G.C.A. Schuit, Proc. XX Internat. Congress, Industrial Catalysis, Italian Chemical Society, Milan (1969).
19. J. Mars and D. W. Van Krevelen, Chem. Eng. Sci. Suppl. 3, 41 (1974).
20. G. W. Keulks, J. Catalysis 19, 232 (1970).
21. R. D. Wragg, P. G. Ashmore, and J. A. Hockey, J. Catalysis 23, 270 (1971).
22. K. M. Sancier, P. Wentrcek, and H. Wise (to be published).

DISCUSSION

R. L. Klimisch *(General Motors Research Laboratories)*

We've looked at a few of these catalysts and found that they weren't selective for nitrogen formation. Two things that we did were different. One was that we were looking at ammonia concentrations of 100 ppm rather than 30,000 ppm, and we also have a lot of water in our system. I would suggest that these two things make a difference. One possibility is that the nitrogen is actually formed in an ammonia-NO reaction and that reaction is certainly much more possible at high concentrations. And, we think, the water interferes with ammonia adsorption on the surface. We haven't looked at all of the materials but I suggest that those two things do tend to change the selectivity.

Wise

To respond to this question, we have not explored the role of water completely. However, in studies of the copper-molybdenum system we have introduced water vapor. We find it plays a very minor role under our experimental conditions. You're right. We're working at fairly high ammonia levels although they're not beyond what you might expect for a catalyst which produces ammonia.

Klimisch

You're much higher, three orders of magnitude.

Wise

All right. Perhaps the much higher ammonia levels used by us make a difference. The water effect, as I have said, has not been explored for all of these catalysts, but since water is formed in the reaction as one of the products of the reaction, I would be very much surprised if water has a very profound effect. There's a concentration effect, obviously in a first order reaction. But here again you would expect that the fractional conversion should remain constant even at low concentrations.

Klimisch

We find conversion, but not to nitrogen.

Wise

What were the temperature and the catalysts?

Klimisch

At 500°C, with copper chromium.

Wise

Yes. Now the copper-chromium catalyst is one of those that indeed showed low selectivity in the presence of oxygen. That's a poor catalyst to use in an oxidizing atmosphere. Perhaps you should try bismuth molybdate.

H. S. Bloch *(UOP, Inc.)*

Can you say anything about the effect of sulfur on this selective behavior?

Wise

No. I have not explored it. UOP knows a great deal about base metal catalysts and their reaction with sulfur containing species. I would say that these are base metals and they will have some of the properties you have experienced in oxidation catalysts.

R. L. Burwell, Jr. *(Northwestern University)*

Did you agree that the lattice oxygen will not be an oxidizing agent?

Wise

The situation is the following. It can be. In studies similar to those by Keulks and coworkers we have observed that at high oxygen levels and moderate temperatures, you don't use lattice oxygen but surface sorbed oxygen from the gas phase.

Burwell

If lattice oxygen is the oxidizing agent, would you tell me what it's reduced to? I missed that.

Wise

Well, we can, for example, write the following reaction involving:

$$NH_3 + O = NH_2^+ + OH^-$$

followed by subsequent hydrogen abstraction and electron transfer to the solid.

Burwell

But the oxygen has not been oxidized or reduced in that reaction.

Wise

You are losing electrons originally attached to the oxygen by injecting electrons into the solid; that is the oxygen step.

Burwell

So then, it's not the oxygen that's the reducing agent, but in essence one of the transition metal ions.

Wise

Well, the problem is much more complicated than that. It appears from conductivity measurements that we do see the electrons injected into the solid when you pulse the reducing agent over the catalyst. So that, indeed, we find that electrons are transferred into the conduction band. We find at the same time, using ESR, that Mo^{+6} is not reduced, which suggests there's an impurity level located here. You see, that's what makes it so complicated. The impurity's energy level is sufficiently below the conduction band so that these electrons are trapped here and not in the conduction band. So Mo^{+6} appears to persist during these pulsing experiments for example. Yet we are losing oxygen from the solid and injecting electrons into the solid.

Burwell

Then I would think you really ought not to say that the lattice oxide is an oxidizing agent. You're saying it as if it were reduced.

Wise

You're removing lattice oxygen.

Burwell

It enters certainly into reaction, but it is not the oxidizing agent at all. There is an old rule that the oxidizing agent is reduced. So, lattice oxygen might enter the products of reaction to form water, but it is not in any sense the oxiding agent.

SESSION IV

CATALYTIC APPLICATIONS

Session Chairman
E. L. HOLT

Exxon Research and Engineering Company
Linden, New Jersey

IRON-CATALYZED REDUCTION OF NO BY CO AND H_2 IN SIMULATED FLUE GAS

D. T. CLAY* and S. LYNN

University of California, Berkeley, California

ABSTRACT

Iron oxide supported on alumina is a promising catalyst/absorbent for use in the simultaneous removal of NO_x and SO_x from power plant stack gases. The process under development would operate under net reducing conditions at temperatures of 370°-540°C. NO_x is converted to H_2 or NH_3, and SO_x is removed as a sulfide or sulfate. The reduction of NO by CO and H_2 in a fixed-bed reactor has been studied to determine the effects of temperature, catalyst composition, and the other components of flue gas (excepting fly ash) on the rate of the reaction and the ratio of N_2/NH_3 produced. Under readily attainable conditions, reaction times of the order of 0.01s are observed. Some of the conditions under which oxygen poisons the reduction of NO have been determined.

INTRODUCTION

The standards proposed by the EPA in 1971 set allowable NO_x emissions for power plant stack gases at 0.2 lb/10⁶ Btu (210 ppm)** for gas-fired units, 0.3 lb/10⁶ Btu (310 ppm) for oil, and 0.7 lb/10⁶ Btu (672 ppm) for coal-fired units. The sulfur emissions allowed were 0.8 lb SO_2/10⁶ Btu (390 ppm) for oil and 1.2 lb SO_2/10⁶ Btu (540 ppm) for coal-fired units. The different standards for the different fuels is a recognition of the different degrees to which emissions can be controlled by fuel pretreatment or combustion modification with current technology. One might thus expect that, as improvements are developed, allowable emissions for oil- and coal-fired units will be reduced.

References pp. 259-260.

*Present address: Weyerhauser Corp., Longview, Washington 98622
**NO_x expressed as NO_2; ppm values based on:
  coal — 0.7 lb C/lb coal, 12000 Btu/lb coal, 14% CO_2
  oil — 0.865 lb C/lb oil, No. 6 fuel oil, 12% CO_2
  gas — 0.97 ft³ CO_2/ft³ gas, 873 Btu/ft³, 9.1% CO_2

The sources of NO in stack gas are the fixation of atmospheric nitrogen and the oxidation of nitrogen-containing compounds in the fuel. The former is quite sensitive to the temperature of combustion, the latter much less so. Bartok et al. have studied combustion modification extensively (1). Turner et al. give a typical range of fuel N conversion to NO of 30-60% for oil-firing (2). Jonke et al. reported 18-25% fuel N conversion for a fluidized bed of coal (3). This means that for heavy fuel oils (1.4% N) and typical coals (1.5% N) an NO level in the flue gas of about 500 ppm or higher results from the fuel N conversion alone. No economically practical method of combustion modification appears likely to result in NO emissions below 300-400 ppm for either fuel.

It therefore seems likely that stack-gas treatment will be required if NO_x emissions from power plants are to be reduced to the order of 100 ppm. Stack-gas treatment also appears to be the most economical method of meeting the limitations on sulfur emissions for large power plants, especially coal-fired units (4). The obvious advantage of combining NO_x removal with SO_x removal in a process which would meet both emissions standards has not yet led to large-scale testing of any device. This work is part of a laboratory study of a process which would meet that goal.

PROCESS DESCRIPTION

The process being developed combines three fundamental techniques for nitric oxide and sulfur oxide removal. These are combustion modification, catalytic reduction, and metal oxide absorption. The process and the function of each technique can best be explained by referring to Fig. 1.

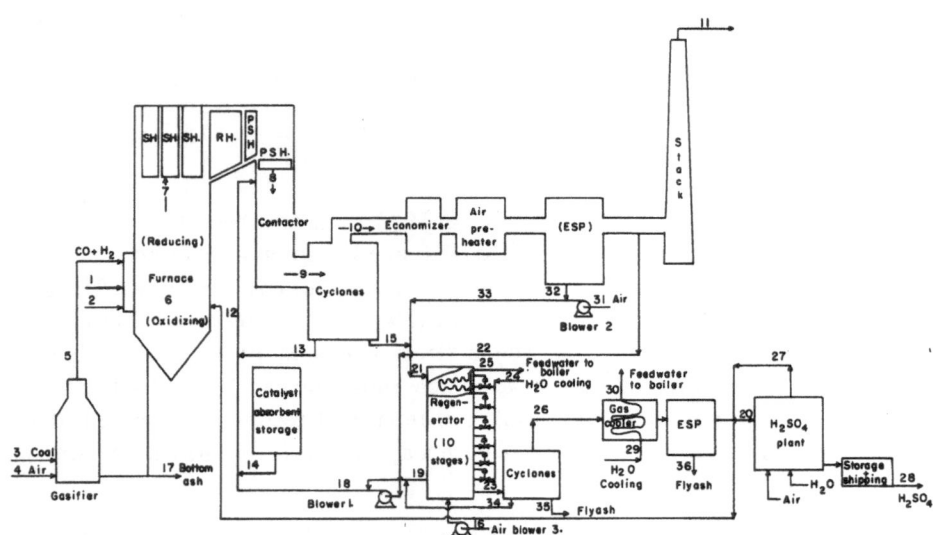

Fig. 1. Conceptual design of NO/SO_2 removal process.

This figure gives the general layout of a power plant furnace and the proposed process. The furnace is operated under oxidizing conditions to assure that all of the coal or oil is oxidized and no soot is formed. The type of fuel and the burner design dictate the lowest level of excess air allowed to meet this requirement. At a point higher up in the furnace a rich stream of CO and H_2 is introduced to shift the flue gas from net oxidizing to net reducing stoichiometry. This rich CO/H_2 stream could be generated by a coal gasification unit adjacent to the power plant. At least enough CO and H_2 is added so that at equilibrium all of the O_2 would be reduced to CO_2 or H_2O, all of the NO to N_2, and all of the SO_2 to H_2S. Since this fuel addition is made in the furnace, its heating value will be recovered either in the superheater or air preheater sections of the power plant.

Between the exit of the superheater and the entrance of the air preheater the flue gas is contacted with the catalyst/absorbent particles. The particles consist of supported Fe_2O_3, which catalyzes the reduction of NO to N_2 or NH_3 and effects the absorption of sulfur as sulfide. The temperatures in this region range from 370°-540°C. Typical flue gas residence times are between 0.30 and 0.50 seconds. After the solids and the flue gas leave the contactor they enter mechanical cyclones. These remove the solids from the gas and segregate the catalyst/absorbent material from the fly ash. After the flue gas passes through the economizer and air preheater, electrostatic precipitators remove the fly ash and catalyst fines. The catalyst/absorbent from the cyclones is sent back for reinjection into the flue gas contactor. A portion of this recycle stream is diverted to the regenerator, where air is added in a fluid bed to convert the iron sulfide back to the oxide. This generates a rich SO_2 stream that can be used as a feed to an H_2SO_4 or sulfur plant. The regenerated catalyst is fed back to the boiler after mixing with the recycle stream.

The primary reactions are listed in Table 1. The catalytic reduction of NO is accomplished by an excess of CO or H_2 in the flue gas. Both N_2 and NH_3 are potential products. It will be shown later that all three of these reactions proceed over either iron oxide or sulfide. The sulfur compounds are removed by reduction with CO or H_2 to form COS or H_2S. These reduced sulfur compounds react with FeO to form FeS. The first two reactions are the most important since more than 95% of the sulfur will be present initially as SO_2.

An excess of reducing agent will be required to achieve high removal efficiencies for both NO and SO_2. The excess reducing agent is oxidized by maintaining an excess of Fe_2O_3 in the recycle solids to avoid emission of CO from the stack. The reactions to remove the excess CO and H_2 are endothermic; the thermal loss associated with their removal is recovered in the regenerator.

In the regenerator FeS and FeO react exothermically with O_2 to produce a self-sustained reaction in a fluidized bed. A rich SO_2 stream and the regenerated ferric oxide are produced. Temperature in the regenerator is maintained at about 675°C by

References pp. 259-260.

TABLE 1

Overall Chemical Reactions for NO/SO_2 Removal Process

Absorption/Reduction

$$2NO + 2CO \rightarrow N_2 + 2CO_2$$

$$2NO + 2H_2 \rightarrow N_2 + 2H_2O$$

$$2NO + 3H_2O + 5CO \rightarrow 2NH_3 + 5CO_2$$

$$FeO + SO_2 + 3CO \rightarrow FeS + 3CO_2$$

$$FeO + SO_2 + 3H_2 \rightarrow FeS + 3H_2O$$

$$FeO + H_2S \rightarrow FeS + H_2O$$

$$Fe_2O_3 + CO \rightarrow 2FeO + CO_2$$

$$Fe_2O_3 + H_2 \rightarrow 2FeO + H_2O$$

Regeneration

$$2FeS + 7/2\ O_2 \rightarrow Fe_2O_3 + 2SO_2$$

$$2FeSO_4 \rightarrow Fe_2O_3 + 2SO_2 + 1/2\ O_2$$

removing heat from the bed. This temperature is high enough to decompose any iron sulfate which may have been formed. The heat is removed with boiler feed water and returned to the power plant heat cycle.

EXPERIMENTAL APPARATUS

The experimental apparatus has been described extensively elsewhere (5). Experiments were carried out in stainless steel fixed-bed reactors having diameters of 6.4, 9.5, and 32 mm. Bed lengths were 30-80 mm. The reactor was placed inside an electric tube furnace for temperature control. The catalyst/absorbent tested was Harshaw Fe - 301-T 1/8, 20% Fe_2O_3, supported on alumina and having a surface area of 41 m^2/g. The material was in the form of extruded pellets 3.2 mm x 3.2 mm and particles of 0.25-0.50 mm diameter. Reactant gas concentrations in the feed stream were generally in the range of 0.5-2.0%, with the carrier gas being helium.

Gas analyses of both inlet and outlet streams were performed by gas chromatography, using two Varian Model 90-P chromatographs in series. Porapak R packing was used in the first column to separate CO_2, N_2O, NH_3, H_2S, COS, SO_2, and H_2O. Molecular sieve 5A was used in the second column to separate H_2, O_2, N_2, NO, and CO. The columns were operated in such a way that the peaks separated in the first did not enter the second. The sensitivity of the analysis for most of these gases was of the order of 100-200 ppm. For H_2, NH_3, and H_2O it was of the order of 2000-2500 ppm. The progress of most reactions was followed by monitoring both the disappearance of reactive species and the formation of products of reaction. Mass balances were generally within 10%.

EXPERIMENTAL RESULTS

Catalysis of NO Reduction — One part of the experimental program was a study of NO reduction to determine reaction products, to obtain an indication of reaction mechanism, and to establish reaction kinetics. In the series of runs presented in Fig. 2, NO/CO mixtures of various concentrations were passed through a bed of 3-mm catalyst pellets at 345°-390°C. The CO/NO ratio appears to be the controlling factor in determining NO removal. Varying degrees of pre-reduction of the Fe_2O_3 with CO had little influence on the results. Later runs in which the iron had been partially converted to FeS correlate well with the curve in this figure. No N_2O was detected in the effluent.

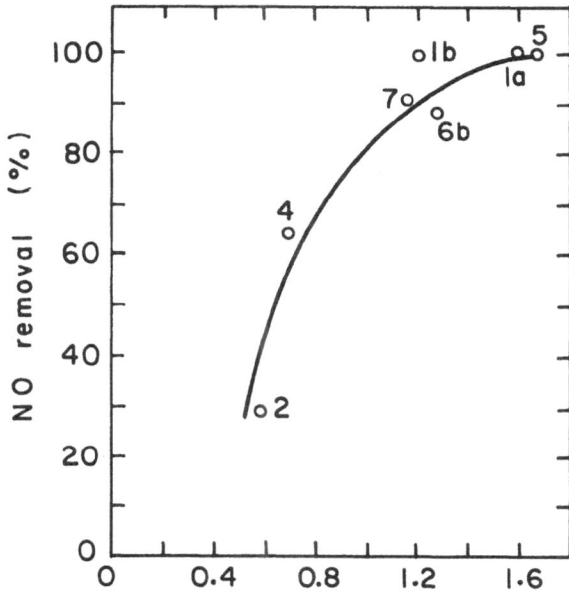

Fig. 2. Percent NO removal versus CO/NO over FeO_x.

Run	1a	1b	2b	4	5	6b	7
[NO] (%)	4.1	3.7	1.7	2.6	2.0	0.29	0.30
[CO] (%)	6.6	4.4	1.0	1.8	3.4	0.37	0.35
FeO_x, x =	1.38	1.01	1.19	1.46	0.92	0.31	⟨0.10
Temp. (°C)	345	347	365	390	390	380	370
Res. time (s)	0.88	0.88	0.37	0.40	0.40	0.39	0.39

Fe-301-T 1/8 (3.2mm x 3.2mm)

References pp. 259-260.

Klimisch and Barnes (6) proposed a redox mechanism for the reduction of NO over iron oxide. Experiments were run to test this mechanism by demonstrating that NO could be reduced with either FeO or FeS. Fig. 3 shows the results of contacting catalyst pellets, pre-reduced to varying degrees, with NO. In run 15b some CO_2 was released, presumably because of CO adsorption during pre-reduction. It corresponded to about 18% of the N_2 formed from NO in this run. It thus appears that reduction of NO can occur both by reaction with reduced Fe and by reaction with adsorbed CO. Fig. 4 shows the results of contacting pellets containing FeS with NO. In contrast to the case of FeO, the initial rate of reaction is slow. Not only N_2 but also SO_2 is formed. The ratio of N_2 to SO_2 concentration is high, especially initially, indicating that $FeSO_3$ may first be formed and then decompose to FeO and SO_2. As was seen in the earlier runs, the presence of CO or H_2 in the gas prevents SO_2 formation when NO is reduced over FeS.

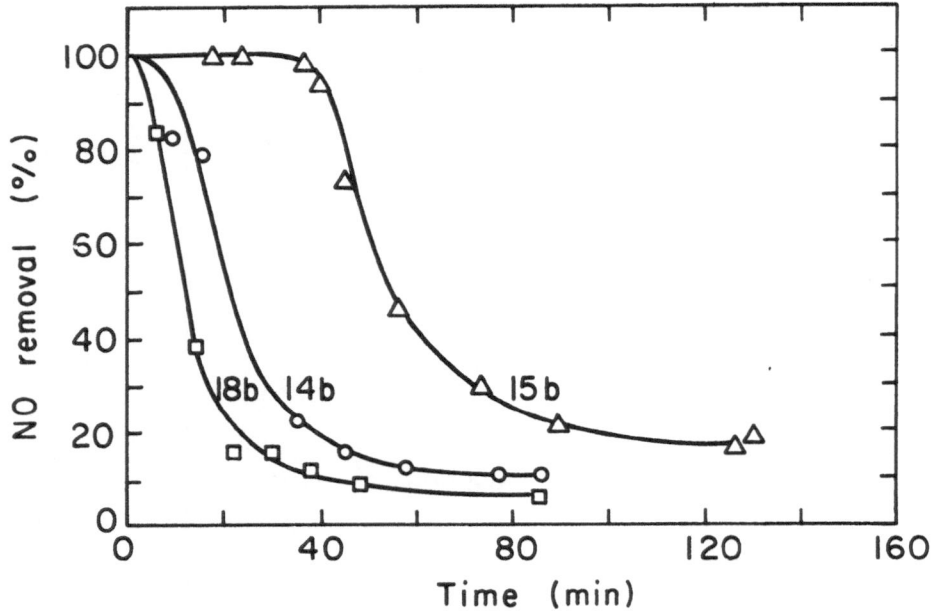

Fig. 3. Oxidation of reduced iron oxide with NO.

Run	14b	15b	18b
T (°C)	374	377	377
θ (s)	0.55	0.51	0.48
[NO] %	2.17	0.98	1.34
Solid (I)	$FeO_{1.34}$	$FeO_{1.19}$	$FeO_{1.42}$
	Fe-301-T 1/8 (3.2mm x 3.2mm)		

NH_3 Formation — In the runs mentioned above, no source of hydrogen was present and N_2 was the only reduction product observed. When mixtures of NO and

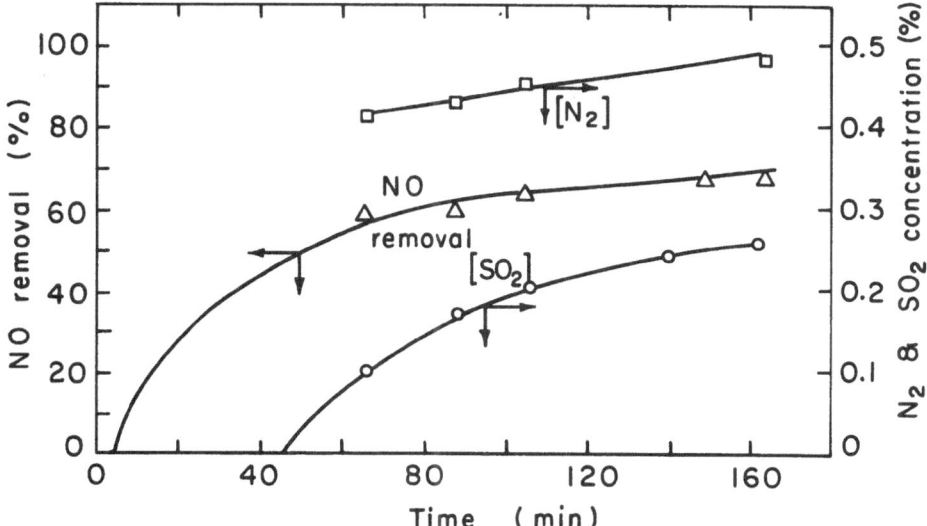

Fig. 4. Oxidation of iron sulfide with NO.

Run	16c
T (°C)	373
θ (s)	0.50
[NO]	1.44%
Solid (I)	$FeO_{.57}S_{.80}$
Fe-301-T 1/8	
(3.2mm x 3.2mm)	

H_2 were passed over catalyst pellets containing FeO at 370°C, under conditions giving complete reduction of NO, it was found that 20-30% of the existing nitrogen was in the form of NH_3. When FeS was present in the catalyst, the formation of NH_3 was significantly higher. It was also found that the reduction of NO by either FeO or FeS in the presence of H_2O led to NH_3 formation, which again was significantly greater for the sulfide. Reduction of NO by CO in the presence of H_2O produced similar results, leading to more than 90% of the nitrogen being converted to NH_3 when a large fraction of the iron was in the form of FeS. Both NH_3 and H_2O were quite strongly adsorbed by the catalyst at 370°C; it required about 20 to 30 minutes to completely flush H_2O from the bed when the residence time of the gas in the bed was about 0.4-0.5 s.

Jones *et al.* (7) and Klimisch and Barnes (6) have shown that the presence of oxygen will decrease ammonia selectivity during NO reduction. A correlation between both this effect and the effect of the sulfur content of the catalyst/absorbent on NH_3 selectivity was found. Fig. 5 shows the effect on the NH_3 selectivity of the product of the ratio of the moles of oxide to moles of sulfide in the solid with the ratio of oxidation to reduction equivalents in the inlet gas. The ratio of the oxide to the sulfide is an average of initial and final conditions. This procedure is only an approximation since initially no sulfide is present and at the end of a run there is a large amount of sulfide. The qualitative trend on this graph is important: the greater the reducing equivalents in the gas or the more sulfide on the solid, the greater will be the NH_3 selectivity. This correlation is insensitive to temperature between 370° and

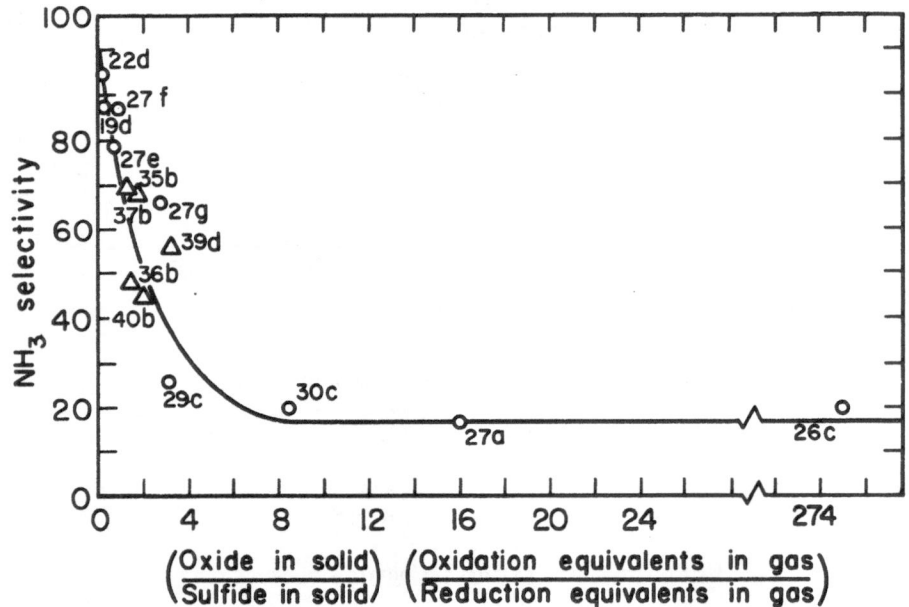

Fig. 5. NH_3 selectivity as a function of solid and gas composition.

Gas composition (%)

Run	[NO]	[SO$_2$]	[CO]	[H$_2$]	[O$_2$]	[H$_2$O]
19d	0.38	0	0	2.5	0	0
22d	0.58	0	1.1	0	0	3.4
26c	0.50	0.95	3.7	3.6	2.0	4.9
27a	0.50	0.41	3.6	3.6	2.0	1.4
27e	0.53	0.44	3.9	3.7	0	0
27f	0.55	0.44	3.8	3.7	0	1.0
27g	0.53	0.45	3.7	3.6	1.3	1.0
29c	0.53	0.50	4.6	0	1.0	1.1
30c	0.61	0.98	4.9	0	1.1	1.0
35b	0.50	0.48	3.7	0	0.49	1.8
36b	0.50	0.50	3.6	0	0.45	1.3
37b	0.49	0.50	3.5	0	0.13-0.48	1.3
39d	0.49	0.56	5.5	0	1.0	1.7
40b	0.53	0.53	5.6	0	1.0	1.7

○ T = 370-380°C, $\theta = 0.5\text{-}0.7$ s
△ T = 538-590°C, $\theta = 0.02\text{-}0.04$ s
(Fe-301-T 1/8)

590°C. NH_3 is unstable relative to N_2 and H_2 at and above 370°C for the concentrations used in this experiment. Shelef (8) reports NH_3 decomposition at the upper temperature limit with an iron oxide catalyst. There is no apparent decomposition in these runs, however, since in all cases the NH_3 selectivity has about the same dependence on the combined factor. Experiments with catalysts containing CuO or NiO in which NO was reduced by CO in the presence of both H_2O and SO_2 at 540°-600°C showed that neither metal oxide was effective in preventing NH_3 formation. For the process being developed, ammonia selectivity is expected to be about 50%.

Effect of O_2 and SO_2 on Catalyst Activity — Ideally, the flue gas entering the contacting zone would contain no oxygen, since the overall composition is to be net reducing. However, oxygen concentrations as high as 1% might result due to poor mixing in the region of the furnace where spontaneous combustion should occur. For this reason, the effect of oxygen on the reaction was studied. It was found that even under net reducing conditions the presence of oxygen led to catalyst deactivation at temperatures of about 370°C. The run presented in Fig. 6 shows the effect

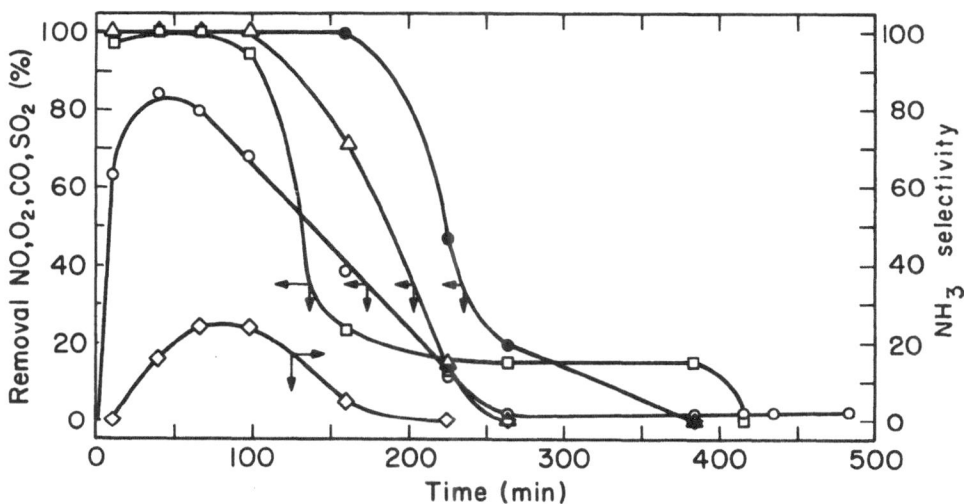

Fig. 6. Simultaneous removal of NO, O_2, SO_2 with CO. Fe-301-T1/8 (0.50mm-0.25mm)

Run	30c	Symbol
T (°C)	371	
θ (s)	0.28	
[CO]	4.95%	○
[NO]	0.61%	△
[O_2]	1.1%	●
[SO_2]	0.98%	□
[H_2O]	0.7-1.2%	
Solid (I)	$FeO_{1.18}$	
Solid (F)	$FeO_{2.2}S_{.29}$	◇

dramatically. The iron is initially present primarily as FeO but the feed gas contains a 16% excess of oxidizing components. Initial removal of NO, SO_2, and O_2 is virtually complete. Then the catalyst becomes poisoned and the removal of all three virtually ceases. Examination of the poisoned catalyst has led to the conclusion that sulfate formation is responsible for the loss of activity. $FeSO_4$ appears to form slowly when O_2 is present in the feed at 370°C even though the net gas composition is reducing. The activity returns after a period of one or two hours if the O_2 is removed from the feed stream.

Investigation of the catalyst deactivation showed that it was much less important at temperatures around 540°C. At the higher temperature no loss of activity occurs if the net composition is reducing, and recovery after exposure to net oxidizing conditions is very rapid. A further advantage of operation at this temperature is the much smaller swing from the 675°C required for regeneration.

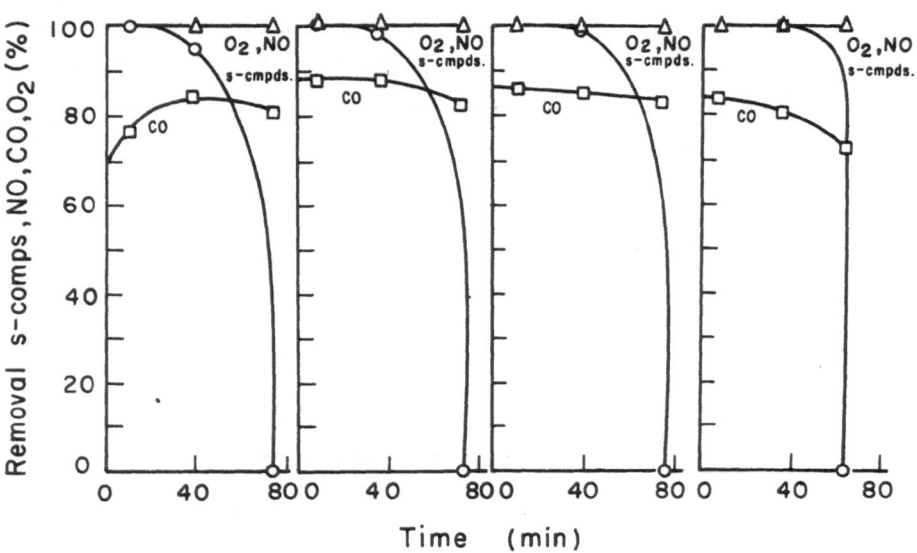

Fig. 7. Percent removal of S-compounds, NO, CO versus run time with four successive removals. Run No. 39.

Removal conditions

[NO] = 0.50% [CO] = 5.5%	Removal at:
[SO$_2$] = 0.56% [O$_2$] = 1.0%	T = 560°C
[H$_2$O] = 1.7.	θ = 0.04 s

FE-301-T1/8 (0.50mm-0.25mm)

Regeneration conditions

1st 4% [O$_2$] for 66 min.	Regeneration at:
2nd 2% [O$_2$] for 129 min.	T = 670°C
3rd 21% [O$_2$] for 63 min.	θ = 0.038 s

The effectiveness of regeneration and an indication of the speed of the reduction/absorption reactions is shown in Fig. 7. The regeneration is carried out in an oxygen-containing atmosphere at a temperature at which both ferrous and ferric sulfates are unstable. All iron compounds should thus be converted to Fe_2O_3. The residence time of the gas in the catalyst bed was about 0.04 s. At the time of breakthrough of sulfur compounds it is estimated that more than 80% of the iron had been converted to FeS. The residence time of the gas in the reaction zone would thus be of the order of 0.01 s. Kinetic measurements of this sort were combined with correlations for mass transfer in packed beds to estimate the catalyst/absorbent flow required in the process.

CONCLUSIONS

The technical feasibility of a process for the simultaneous removal of NO and SO_2 from power plant stack gases has been demonstrated. The mechanism of the reduction of NO by CO or H_2 over an iron oxide or iron sulfide catalyst is probably primarily through reaction between adsorbed molecules, but reduction of NO directly by FeO or FeS can also take place. Ammonia formation occurs during reduction of NO in the presence of hydrogen-containing species. Since NH_3 is not photochemically reactive (9), and the expected level of 300-400 ppm is well below the 2500 ppm limitation which has been set (10), its presence in the stack gas should not be objectionable. Catalyst deactivation occurs if excessive oxygen is present in the gas, but at 540°C reactivation is rapid when the net gas composition becomes reducing. Regeneration of the catalyst/absorbent has been demonstrated, and the kinetics of the reduction/absorption reactions have been shown to be quite rapid.

ACKNOWLEDGMENT

Fellowship support for David T. Clay was provided by the National Science Foundation, the Environmental Protection Agency, and the Atomic Energy Commission during the course of this project.

REFERENCES

1. W. Bartok, et al., "Systems Study of Nitrogen Oxide Control Methods for Stationary Sources – Volume II", NAPCA PB 192789, November, 1969.
2. D. W. Turner, et al., "Influence of Combustion Modification and Fuel Nitrogen Content on Nitrogen Oxides Emissions from Fuel Oil Combustion", from Air Pollution and its Control by R. W. Coughlin, et al., AIChE Symposium Series No. 126, Volume 68, p. 55, 1972.
3. A. A. Jonke, et al., Argonne National Lab. Monthly Report No. 8, March 1969; Monthly Report No. 16, January 1970.
4. F. R. Princiotta, W. H. Ponder, "Current Status of SO_2 Control Technology," Presented at the Lawrence Berkeley Laboratory Seminar entitled Sulfur, Energy, and Environment, April 4, 1974.

5. *D. T. Clay, Ph.D. Thesis "Development of a Feasible Process for the Simultaneous Removal of Nitrogen Oxides and Sulfur Oxides from Fossil Fuel Burning Power Plants", Department of Chemical Engineering, University of California, Berkeley, 1974.*

6. *R. L. Klimisch, G. J. Barnes, "Chemistry of Catalytic Nitrogen Oxide Reduction in Automotive Exhaust Gas", Environmental Science and Technology, 6, 543, 1972.*

7. *J. H. Jones, et al., "Selective Catalytic Reaction of Hydrogen with Nitric Oxide in the Presence of Oxygen", Environmental Science & Technology, 5, (9), 790, 1971.*

8. *M. Shelef and H. S. Gandhi, "Ammonia Formation in Catalytic Reduction of Nitric Oxide by Molecular Hydrogen. I. Base Metal Oxide Catalysts", Ind. Eng. Chem. Prod. Res. Develop., 11, 2, 1972.*

9. *E. Robinson and R. C. Robbins, "Emissions, Concentrations, and Fate of Gaseous Atmospheric Pollutants" in "Air Pollution Control", W. Strauss, Ed., Wiley-Interscience, New York, 1972.*

10. *BAAPCD, Amendment to Regulation 2, Division 15-Odorous Substances, Bay Area Air Pollution Control District, December 6, 1972.*

DISCUSSION

L. Hegedus *(General Motors Research Laboratories)*

Did you say you have a continuous regeneration process in a moving bed or is it a batch-wise process?

Lynn

We propose a continuous process in a fluidized bed.

K. C. Youn *(Shell Development Co.)*

I'd like to get your comment on the technical feasibility of getting about two percent carbon monoxide in the boiler when you burn fuel oil or coal.

Lynn

What we're proposing doing is to have a separate gasification unit . . . adiabatic gasification which would, in essence, generate a low BTU gas. The gas would be injected into the boiler in the upper regions rather then trying to burn the primary fuel with a deficiency of oxygen.

Youn

If you have a separate gasification process, wouldn't it be more economical to remove most of the fuel nitrogen and sulfur in the gasification process rather than burning the gasified fuel in the boiler.

Lynn

In the first place, only about 15% of the powerplant fuel would go through the gasification step, so that 85% would be burned in the normal manner. In the second place, I don't propose treating that gas at all . . . I propose having it as an adiabatic gasification step which is close-coupled with the furnace, so that the CO-hydrogen mixture would be injected directly into the furnace without treatment at a temperature of the order of 1800°F (the adiabatic temperature at which air reacts with coal to give CO and hydrogen).

T. Dingo *(General Motors Manufacturing Staff)*

Just a couple of comments on your conceptual design. On most utility and industrial powerhouse applications, you have quite a bit of excess air. You have, in industrial applications, up to 10% oxygen in your flue gas and in utility applications the tightest that they can control is perhaps 2 or 3% oxygen. So you have to use quite a bit of carbon monoxide to get a net reducing atmosphere.

Lynn

That's right. On the other hand, the heat which is released when you inject that carbon monoxide is going to be recovered in your powerplant.

Dingo

Well, what the other gentleman I think was getting at is, if you're going to have to have a gasifier to get a net reducing atmosphere, you have to have a small gasifier . . . you may as well have a big gasifier and just run your powerhouse on the gasifier to avoid any SO_2 or NO_x problem.

Lynn

Your gasifier will still release sulfur when you have sulfur in the fuel which you're burning. It depends on whether it would be cheaper to remove the sulfur from the CO and hydrogen before you put it into your boiler and add more air . . . I think that when you add the air to this hot CO-hydrogen mixture, you would still get thermal NO. I don't believe that you would solve the NO or the SO_2 problem more economically by gasifying all of the fuel ahead of the boiler.

Dingo

Another comment if you have to have very high CO, and I notice some of your efficiencies for CO conversion were around 80%, I don't know what the limits are on CO coming out of the stack gases but that's something that should be checked. Now, what are you going to do with the CO going out of the stack gases?

Lynn

The CO level that we anticipate, assuming that we can use first order reaction kinetics, is of the order of 100-500 ppm. That is well below the levels of about 1% CO currently allowable in exhaust in California.

Dingo

One more quick comment about your conceptual design. You're taking out sulfur simultaneously with your NO. You then go through your cycle in the electrostatic precipitator and the efficiency of the electrostatic precipitator is usually based on how much sulfur is left in the flue gas. When you're burning low sulfur fuel . . . less than 1% sulfur . . . the electrostatic precipitator isn't as efficient to deal with the particulate problem. So this is something you should look at too.

Lynn

I think the ammonia will have a beneficial effect.

Dingo

It has benefits but I don't think it's as good as the sulfur.

R. Voorhoeve *(Bell Laboratories)*

Did you form COS? Similar catalysts have been known to form COS.

Lynn

What we did was to demonstrate that FeO will adsorb COS, forming CO_2 and FeS.

NO_x CATALYST DEGRADATION
BY CONTAMINANT POISONING

D. P. McARTHUR

Union Oil Research, Brea, California

ABSTRACT

The main objective of earlier studies dealing with contaminant poisoning of auto exhaust catalysts was to develop catalysts with superior tolerance to contaminant poisons. The added objectives of our present research efforts are: to determine the "optimum permissible" operating conditions for minimizing contaminant poisoning, and to provide information for the development of methods for the rejuvenation of catalysts which have been severely deactivated, predominantly by contaminant poisoning.

Work is being conducted to determine the following: a) the form, physical and chemical, in which the contaminants are initially deposited on the catalyst, b) the subsequent gas-solid and solid state reactions which these deposits participate in once on the surface, c) the temperature dependence of these reactions, d) the various possible reaction products and their relative thermal stabilities, and e) the relative "toxicities" of the various contaminant poisons.

Results from preliminary studies and work in progress will be presented.

INTRODUCTION

The majority of the 1975 model year production of automobiles will be equipped with oxidation catalysts for the control of HC/CO emissions. These catalysts will be exposed to contaminant poisons, principally lead, sulfur, and phosphorus, originating from motor oil and contaminated fuel. The severity of contaminant exposure will vary with the individual motorist, depending largely on the frequency of occasional use (unavoidable or otherwise) of leaded gasoline.

References pp. 278-279.

Earlier in our study of catalyst deactivation by contaminant poisoning, the main objective was the development of catalysts with superior tolerance (and/or resistance) to contaminant poisoning. When it became apparent that the manufacture and use of auto exhaust catalysts would become a reality, the emphasis of our poisoning studies changed. Presently we are trying to determine the "optimum permissible" operating conditions for minimizing contaminant poisoning, and obtain information useful for the development of methods for the rejuvenation (regeneration) of auto exhaust catalysts which have been severely deactivated, principally due to contaminant poisoning vis-à-vis thermal degradation.

The knowledge we seek to obtain concerns the following:

a. the initial chemical and physical form of the contaminant compounds upon deposition onto the catalyst surface,

b. the subsequent gas-solid and solid state reactions in which these compounds participate once on the surface,

c. the effects of temperature on these reactions,

d. the identity and relative stability of the various reaction products.

e. the relative toxicity of the various contaminant compounds.

Fortunately for the catalyst researcher, the automotive engineering literature contains numerous excellent papers dealing with the subject of engine combustion chamber deposits resulting from the use of gasoline containing various antiknock and deposit modifying additives, e.g., motor mix, TCP, etc. Most of the literature concerning the identity, composition and mechanism of formation of combustion chamber deposits was published during the years 1950-1960. It is our feeling that a good analogy can be drawn between combustion chamber deposits and contaminant poisons on auto exhaust catalysts. A review of the literature dealing with engine deposits provides a very good and comprehensive perspective regarding deposition and reaction phenomena which are most likely involved in the contaminant poisoning of auto exhaust catalysts.

It should be pointed out that the material to be presented here primarily concerns monolithic NO_x reduction catalysts. However, it is felt that the important principles and conclusions discussed in this paper are generally applicable to HC/CO oxidation catalysts as well.

CATALYST CONTAMINANT POISONS – ANALOGY WITH ENGINE DEPOSITS

The contaminant poison precursors which originate from the combustion of leaded gasoline include lead oxide (PbO), hydrogen halides (HX), lead halides (PbX_2), SO_2, and probably P_2O_5. Newby and Dumont (6) have shown that gas phase reactions resulting in the formation of lead sulfate, lead halides, or lead oxyhalides can proceed only to a very limited extent at the high temperature conditions present in the combustion chamber. However, the temperature of the combustion gases decreases considerably from the combustion chamber to the catalyst, and these reactions become thermodynamically favorable. For example, Newby and Dumont have reported thermodynamic calculations for gas phase reactions between lead oxide and hydrogen halides for reactant concentrations equal to those present in combustion gases.

$$PbO + 2HCl \;\rightleftharpoons\; PbCl_2 + H_2O \tag{A}$$

$$PbO + 2HBr \;\rightleftharpoons\; PbBr_2 + H_2O \tag{B}$$

Under conditions for which thermodynamic equilibrium is reached, the major lead compounds present in the gas phase will be $PbCl_2$ for T $\langle$ 700°C (reaction A), $PbBr_2$ for T $\rangle$ 800°C (reaction B), and PbO for T $\rangle$ 900°C.

Thus, the nature of the lead compounds present in the exhaust gas as it encounters the catalyst can be expected to be strongly dependent on the exhaust gas temperature. For the expected "normal" operating temperature range for auto exhaust catalysts, viz., 600-800°C, it is reasonable to expect that the exhaust gas will contain lead oxide (PbO) and lead halides (PbX_2). The presence of $PbSO_4$ in the gas phase seems unlikely since it has been reported (14) that SO_2, rather than SO_3, is thermodynamically favored in the hot exhaust gases produced in an engine. The catalyst may also encounter extremely small particles of PbO resulting from gas phase agglomeration of this species during transit from the combustion chamber to the catalyst.

A "dusty gas" model for the encounter between the contaminant poison precursors and the catalyst is depicted in Fig. 1. Included in Fig. 1 is a representation of the flow

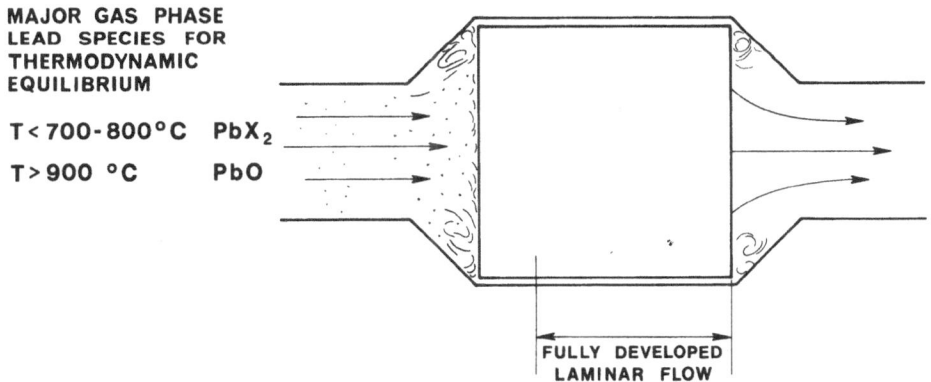

MAJOR GAS PHASE
LEAD SPECIES FOR
THERMODYNAMIC
EQUILIBRIUM

T < 700-800°C PbX_2

T > 900 °C   PbO

FULLY DEVELOPED
LAMINAR FLOW

Fig. 1. "Dusty gas" model for lead deposition on auto exhaust emission control catalysts.

References pp. 278-279.

disturbances which are present at the inlet and outlet of the catalyst. The work of Shelef et al. (12) with monolithic HC/CO oxidation catalysts, and our own work with NO_x catalysts (5), show that the concentration of contaminant poisons on the catalyst decreases non-linearly from the inlet to the outlet. The disproportionately high contaminant concentrations at the catalyst inlet are associated with this transitional flow condition (disturbance) and the enhanced mass transfer between the gas phase and the catalyst surface which results. Schoenherr et al. (11) have determined that fully developed laminar flow does not occur in monolith flow channels until a point $\sim 3/4$" downstream of the inlet.

Deposition — For temperatures below about 900°C, lead oxide will be deposited on the catalyst by condensation — adsorption. Deposition of PbX_2 by condensation is precluded at normal operating temperatures, but reversible and irreversible adsorption of this species is possible.

Gas-Solid Reactions — The lead oxide deposited onto the catalyst surface is very reactive. Lead oxide can react with the gas phase hydrogen halides (HX) to form lead halides (PbX_2). The PbX_2 thus formed can either volatize or react with additional lead oxide to form the more stable lead oxyhalides ($nPbO \bullet PbX_2$, n = 1,2,4,6). Of course, the halogen scavengers are purposely added to leaded gasoline because they do react with PbO to form volatile lead halides and thus minimize the accumulation of combustion chamber lead deposits. We can expect that these lead scavenging phenomena will occur on the catalyst surface.

Alternatively, the lead oxide can react with SO_2, presumably through a chemisorbed SO_3 like species to form very stable lead sulfate ($PbSO_4$). If it were not for the formation of highly stable $PbSO_4$, and the relatively stable lead oxyhalides (n $PbO \bullet PbX_2$), the halide scavenging reactions might effectively keep the catalyst surface free of lead deposits. To emphasize the reactivity of lead oxide (PbO), it can be pointed out that we have never detected the presence of free lead oxide on either dynamometer tested or laboratory poisoned catalyst samples.

Solid State Reactions — The simple lead salts, viz., PbO, PbX_2, and $PbSO_4$, can participate in subsequent solid state addition and replacement reactions. The addition reaction between PbO and PbX_2 to form lead oxyhalides ($nPbO \bullet PbX_2$) has already been mentioned. A second important addition reaction is the formation of lead oxysulfates ($nPbO \bullet PbSO_4$, n = 1,2,4). The lead oxysulfates are less thermally stable than is $PbSO_4$. These important solid state addition reactions and their products are summarized in Fig. 2. The "reaction" temperatures listed in Fig. 2 represent the approximate temperatures at which the respective reactions proceed to a significant extent upon heating of the reactants for a period of 2 hours. The direction of changes in the composition of the lead deposits with changes in catalyst temperature and engine air/fuel ratio is also indicated in Fig. 2.

		REACTANTS	COMPOUND FORMED	REACTION TEMPERATURE, °C
INCREASING TEMPERATURE	INCREASING AIR-FUEL RATIO	PbO + PbX$_2$	PbO · PbX$_2$ PbO · PbX'X"	250 - 300
		2PbO + PbX$_2$	2PbO · PbX$_2$	350 - 400
		PbO + PbSO$_4$	PbO · PbSO$_4$	500
		4PbO + PbSO$_4$	4PbO · PbSO$_4$	600

Fig. 2. Reactants and reaction temperatures for compounds formed by solid state reactions involving lead (after Lamb and Niebylski, 1951).

The complex lead salts (oxyhalides and oxysulfates) and the simple lead salts can also participate in replacement reactions. For example, PbSO$_4$ can react with PbO • PbX$_2$ to form PbO • PbSO$_4$ and PbX$_2$. These replacement reactions are reversible. If PbO • PbX$_2$ is heated to a temperature of ∿ 500°C in the presence of PbSO$_4$, the replacement product PbO • PbSO$_4$ will be formed. If the PbO • PbSO$_4$ is then cooled to a temperature of ∿ 250°C in the presence of PbX$_2$, the replacement product PbO • PbX$_2$ will be formed. The reversibility of these solid state replacement reactions is pointed out in Fig. 3.

The various phenomena involved in the contaminant poisoning of auto exhaust catalysts, viz., deposition, gas-solid, and solid state reactions, are summarized in Fig. 4. At the lower end of the normal operation temperature range, i.e., 900-1300°F, the

TEMPERATURE, °C

250 - 400 $\quad$ nPbO + PbX$_2$ $\longrightarrow$ nPbO · PbX$_2$

500 $\quad$ PbO + PbSO$_4$ $\longrightarrow$ PbO · PbSO$_4$

$$\text{PbO} \cdot \text{PbX}_2 + \text{PbSO}_4 \underset{250}{\overset{500}{\rightleftarrows}} \text{PbO} \cdot \text{PbSO}_4 + \text{PbX}_2$$

600 - 700 $\quad$ 4PbO + PbSO$_4$ $\longrightarrow$ 4PbO · PbSO$_4$

Fig. 3. Representative solid state addition and replacement reactions involving lead compounds.

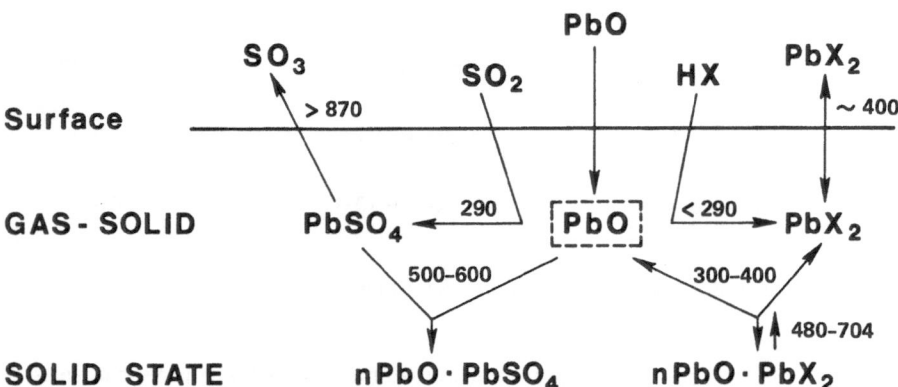

Fig. 4. Model for the mechanism of the degradation (poisoning) of auto exhaust catalysts by contaminant deposition (after Newby and Dumont).

contaminant deposits would be expected to consist of lead oxyhalides, the lead oxysulfates, and lead sulfate. When the catalyst temperature exceeds $\sim$ 700°C, the oxyhalides will decompose into lead oxide and volatile lead halides. The free lead oxide thus generated can either react with SO_2 to form lead sulfate, react with HX to form volatile halides, or participate in an addition reaction with existing $PbSO_4$ to form lead oxysulfates. When the operating temperature of the catalyst exceeds $\sim$ 870°C, decomposition of lead oxysulfates and lead sulfate can occur, the products being PbO and SO_3. The free lead oxide formed by the sulfate decomposition can react with HX to form volatile lead halides.

This proposed model for the mechanism of contaminant poisoning predicts that the composition of the contaminants present on the catalyst will vary with temperature. Furthermore, for catalyst operating temperatures greater than 700-760°C, the total amount of contaminants present on the catalyst will decrease as the operating temperature increases. Accordingly, this model suggests that in-situ thermal rejuvenation of contaminant poisoned catalysts can occur. This would be analogous to the phenomena encountered with engine deposits. Combustion chamber deposits formed mainly by light load and stop and go service can be largely removed by acceleration at wide open throttle (7). However, it should be kept in mind that an automobile engine can be lightly loaded during travel on level freeways, even at 60 mph.

The proposed catalyst poisoning mechanism (Fig. 4) predicts that contaminant retention will decrease with increasing operating temperature for temperatures *greater* than 700-760°C. At this point, it is appropriate to make comment on the effect of catalyst operating temperature on contaminant retention for temperatures *below* 700-760°C. It has already been mentioned that lead is present in the exhaust gas primarily as PbX$_2$ and PbO. Under conditions of thermodynamic equilibrium, the ratio of PbO/PbX$_2$ increases with increasing temperature. Since PbX$_2$ will not readily "condense" on surfaces with temperatures greater than ~ 200°C, it can be expected that the sticking probability of PbX$_2$ will be very low at normal catalyst operating temperatures, i.e., T $\rangle$ 425°C. For temperatures greater than ~ 425°C, the sticking probability of PbO will be very much greater than that of PbX$_2$. Therefore, because the ratio of PbO/PbX$_2$ in the gas phase increases with increasing temperature, it is expected that lead deposition will increase with increasing temperature for temperatures *below* 700-760°C. This conclusion is supported by the experimental results of Shelef et al. (12) which showed that contaminant retention (Pb, Zn, P), with the exception of sulfur, increased with increasing catalyst operating temperature in the temperature range 400-800°C.

Phosphorus Containing Contaminant Compounds — So far we have focused our attention on the lead oxyhalides, oxysulfates, and sulfates. Unfortunately, to the author's knowledge, the literature is not an abundant source of information regarding the possible mechanisms involved in the formation of phosphorus compounds in the auto exhaust environment. However, we can speculate on this matter, and draw analogies with the contaminant poisoning mechanisms discussed previously. Phosphorus is most likely present in the combustion gases in the form of P$_2$O$_5$ (P$_4$O$_{10}$). We can expect that gas phase P$_2$O$_5$ will react with lead oxide, and also with the alumina wash-coat catalyst support material

$$P_2O_5 + Al_2O_3 \rightarrow 2AlPO_4 \tag{C}$$

$$P_2O_5 + 3PbO \rightarrow Pb_3(PO_4)_2 \tag{D}$$

$$2PbO + P_2O_5 \rightarrow Pb_2P_2O_7 \tag{E}$$

The lead phosphates formed by reactions (D&E) can participate in the following solid state addition reactions.

$$PbO + Pb_3(PO_4)_2 \rightarrow PbO \bullet Pb_3(PO_4)_2 \tag{F}$$

$$3Pb_3(PO_4)_2 + PbBr_2 \rightarrow 3Pb_3(PO_4)_2 \bullet PbBr_2 \tag{G}$$

$$PbSO_4 + Pb_3(PO_4)_2 \rightarrow PbSO_4 \bullet Pb_3(PO_4)_2 \tag{H}$$

$$2PbBr_2 + Pb_2P_2O_7 \rightarrow 2PbBr_2 \bullet Pb_2P_2O_7 \tag{I}$$

It is likely that the lead phosphates can also participate in solid state replacement reactions of the type already discussed.

For example,

$$PbO \bullet PbBr_2 + Pb_3(PO_4)_2 \rightarrow PbO \bullet Pb_3(PO_4)_2 + PbBr_2 \tag{J}$$

The number of possible replacement reactions involving the oxyhalides, oxysulfates and phosphates is indeed large. It is sufficient for the purposes of this paper merely to point out the possibility of the occurence of these reactions on the catalyst surface. The work of Street (13) has shown that all of the complex lead phosphate compounds shown above (the products of reactions F through I) can be found present in combustion chamber deposits. A list of the types of contaminant compounds, *including those containing phosphates*, that one may expect to find present on auto exhaust catalysts is presented in Fig. 5. The compounds that appear in Fig. 5 with a check mark are contaminants which we have found to be present on used catalysts and on the surfaces of the catalyst container. These compounds have been positively identified by x-ray diffraction. It is our experience that used auto exhaust catalysts typically contain from a few tenths to ~ 10 wt% lead in the form of contaminant poisons. This amount of lead is associated with several possible lead compounds, i.e., lead oxyhalides, oxysulfates, sulfates, phosphates, etc. Therefore, the amount of lead present as any one particular compound can be quite small ($\langle 1\text{-}2$ wt%). Furthermore, these compounds are fairly well dispersed over a rather large surface area. For these reasons, it is difficult to identify all of the lead compounds which may be present on the catalyst by ordinary x-ray diffraction techniques.

METALLIC LEAD

Greenshields (2) has reported that under certain engine operating conditions metallic lead is found to be present in combustion chamber deposits. The metallic

	COMPOUND	M. P., °C
	$Pb\ Br_2$	370
	$Pb\ Cl_2$	496
$\underline{710\ °C}$ $(\sim 1300\ °F)$	✓ $nPbO \cdot PbX_2$	495 - 710
	✓✓ $3Pb_3(PO_4)_2 \cdot PbX_2$	
	PbO	888
	✓ $PbO \cdot Pb_3(PO_4)_2$	895 - 975
	$nPbO \cdot PbSO_4$	
	$Pb_3(PO_4)_2$	1014
	✓ $PbSO_4$	1170
	$AlPO_4$	>1500

Fig. 5. Important lead and phosphorus compounds likely to be present on poisoned auto exhaust catalysts (Street, 1953).

lead is thought to be formed as a result of reactions between lead compounds and carbonaceous deposits at elevated temperatures. We have observed the following. X-Ray analysis of a sample of used catalyst could *not* detect the presence of metallic lead. The catalyst sample was then calcined in air for ~ 16 hours at 1000°C. Subsequent x-ray analysis of this catalyst sample did reveal the presence of metallic lead. The reasons for this phenomenon are not well understood. Perhaps metallic lead can be formed on the catalyst surface under certain special operating conditions. One would expect that any metallic lead formed would not exist in that form for very long.

EXPERIMENTAL RESULTS FOR CATALYSTS SUBJECTED TO ENGINE DYNAMOMETER DURABILITY TESTS

Of the contaminant lead compounds discussed in this paper, the most thermally stable are the oxysulfates, sulfates and phosphates. For these compounds the weight ratios of Pb/S and Pb/P vary between 6.5-32.3 and 6.7 to 13.4, respectively. We would expect, therefore, that the weight ratios of Pb/S and Pb/P determined from analysis of used catalyst would fall in these ranges. Our experimental data show that the Pb/S weight ratio for used catalysts varies between 2-45. This result is consistent with the expected Pb/S weight ratio. The experimental data also show that the Pb/P weight ratio for used catalysts can vary between 0.05-20. However, for the majority of catalysts analyzed the Pb/P weight ratio is less than ~ 2. This result shows that phorphorus is present on the catalyst in the amounts greater than that which can be accounted for by lead phosphate compounds. This suggests that a large fraction of the phorphorus which is present on catalysts is in the form of compounds resulting from the reactions between P_2O_5 and the wash-coat catalyst support material, e.g., $AlPO_4$.

Fig. 6 summarizes data which we have accumulated regarding the percent contaminant retention on catalysts subjected to engine dynamometer testing. Contaminant retention values are listed for various modes of catalyst operating conditions (constant reducing, cyclic oxidizing/reducing) and catalyst operating temperatures. Contaminant retention is defined here as the amount of a particular contaminant found to be present on the catalyst, expressed as a fraction of the total amount of that contaminant which the catalyst was exposed to by the engine consumption of engine oil and contaminated gasoline. The retention values for calcium and zinc are relatively small and vary between 2-8% and 3-11%, respectively. Sulfur retention is extremely small, viz, 0.002-0.015%. Lead retention varies between $\langle 0.5$ to $\sim 15\%$. The largest retention values observed are those for phosphorus which usually vary between ~ 15-40%. The low phosphorus retention value of 4% is a special case. This result will be discussed shortly. It should be noticed that the sum of the lead and phosphorus retention is reasonably constant for catalyst operating temperatures in the range 550-750°C. This is so even with varying mode of catalyst operation. The sum of the lead and phosphorus retention decreases greatly when the

References pp. 278-279.

CATALYST OPERATING MODE	CATALYST OPERATING TEMPERATURE °C	PERCENT CONTAMINANT RETENTION ON CATALYST				
		Ca	Pb	P	S ppm	Zn
REDUCING	565–600	2	13	4–28	40–60	4
OX – RED	620–680	2–3		32–40	60	9–11
RED + OX – RED	565–730	2–3	14–16	30	150	5–6
OX – RED	790–870	3–8	0.3–3	13–19	20–100	3–5

Fig. 6. Contaminant poisons retention on auto exhaust catalysts — effect of catalyst operating temperature.

catalyst operating temperature is increased to 790-870°C. This result is consistent with the contaminant deposition and reaction processes which are thought to occur on the catalyst surface as presented in Fig. 4. These experimental data support our earlier hypothesis that catalyst contaminant retention can be minimized by catalyst operation at temperatures above 700-760°C.

A dramatic demonstration of in-situ thermal rejuvenation of contaminant poisoned catalysts is presented by the results of a special engine dynamometer test which we performed. Two identical 3-21/32" diameter by 3" long mixed metal monolithic catalysts were tested in the dual exhaust system from a 350 CID V-8 engine. Certification fuel (0.03 g/gal Pb + 0.005 g/gal P) and ashless oil were used. The catalyst operating mode was maintained at constant reducing condition, and the normal catalyst operating temperature was 550-600°C. Catalyst activity was determined once daily at a catalyst operating temperature of approximately 590°C. The two catalyst cannisters were switched from one side of the engine to the other at two day intervals to ensure that the catalysts each sustained the same service exposure.

The results of the dynamometer test are presented in Fig. 7. The activity of both catalysts decreased, rapidly initially and then gradually, with accumulated mileage. During the first eleven days of the test, gross NO_x conversion decreased from ∿85% to ∿35%. At this point in the test, catalyst A was removed from the test unit. Catalyst B was then operated for a period of 21 hours under cyclic oxidizing/reducing conditions which resulted in catalyst operating temperatures of 790-870°C. At the end of this period the activity of catalyst B was again measured at ∿600°C under constant reducing conditions. As a result of this high temperature mode of operation the catalyst activity had increased substantially to ∿75% gross NO_x conversion. The constant reducing 600°C mode of catalyst operation was then resumed for seven more days. Again, the activity of catalyst B steadily decreased, this time from ∿ 75% to ∿ 50% NO_x conversion. Catalyst B was then once again subjected to the high temperature cyclic oxidizing/reducing mode of operation, this time for a period of 16 hours. As before, the activity of catalyst B was improved substantially, i.e., gross NO_x conversion at 600°C increased from ∿ 50% to ∿ 85%. At this point, the dynamometer test was terminated.

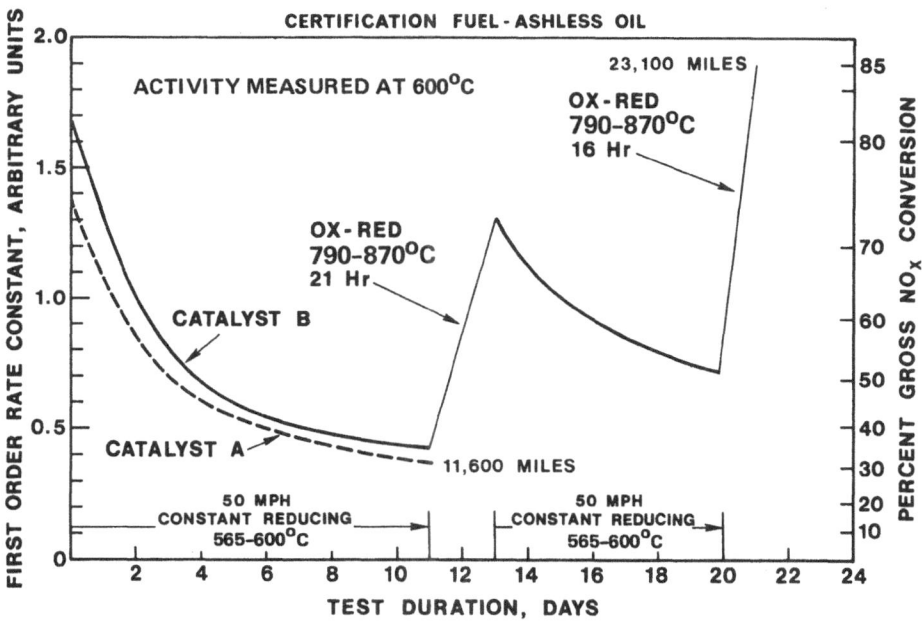

Fig. 7. Demonstration of in-situ "thermal rejuvenation" (regeneration) of a monolithic NO_x catalyst during engine dynamometer testing.

Core samples (1" diameter) were then taken of both catalyst A and catalyst B for subsequent laboratory tests, viz., contaminant analysis and bench reactor activity tests. Contaminant concentrations for the used catalyst samples, as determined by x-ray fluorescence, are reported in Fig. 8. As is usually the case, contaminant concentrations decreased in the axial direction from inlet to outlet. Even though catalyst B was exposed to twice the service mileage as was catalyst A, the average contaminant lead concentration on the former was twenty times lower than that of the latter, i.e., 0.04 wt% versus 0.81 wt%, respectively. The average phosphorus concentrations on both catalysts were rather low, viz., 0.04 and 0.08 wt% for catalysts A and B, respectively. The higher contaminant phosphorus concentration on catalyst B is associated with the fact that catalyst B was exposed to twice the service mileage as was catalyst A. It should also be remembered that reactions between P_2O_5 and the wash-coat support material (e.g., Al_2O_3) contribute significantly to the over-all

References pp. 278-279.

CATALYST	CATALYST OPERATION MODE	CATALYST MILEAGE	AVE. SPEED MPH	CATALYST OPERATING TEMPERATURE °C	WEIGHT PERCENT CONTAMINANT ON CATALYST					
					LEAD			PHOSPHOROUS		
					INLET	OUTLET	AVG.	INLET	OUTLET	AVG.
A	RED	11,600	51	565–600	1.56	0.35	0.81	0.07	0.01	0.04
B	RED	20,450	51	565–600	0.07	0.02	0.04	0.18	0.03	0.08
	OX + RED	2,650	72	840–900						

FUEL: g/gal
LEAD 0.03
PHOS. 0.005
SULF. 0.850

ASHLESS OIL

Fig. 8. Demonstration of the removal of contaminants from a NO_x catalyst by in-situ "thermal rejuvenation."

phosphorus retention on the catalyst. Thus, the concentration of contaminant phosphorus present on the catalyst can be largely independent of the contaminant lead concentration.

Four core-sample quarter sections (inlet quarter, 2nd, 3rd, outlet quarter) from each of the two catalysts were activity tested in our bench reactor using synthetic exhaust gas containing 2% CO and 45 ppm SO_2. The results of these tests are shown in Fig. 9. Activity data are presented for the inlet and 3rd quarter core-sample sections of each of the two catalysts. For purposes of comparison, data describing the activity of a fresh catalyst sample are also presented in Fig. 9. For both catalysts A and B the activity of the 3rd quarter sections is substantially better than that of the inlet quarter sections. This result is typical, and is expected on the basis of the known distribution of contaminant concentrations, i.e., greatest at the inlet and decreasing in the axial direction towards the outlet. The activity of the catalyst B samples is much better than that of the catalyst A samples. Again, this result is consistent with the amounts of contaminants present on the two catalysts. See Fig. 8. It should be pointed out that the activity of the outlet portion of catalyst B is almost equal to that of a fresh catalyst.

The results of this dynamometer test indicate quite clearly that a catalyst which has become severely deactivated due to contaminant poisoning can be rejuvenated thermally, in-situ. Operation of the catalyst at temperatures above $\sim$ 760°C will result in the removal of a substantial amount of contaminant poisons which were deposited and retained on the catalyst at lower operating temperatures. Concomitant with contaminant removal will be an increase in catalyst activity.

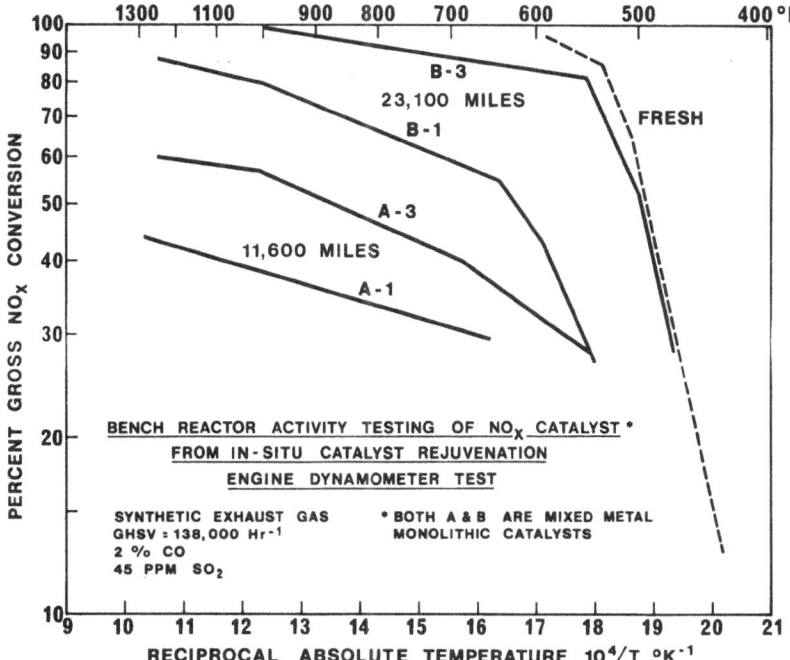

Fig. 9. Bench reactor activity testing of catalyst samples from the in-situ "thermal rejuvenation" dynamometer run.

PHOSPHORUS RETENTION ON CATALYSTS

The federal government has passed legislation which will ensure the availability of unleaded gasoline for use with the 1975 model year catalyst equipped automobiles. Maximum permissible levels for *contaminant* lead (0.05 g/gal) and phosphorus (0.005 g/gal) in unleaded gasoline have been set by law. However, most engine oils contain phosphorus, in the form of additives, in amounts equal to approximately 0.15 wt%. To our knowledge, there are no current plans for the removal of phosphorus additives from engine oil.

It has been well established that phosphorus is a potent contaminant poison for auto exhaust catalysts. We have been trying to determine the relative values for phosphorus retention on catalysts from the two potential sources of this contaminant, i.e., contaminated gasoline and engine oil. Fig. 10 presents the results of two consecutive engine dynamometer tests which were performed to study phosphorus retention. A different fresh sample (3-21/32" diameter by 3" long monolith) of the same mixed metal NO$_x$ catalyst was used in each of the dynamometer runs. Catalyst AA-4611 was subjected to the equivalent of 31,300 miles of testing using sterile fuel and a fully formulated SAE 10W-40 engine oil containing 0.15 wt% phosphorus. The catalyst was operated under constant reducing conditions, and the normal operating temperature range of the catalyst was 565-600°C. This catalyst was removed from the dynamometer test unit and the engine was thoroughly cleaned (by repeated flushings with fresh ashless oil) to remove all traces of the contaminant containing SAE 10W-40 engine oil. A second fresh catalyst (AA-4636-3) was then subjected to the equivalent of 11,600 miles of testing using ashless engine oil and contaminated (certification) gasoline containing 0.05 g/gal lead and 0.005 g/gal phosphorus. The catalyst operating mode and operating temperature were the same for this test as they were for the first dynamometer test.

Subsequent to the two dynamometer tests, the two catalysts were analyzed for contaminant phosphorus content by x-ray fluorescence. These data together with the known consumption of engine oil and gasoline for the two tests permitted determination of the percent phosphorus retention on the two catalysts. The contaminant phosphorus retention when the contaminant source was the engine oil (first test) was 6-1/2 times greater than when the contaminant source was the gasoline, viz., 26% versus 4%, respectively. The much lower catalyst phosphorus retention from gasoline may be due to the presence of lead. It is quite likely that lead contamination of the catalyst surface would greatly reduce the extent to which reaction would occur between gas phase P_2O_5 and the wash-coat support material (Al_2O_3). Another possible explanation for the large difference in catalyst phosphorus retention from fuel and engine oil contaminant sources may be that the catalyst encounters the phosphorus in different forms in each case. The combustion of gasoline most likely results in generation of gaseous P_2O_5. On the other hand, the phosphorus compounds (additives) in the engine oil may or may not be combusted. Each stroke of the piston (including the exhaust stroke) spreads a very thin film of engine oil on the combustion chamber wall. Some of this oil may be entrained in the combustion gases and transported to the catalyst in some form other than

CATALYST	CATALYST MILEAGE	FUEL	ENGINE OIL	PERCENT PHOSPHORUS RETENTION ON CATALYST
AA-4611	31,300	STERILE	10W-40 0.15 wt. % P	26
AA-4636-3	11,600	CERTIFICATION 5 mg/gal P	ASHLESS	4

FUEL SULFUR; g/gal   CATALYST OPERATING MODE: CONSTANT REDUCING
CERTIFICATION: 0.85   CATALYST OPERATING TEMPERATURE: 565-600°C
STERILE 1.27

Fig. 10. Dependence of catalyst phosphorus retention on contaminant source: engine oil phosphorus versus fuel phosphorus.

combustion products; i.e., phosphorus would be present in some form other than P$_2$O$_5$. For example, condensation and decomposition of an oil vapor (mist) on the catalyst surface could be the reason for the high values for contaminant phosphorus retention on catalysts resulting from the consumption of engine oil.

THE UNION OIL Z-3500 MONOLITHIC NO$_x$ CATALYST

In this paper it has been argued that catalyst deactivation by contaminant poisoning can be minimized by operating the catalyst at elevated temperatures, viz.⟩ 1400°F. Prolonged operation of a catalyst at such high temperatures requires an active metal(s)/wash-coat combination possessing exceptional thermal stability. Presented in Fig. 11 are data which compare the activity of the Union Oil Z-3500 NO$_x$ catalyst when fresh, and after accelerated engine dynamometer aging. This catalyst consists of mixed metals supported on a proprietary wash-coat catalyst support material. For this particular dynamometer run, the catalyst was operated under cyclic oxidizing/reducing conditions. The normal catalyst operating temperature varied between 790-870°C. Certification fuel and a fully formulated engine oil were used. The catalyst was dynamometer aged for 550 hours, or the equivalent of 39,700 miles.

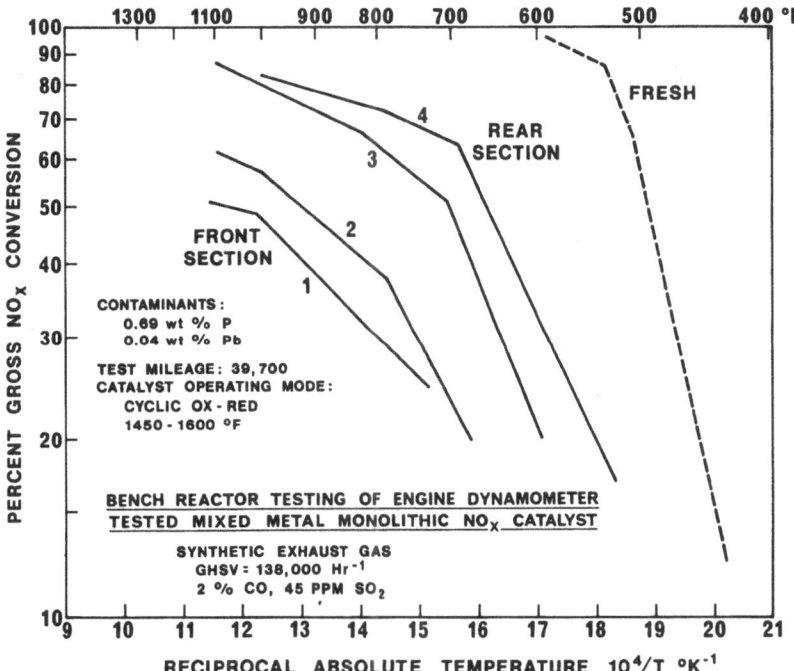

Fig. 11. Bench reactor activity testing of the most durable Union Oil NO$_x$ catalyst developed to date after 40,000 miles of accelerated engine dynamometer testing.

References pp. 278-279.

Data are shown in Fig. 11 which describe the activity of each of the four (inlet, 2nd, 3rd, outlet) quarter sections of a 1" diameter core-sample. The front half of the monolithic catalyst has been severely deactivated as a result of the dynamometer aging. The rear half of the aged catalyst, on the other hand, exhibits reasonably good activity.

Analysis of the aged catalyst by x-ray fluorescence revealed that the contaminant lead and phosphorus concentrations were 0.04 and 0.69 wt%, respectively. From this result we conclude that catalyst degradation due to contaminants (vis-à-vis thermal degradation) was primarily due to phosphorus poisoning.

SUMMARY

A useful analogy drawn between combustion chamber deposits and catalyst poisoning provides insight into the various phenomena which are involved in this complex catalyst degradation mechanism.

For catalyst operating temperatures *below* 700-760°C, the retention of contaminant poisons *increases* with increasing temperature because the ratio of PbO/PbX_2 in the exhaust gas increases with increasing temperature. For operating temperatures *above* 700-760°C, contaminant compounds which were formed at lower temperatures begin to decompose and volatilize, and contaminant retention *decreases* with increasing temperature.

The amounts and composition of contaminant poisons present on the catalyst at any time are dependent on the *immediate* past history (operating conditions) of the catalyst, and are very strongly temperature dependent.

The ratio of Pb/P determined for poisoned NO_x catalysts is usually much smaller than the stoichiometric Pb/P ratio calculated for known lead phosphate compounds. This result indicates that much of the phosphorus retained on the catalyst is in the form of compounds resulting from the interaction of phosphorus with the catalyst support material.

Catalyst phosphorus retention is much greater for phosphorus originating from engine oil than it is for phosphorus originating from gasoline.

Catalyst degradation due to contaminant poisoning can be minimized by maintaining the catalyst operating temperature in the range 730-870°C.

REFERENCES

1. F. J. Cordera, H. J. Foster, B. M. Henderson, and R. L. Woodruff, "TEL Scavengers in Fuel Affect Engine Performance and Durability," *SAE Trans, 73, 1528 (1965).*
2. R. J. Greenshields, "Spark-Plug Fouling Studies," *SAE Trans, 61, 3 (1953).*

3. F. W. Lamb, "Postulated Mechanism of Spark-Plug Fouling," Aircraft Spark-Plug and Ignition Conference, Champion Spark-Plug Company, Toledo, Ohio, October 5-7 (1954).

4. F. W. Lamb and L. M. Niebylski, "Formation of Engine-Deposit Compounds by Solid-State Reactions, An X-Ray Diffraction Study," API, Division of Refining Meeting, Tulsa, Oklahoma, April 30-May 3, 1951.

5. D. P. McArthur, "The Deposition and Distribution of Lead, Phosphorous, Calcium, Zinc, and Sulfur Poisons on Automobile Exhaust NO$_x$ Catalysts," Symposium on Catalysts for Removal of Automobile Pollutants, 167th National ACS Meeting, Los Angeles, California, April 1-5 (1974).

6. W. E. Newby, L. F. Dumont, "Mechanisms of Combustion Chamber Deposit Formation with Leaded Fuels," Ind & Engr Chem 45 (6), 1336 (1953).

7. E. F. Obert, "Internal Combustion Engines," 3rd ed, Inter Textbook Company, Scranton, Pa., (1968), chap. 9.

8. A. Ross, E. B. Rifkin, "Theory of Tetraethyllead Actions," Ind & Engr Chem 48 (9) 1528 (1956).

9. A. Ross, C. Walcutt, "Decomposition of TEL in an Engine," Ind & Engr Chem 48 (9) 1532 (1956).

10. E. B. Rifkin, "The Role of Physical Factors in Knock," Div. of Refining, API Meeting, Los Angeles, California, May 12, 1958.

11. R. Schoenherr, J. A. Woolley, and Wing Chow, "Ceramic Monoliths as Catalyst Supports for Emission Control Applications," Symposium on Catalysts for Removal of Automobile Pollutants, 167th ACS National Meeting, Los Angeles, California, March 31-April 5, 1974.

12. M. Shelef, R. A. Dalla Betta, J. A. Larson, K. Otto, and H. C. Yao, "Poisoning of Monolithic Noble Metal Oxidation Catalysts in Automobile Exhaust Environment," Symposium on Catalyst Poisoning, 74th National AIChE Meeting, New Orleans, March 11-15 (1973).

13. J. C. Street, "Mode of Formation of Lead Deposits in Gasoline Engines," SAE Trans, 61, 443 (1953).

14. M. J. van der Zijden, J. E. van Hinte, and J. C. van den Ende, J. Inst Petrol 36, 561 (1950).

DISCUSSION

L. L. Hegedus *(General Motors Research Laboratories)*

I have three short questions. Why do you propose P_2O_5? Don't you have 10% water?

McArthur

Yes, you do. The question whether or not the phosphorus species in exhaust gas is P_2O_5 or phosphoric acid, I can't answer. It would have to do with the relative stability of the two species. Certainly you could form phosphoric acid on the catalyst. Under strongly reducing conditions the species might even be phosphine (PH_3).

Hegedus

We have done some work with phosphorus poisoning. The assumption of ortho-phosphoric acid proved to be rather descriptive in predicting the rate at which the phosphorus front propagates into the catalyst by diffusion. The second question is, you had a reaction there:

$$SO_2 + PbO = PbSO_4$$

I thought you needed one more oxygen.

McArthur

Yes. That was not a stoichiometric equation. I was just indicating that lead sulfate is formed by the interaction between PbO and SO_2. Thermodynamic calculations of Newby and Dumont[*] show that at the high temperatures present in the combustion chamber, SO_2 is the stable species relative to SO_3. Now as the exhaust gas moves from the combustion chamber to the catalyst, it is cooled considerably and there are several possible reactions including the formation of lead sulfate in the gas phase. But what I think happens is that the formation of SO_3 occurs primarily on the catalyst surface (catalytically), vis-à-vis homogeneously in the gas phase.

Hegedus

Quickly, the third one. You plotted on your thermal regeneration curve a first order rate constant at 600°C. I thought you are diffusion controlled there . . .

McArthur

Yes, you are. That's why the right side ordinate of Fig. 7 indicates gross NO_x conversion. We generally correlate catalyst activity in terms of first order rate constants. Of course, diffusion is a first order process.

G. H. Meguerian   (Amoco)

Could you see a difference in the poisoning capability of the oil phosphorus versus the fuel phosphorus?

McArthur

No. The only thing we've studied so far is the relative retention. We assume the final product is the same.

G. L. Haller   (Yale University)

Dennis, on that measurement of retention of phosphorus from oil versus gasoline, were those quoted as percent of catalyst weight or percent of exposure to phosphorus?

[*]W. E. Newby, L. F. Dumont, "Mechanisms of Combustion Chamber Deposit Formation with Leaded Fuels." Ind. & Engr. Chem. 45 (6), 1336 (1953).

McArthur

That was the amount of phosphorus retained on the catalyst expressed as a percent of the total exposure. Phosphorus retention remains essentially constant with aging value, but you get a steady accumulation of phosphorus with mileage. Thus, the catalyst that was run for the longer mileage had about twice as much phosphorus on it.

H. C. Yao *(Ford Motor Co.)*

In order to minimize the contamination you have tried higher temperatures; did you try other methods of minimizing the contamination besides increasing temperature, for example did you put in scavengers?

McArthur

No, we haven't. We considered that or some kind of a trap in front of the catalyst, maybe an inert zone or a guard chamber in front of the catalyst.

K. Otto *(Ford Motor Co.)*

You've shown in your rejuvenation experiment (Fig. 11) that you removed quite a bit of lead from your catalysts, but on the other hand the activity which you see is relatively low. Do you expect that your heat treatment has affected your catalytic surface in such a way that sintering has taken place, perhaps promoted by the presence of lead?

McArthur

Even when the lead retention on a catalyst is minimized by operating it at elevated temperatures, we observe a very significant decrease in catalyst activity. We attribute a large portion of this decrease to phosphorus poisoning. The loss of metal surface area by sintering and agglomeration also contributes to the observed activity decline. Your hypothesis that lead (even very small amounts) may promote sintering perhaps is correct.

R. L. Klimisch *(General Motors Research Laboratories)*

You talk in your abstract about optimum permissible operating conditions for minimizing poisoning. I was wondering about retention as a function of poison level, in particular for lead. I wonder if you'd comment on what level of lead below which we might minimize poisoning effects.

McArthur

The only way I can answer that is by the following. We are also doing catalyst rejuvenation and we find that you do not have to remove all the lead from an oxidation catalyst to substantially restore its activity. In fact we've had cases where we restored most of the activity by removing only 20-60% of the lead which was present on the catalyst. The complete removal of the contaminants is very difficult because the contaminants are present in the form of many different compounds. Some compounds are easy to remove. We have found that catalysts that are operated using fully formulated engine oil containing phosphorus ... are much more difficult to rejuvenate, presumably due to the fact that the phosphorus compounds ... phosphates ... are much more difficult to remove without destroying the catalyst. But the amounts of contaminants which must be removed in order to restore the activity to an acceptable level depend on the form in which the contaminants are there.

J. M. Colucci *(General Motors Research Laboratories)*

Would you care to speculate on the possible effects of manganese on the reducing catalysts?

McArthur

Well, I'll try. What I have described here, I hesitate to call it degradation by poisoning since we normally think of poisoning as a one to one chemical relationship. For example, sulfur poisoning of nickel. The type of contamination we're talking about here is a deposition. It's really just "crudding" up the catalyst, i.e., covering up the active sites in the region of the pore mouth or blocking the pore mouths altogether. And I believe that you'd have the same problem with manganese; you'll have manganese deposits on catalysts.

APPLICATION OF METALLIC CATALYSTS
FOR AUTOMOTIVE NO$_x$ CONTROL

R. J. FEDOR

Gould, Inc., Cleveland, Ohio

ABSTRACT

With the delay in the U.S. 0.4 gram per mile NO$_x$ standard from 1976 to 1978 coupled with EPA's long term 2.0 gram per mile recommendation, Gould has emphasized the Japanese market as the initial target for its NO$_x$ catalyst development program.

This report describes Gould's progress toward qualifying a product for the 1976 Japanese market. Development of a new catalyst system, its adaption to a prototype small size vehicle, and results of durability tests are presented.

The data indicate compliance with the Japanese 10-mode test standards through the entire 60,000 kilometer road durability test without catalyst replacement or the use of exhaust gas recirculation.

Application of this concept to full size vehicles is also discussed with respect to both U.S. and Japanese emission standards.

INTRODUCTION

Gould's effort to develop a nitrogen oxide reduction catalyst began in 1970. For the first two and a half years, this work was carried out in cooperation with Exxon Research and Engineering Company. Reports published by Exxon on this early work can be found in SAE papers published in 1971 (1), 1972 (2), and 1973 (3). Since early 1973, Gould has conducted its own engine dynamometer and vehicle test programs. Our total vehicle durability experience now exceeds one-half million road miles.

References p. 290.

In July, 1973, the Environmental Protection Agency held hearings to review the state of the art of NO_x control technology and to decide whether to impose the stringent 0.4 gram per mile NO_x standard for the 1976 model year. At this hearing, Gould, various auto companies, and Exxon presented data indicating that the 0.4 gram per mile NO_x standard could be achieved at low mileage; however, it was apparent from the data submitted that the sensitivity of the catalyst to carburetor transients and common engine malfunctions severely limited the practical durability of the Gould NO_x catalyst.

As a result of these and other deficiencies with respect to other candidate catalysts, the 0.4 gram per mile standard was delayed until 1977. A further delay, until 1978, emanating from the "energy crisis" was inacted in June of this year. This delay in the U.S. NO_x standard until at least 1978 led Gould to concentrate on the Japanese market which still has a 0.4 gram per mile equivalent standard scheduled for their 1976 model year.

It is the intent of this report to present Gould's progress toward developing a practical catalyst capable of meeting the Japanese NO_x requirement and to comment on its applicability toward meeting the U.S. 1978 NO_x standard.

PROTOTYPE SYSTEM FOR THE JAPANESE MARKET

Shown in Fig. 1 is a 1973, 1.8 liter, manual transmission, Datsun 610 which was equipped with our preferred catalytic control system. The dual bed system for

Fig. 1. Prototype vehicle for the Japanese Market. 1.8 liter, Manual Transmission, Datsun 610.

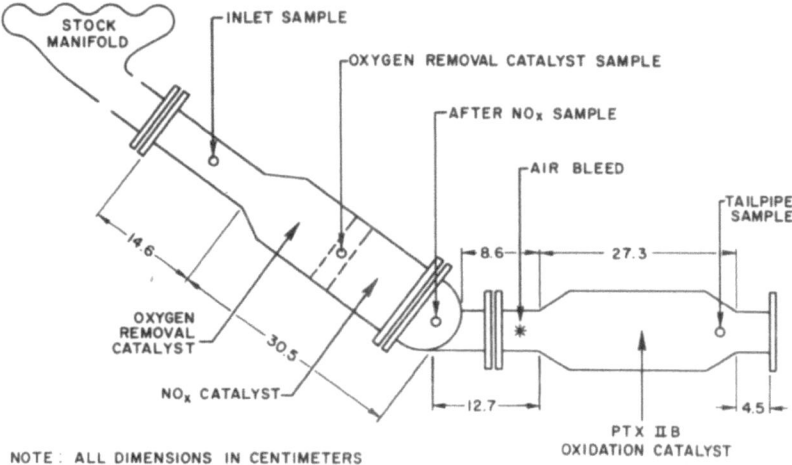

Fig. 2. Schematic of Datsun 610 equipped with catalyst system.

emission control is illustrated in Fig. 2. It should be noted that the vehicle does not employ EGR. The NO$_x$ control package features a Gould GEM 68 catalyst and a small oxidation catalyst in front of it. Carbon monoxide and unburned hydrocarbon control is achieved by means of an Englehard IIB oxidation catalyst. The purpose of the first catalyst in the NO$_x$ bed is to remove the majority of the oxygen in the exhaust before it contacts the NO$_x$ catalyst. This element is referred to an oxygen removal catalyst or "getter." It is utilized because previous work at Gould has identified oxidation of the NO$_x$ catalyst as the main deterrent to durability. Fig. 3 shows a photo of the actual cross-section of the NO$_x$ control package, illustrating the relative sizes and position of the NO$_x$ control elements.

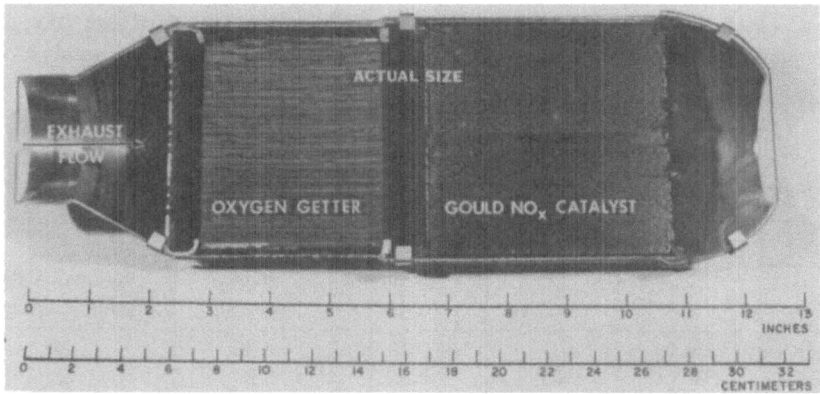

Fig. 3. Cross section of NO$_x$ catalyst bed illustrating oxygen removal catalyst in front of NO$_x$ reduction catalyst.

References p. 290.

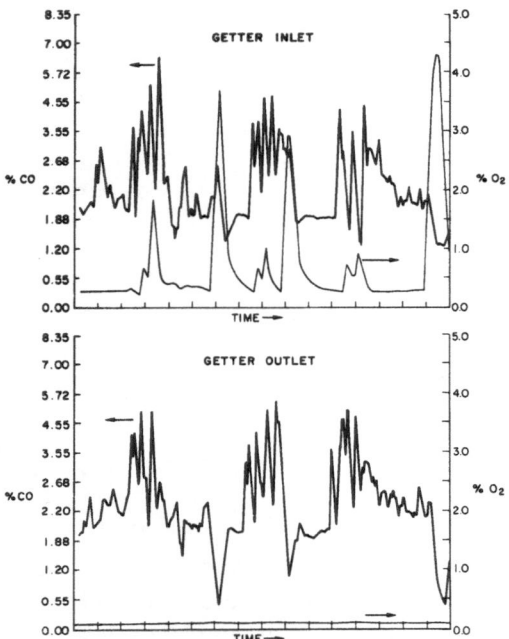

Fig. 4. Effect of getter on inlet CO and O_2 during portion of CVS test. Test conducted on 1800 cm^3 Datsun, manual transmission.

The effectiveness of the oxygen removal catalyst on this vehicle can be seen in Fig. 4. The top portion of the figure illustrates the input CO and O_2 to the NO_x reduction system during the last three hills of Bag 1 of the Federal Test Procedure. It can be seen that oxygen spikes in excess of 2% occur. These spikes have been correlated with both shift points and decelerations on this manual transmission vehicle. In the lower portion of Fig. 4, the O_2 and CO trace is illustrated for the exhaust after the oxygen removal catalyst. This is the input condition for the NO_x catalyst. Thus, the first catalyst is effective in removing most of the oxygen and permits an exhaust with very low oxidizing potential to enter the NO_x catalyst.

JAPANESE EMISSION CONTROL STANDARDS

Fig. 5 presents the proposed 1976 exhaust emission standards for Japan. Numerically the mean standards are equivalent to the U.S. 1978 standards, but are

Vehicle	EM	10-Mode		
		Mean (g/km)	Max (g/km)	Reduction Rate From '73 (Mean)
Light Duty Vehicle	CO	2.10	2.70	89
2.5 Ton GVW	HC	0.25	0.39	91
10 Persons	NO_x	0.25	0.33	91

Fig. 5. Proposed 1976 Japanese exhaust emission standard 4-cycle engine.

presented in gram per kilometer units instead of gram per mile as in the U.S. However, the test procedure is quite different from the U.S. Fig. 6 illustrates the details of the 10-mode driving cycle. The test does not employ a cold start and is merely a repetition of three low speed hills. The top speed is approximately 25 miles per hour. These conditions are easier on an NO$_x$ catalyst from the standpoint that only a minimum amount of NO$_x$ is produced by the engine; however, this test is difficult for a metallic catalyst such as Gould's because of a minimal exhaust temperature during the low speed driving and idle periods. The Japanese durability limit is 60,000 kilometers but a replacement is permitted after 30,000 kilometers.

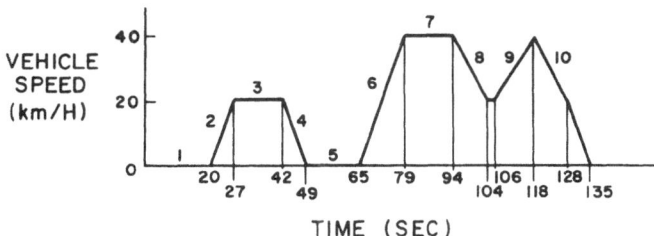

Fig. 6. Japanese emission test driving cycle.

VEHICLE TEST RESULTS

Results for the Datsun on Japanese 10-mode tests are given in Fig. 7. It can be seen that the 0.25 gram per kilometer standard is attainable, not just after 30,000 kilometers, which is the minimum requirement, but even at approximately the full 60,000 kilometer test limit. Also, it should be noted that the HC and CO standards

NO$_x$ (g/km)	CO (g/km)	HC (g/km)
0.22	2.66	0.36
0.19	1.41	0.19
0.23	2.30	0.23
0.21	2.60	0.19
	Average	
0.21	2.25	0.24
	Standard	
0.25	2.10	0.25

Fig. 7. 10 Mode Test – Japan Vehicle Inspection Association at $\sim$ 60,000 kilometers.

References p. 290.

have been approximated without catalyst replacement. Mileage was accumulated using the AMA test procedure on public roads in Ohio. The fuel used was commercially available, unleaded gas purchased from the pump. These tests, described in Fig. 7, were conducted in Japan by the Japan Vehicle Inspection Association at approximately 60,000 kilometers.

Perhaps a more meaningful representation of the performance of the NO_x control system is given in Fig. 8, which depicts the NO_x conversion efficiency as a function of miles at 50 miles per hour for both AMA mileage accumulation and 60 mile per hour steady state dynamometer mileage accumulation. The near constant 80% performance for both types of tests tends to verify the stabilizing effect of the oxygen removal catalyst through at least 60,000 kilometers.

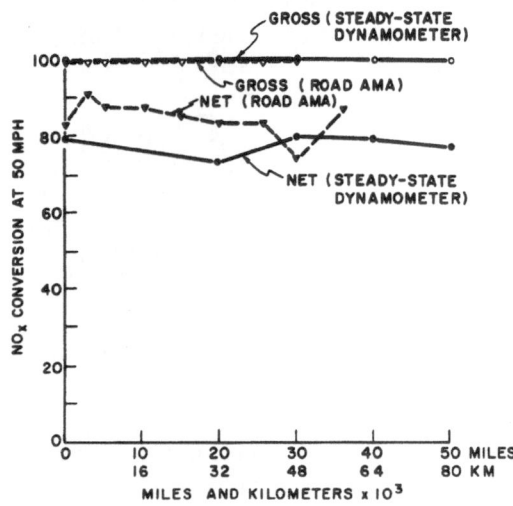

Fig. 8. Comparison of steady state dynamometer to road AMA activity.

LARGER CAR 10-MODE PERFORMANCE

A full sized V-8, 1973 Plymouth and a 1973 Ford have also been equipped with a similar dual bed system as the Datsun; however, a separate system is employed for each four cylinder bank. Fig. 9 presents results on the 10-mode test after 26,000 miles on the Ford at a low mileage on the Plymouth. Ability to meet the 0.25 gram per kilometer NO_x standard is demonstrated. These vehicles, however, did utilize the EGR systems with which they were originally equipped.

APPLICATION TO U.S. STANDARD

While Gould has not recently emphasized the testing of systems capable of meeting the U.S. 1978 NO_x standard, it may be interesting to note the data obtained to date.

1973 Ford 351 Cubic Inch V-8
With EGR 2,000 kg. Inertia Wt.
Test Mileage: 26,000 Miles

10 Mode		
NO$_x$ (g/km)	HC (g/km)	CO (g/km)
0.16	0.14	1.54
0.19	0.29	0.028
0.18	0.27	0.028
0.34	0.34	0.49
0.21	0.29	0.028
0.21	0.30	0.27
0.24	0.32	0.027

1973 Plymouth 318 Cubic Inch V-8
With EGR 2,000 kg. Inertia Wt.
Test Mileage: Fresh

NO$_x$ (g/km)	HC (g/km)	CO (g/km)
0.089	0.097	0.028

Fig. 9. Preliminary Japanese emission test results on large size U.S. vehicles.

Fig. 10 shows the Datsun, still without EGR, was capable of achieving the 0.4 gram per mile standard through at least 36,000 miles of AMA mileage accumulation. The full size Ford, without EGR, was also durability tested and indicated a net 80% NO$_x$ conversion efficiency performance after 26,000 miles. Fig. 11 shows that with stock EGR both this Ford and a Plymouth are capable of achieving the 0.4 gram per mile standard. It should be noted, however, that both the oxygen removal catalyst and the oxidation catalyst were replaced during the 26,000 mile test on this Ford.

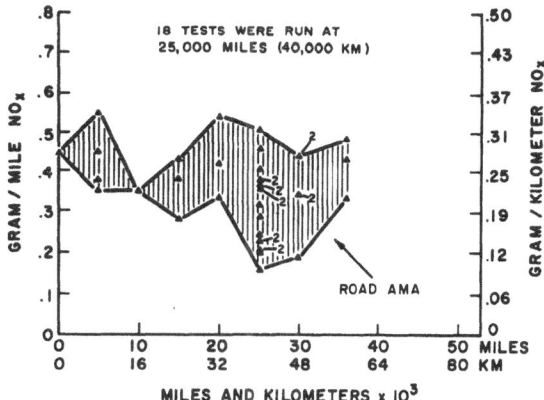

Fig. 10. Datsun 610 — U.S. Federal Test Procedures results as a function of durability mileage.

References p. 294.

1973 Ford 351 Cubic Inch V-8
With EGR — 4,500 lb. Inertia Wt.
Test Mileage: 26,000 Miles

1975 CVS		
NO$_x$ (g/mile)	HC (g/mile)	CO (g/mile)
0.27	0.39	2.04
0.30	0.46	2.36
0.36	0.51	2.40

1973 Plymouth 318 Cubic Inch V-8
With EGR — 4,500 lb. Inertia Wt.
Test Mileage: 6,000 Miles

1975 CVS		
NO$_x$ (g/mile)	HC (g/mile)	CO (g/mile)
0.18	0.33	2.4
0.38	0.33	3.4
0.24	0.25	1.8

Fig. 11. Large car U.S. Federal Test Procedure results.

SUMMARY

The test results presented on the Gould NO$_x$ catalyst with an oxygen removal catalyst ahead of it indicate the ability to achieve the 1976 Japanese NO$_x$ standard. Gould's experience indicates that control of input oxygen is the critical factor controlling the stability of the Gould NO$_x$ catalyst. Preliminary data also indicate that the 1978 U.S. NO$_x$ standard is obtainable in systems employing EGR. However, much more work is required to define the limit of NO$_x$ catalyst durability, total system performance, and fuel economy factors.

REFERENCES

1. L. S. Bernstein, et al., "Application of Catalysts to Automotive NO$_x$ Emission Control," SAE Paper No. 710014, January, 1971.
2. R. S. Lunt, et al., "Application of a Monel-Platinum Dual-Catalyst System to Automotive Emission Control," SAE Paper No. 720209, January, 1972.
3. L. S. Bernstein, et al., "Nickel-Copper Alloy NO$_x$ Reduction Catalysts for Dual Catalyst Systems," SAE Paper No. 730567, May, 1973.

DISCUSSION

J. Hightower *(Rice University)*

I found your paper to be very interesting. But I wanted to get your comments concerning fuel economy. Do you have anything you can say about that at this time? That's always been a big worry about this kind of system.

Fedor

Well, I would say that in our previous experience in NO$_x$ catalysis we had to set the carburetor very rich, around 3% CO. With this system we've been able to set it as lean as 1-1.25% CO. We normally tested the Datsun at 1-1.5% CO. That should be some help. In addition, we feel that there is evolving technology that will allow cars to run near 1-1.5% CO and not suffer a fuel economy penalty by changing the air/fuel ratio. This is basically the EGR and advanced spark timing work of General Motors*. Our data is certainly inconclusive. We've seen fuel economy gains from test to test and have seen fuel economy losses. But basically we can't optimize the system for fuel economy. We will try to do our best in the next few months but that question can't be answered by us.

R. L. Klimisch *(General Motors Research Laboratories)*

Will you comment on the problem of running out of fuel with this catalyst system?

Fedor

Yes. First of all at the SAE Meeting in Toronto on Oct. 25th, 1974**, we'll go into that factor in depth. But we have tests on running out of gas through 60 miles per hour and no damage to the catalyst occurs, because of the use of the getter. Now before the use of this oxygen removal catalyst that certainly was a problem above about 20 or 30 miles per hour. In other words, if the oxygen is removed ahead of the NO$_x$ catalyst, there is little reason for it to oxidize.

Klimisch

How did the getter fare in this experiment?

Fedor

No damage to the *system* through 60 miles per hour.

G. Barnes *(General Motors Research Laboratories)*

That was my question also, Bob. It seems like the getter is just another oxidation catalyst. Whatever you've done to prevent damage to the getter during these

---

*J. J. Gumbleton et al. "Optimizing Engine Parameters with Exhaust Gas Recirculation", SAE Paper No. 740104, February, 1974.

**R. J. Fedor, F. M. Dunleavy, R. W. Henry and R. R. Steiner, "Durability Experience with Metallic NO$_x$ Catalysts", SAE Paper No. 741081, October, 1974.

malfunctions would seem to be technology applicable to preventing damage to the oxidations catalysts and to the metallic catalyst itself. Why isn't it destroyed?

Fedor

Well, we're not running it as a classic oxidation catalyst, we're not putting oxygen over it.

Barnes

You are when you run out of fuel.

Fedor

Yes, at that instant, but you're starting from a lower temperature. I can just present the data and you can draw the conclusions.

D. E. Achey *(GM AC Spark Plug Division)*

It seems to me from earlier tests of these catalyst that they were kept very warm, or extremely hot, to prevent ammonia formation. You said it wouldn't fail as much because it operates cooler. Are you still hot enough in your system to prevent ammonia or are you getting ammonia now?

Fedor

Well, the key to that is the fact that when we take oxygen out, the catalyst has to form a relatively small amount of ammonia in that low oxygen environment. We have altered our catalyst composition and configuration for that. This is a new catalyst and I think your prior experience is not applicable. It does run at lower temperatures. It forms a maximum 20% ammonia under those conditions and we believe we can get it down to 10% ammonia in the future. So that is the key to the performance.

Achey

Do the results show an accounting for the ammonia formation so that they are net results rather than gross results.

Fedor

Yes. We gave 80% net conversion. There will be more details on that in the SAE paper.

K. Hellman *(Environmental Protection Agency)*

Bob, in the past work with your other catalysts which we discussed in somewhat more formal proceedings, it seems to have the same type of behavior that was discussed here yesterday by the gentleman who discussed the reactor that turned into a catalyst. Is the performance now superior to the old suicidal catalyst? In view of the fact that the oxygen spikes are eliminated, are the grain boundaries attacked just by thermal cycles without going through redox cycles?

Fedor

Basically, we have found oxidation as the reason for the "catastrophic" failures and also as the reason for the gradual deterioration. Our laboratory results show that it is not thermal cycling, per se, it is a function of the oxygen over the catalyst, and also a function of time and temperature. If we control the oxygen input, we do control the durability that way. And to answer your first question, yes. And I think the data will prove it in the following months, that we have made significant strides on that problem.

R. Mondt *(General Motors Research Laboratories)*

Do you have an air pump on this system to accomplish the oxidizing reaction?

Fedor

Certainly.

Mondt

Is there a control system of some sort to modulate the air supply to the oxidizing section?

Fedor

No. It's rather crude.

J. J. Carberry *(University of Notre Dame)*

Will this system survive 50,000 miles?

Fedor

Well, we've only gone 36,000 miles in vehicle tests. With engine dynamometers we've gone 50,000, but those were steady state conditions.

Carberry

Can you comment on the role that sintering plays in the ultimate deterioration of the catalyst? I presume it doesn't last forever.

Fedor

This system is totally different from a high surface area catalyst deposited on ceramic. We start with low surface area. So it's a question which Karl was getting into, to prevent the surface area from generating with time. We have a stable catalyst as long as the surface area doesn't increase with time.

W. K. Hall *(University of Wisconsin)*

Can you say something about its form, is it on a support?

Fedor

I tried to explain it before. It's a coil of expanded metal with a special processing in which we deposit the catalyst and form a catalytic surface and a corrosion resistant substrate. I'm sure that doesn't explain it, but that's a cross-section through it. It has a relatively high amount of geometric surface, because of the number of cuts into the strip. But it's low surface area compared to the traditional deposited ceramic catalyst.

W. E. Bernhardt *(Volkswagenwerk)*

What is the cold start performance of your catalyst in this system? What can you say about the low temperature performance of the catalyst?

Fedor

Well, we gave some of the CVS data. Until the hydrocarbon catalyst starts degrading significantly, we can meet HC, CO and NO_x. There is a Japanese cold start test, and we can reach the HC and CO very readily, through 60,000 kilometers. There is no NO_x standard for this test yet.

E. E. Weaver *(Ford Motor Co.)*

Maybe his question is the same as mine. During the cold start do you put air before the NO_x catalyst or the getter catalyst?

Fedor

To date, this is a very crude system where we use stock carburetors with about a 90-second choke. The choke, for example, is 6% CO and we do add oxygen for about 30 seconds, approximately 3% oxygen. But it's still not reducing in this system, and that does tend to preheat the thing. That's all we really use it for. Despite that crude technique we come pretty close to HC and CO.

M. Shelef *(Ford Motor Co.)*

You have put the onus of catching the poisons on your little getter in front because it is first in the gas stream. How does it then affect the catalyst after 36,000 miles? Does it still remove the O$_2$ spikes?

Fedor

Yes. We do that test. We take three hills out of the CVS and calculate the percentage oxygen removed during those three hills. It started out around 90-92% and it's down to 86% after 36,000 miles. We have this data plotted out, with no significant deterioration yet, I would say.

H. Gandhi *(Ford Motor Co.)*

I have a question about the slide which shows that the inlet oxygen level to the getter catalyst is about 4%, but the CO level is only 1%. How do you account for the removal of all the 4% O$_2$ down to .03%?

Fedor

That has been asked often. I don't know what the HC is at that time. During shifts there are tremendous HC spikes. They're basically instantaneous oxygen peaks. That's how the measurement is but I don't really know the time sequence. I really don't know where that occurs in real time. I can show you all the traces and show you the output but I can't really account for that.

J. B. Butt *(Northwestern University)*

Bob, why do you use half of a PTX?

Fedor

That was the only space available at that time. That's the only size we've ever tried.

STUDIES ON NO$_x$ REDUCTION AND THREE-COMPONENT CATALYSTS

W. E. BERNHARDT

Volkswagenwerk AG, Wolfsburg, Germany

ABSTRACT

Nearly two dozen promising NO$_x$ reduction and three-component catalysts have been evaluated in laboratory activity tests, using actual engine exhaust gas to determine NO$_x$ conversion rates. Furthermore, many of these catalysts were exposed up to 50 hours, mainly under stoichiometric conditions, to engine exhaust in dynamometer tests, and then re-evaluated in the laboratory test apparatus for aging and deactivation behavior. For supported pelleted and monolithic catalysts, it was found that the primary causes of deactivation were sulfur, phosphorus and lead poisoning. Some noble-metal catalysts showed rapid deactivation because of unsatisfactory chemical stability of specific active components, e.g. ruthenium.

Typical NO$_x$ reduction and three-component catalyst characteristics will be presented in the paper. For three-component catalysts the variation of the operation "window" by aging is discussed.

INTRODUCTION

Reactions of various nitrogen oxides (NO$_x$) from automotive and stationary sources with unburned hydrocarbons to produce photochemical smog have been well examined. The U. S. Government has proposed standards and goals that will further reduce emission of NO$_x$ from automobiles in the forthcoming years. Hence, the internal combustion engine will perhaps require a modified combustion process (stratified charge engine) or will at least require a combination of engine modifications to minimize NO$_x$ formation and an efficient catalytic system for reducing NO$_x$ in the exhaust gases. The development of a catalyst technology, however, depends on a thorough understanding of how NO$_x$ can be reduced

References p. 304.

catalytically without having so-called secondary emissions (e.g., sulphur and ruthenium compounds) and pronounced aging effects.

EXPERIMENTAL

Some operating parameters are very important to the overall effectiveness of a catalytic system, such as: catalyst inlet concentrations, inlet temperature, space velocity, and $O_2/(CO + H_2)$ ratio. Therefore, it is necessary to make laboratory conversion tests with the catalysts to find out under which conditions an optimal catalytic efficiency can be achieved. It was concluded that especially the ignition temperature* and the low temperature performance of the catalyst are important items of information for the development of emission controlled vehicles. Furthermore, in connection with a closed-loop electronic fuel injection control which includes an oxygen-sensor, it is important to investigate whether the catalyst is able to operate as a three-component catalyst at stoichiometric mixtures.

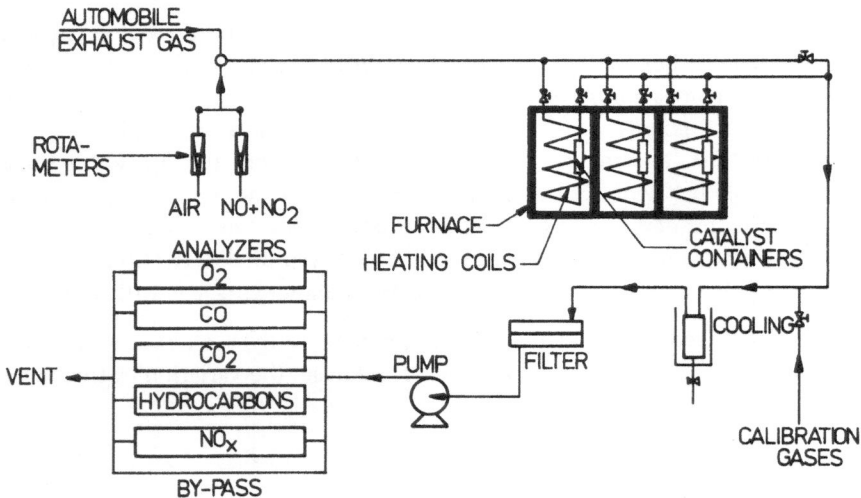

Fig. 1. Apparatus for measuring catalyst activity.

In order to get all this important information about the catalysts, the equipment used in our laboratory is shown schematically in Fig. 1 (1). The catalyst test apparatus consists essentially of three electrically heated furnaces. In this test method, a small quantity of catalyst is placed in a container. Vacuum pumps are used to draw a portion of exhaust gas through the sample. This system enables the catalyst to be tested at different gas concentrations and at different space velocities. The

*Ignition temperature is defined as the temperature at which 50% conversion has occurred.

composition of the engine exhaust gas used for the evaluation of reduction catalysts is summarized below (dry basis):

- 250 ppm HC (as C$_6$ FID)
- 1.5 vol % CO
- 0.35 vol % O$_2$
- 13 vol % CO$_2$
- $\sim$3 vol % H$_2$O (wet basis)
- 1500 ppm NO$_x$
- remainder N$_2$ + H$_2$

The above mentioned gas composition is used for testing three-component catalysts. The oxygen concentration, however, varies between 0.4 and 1.65%.

The space velocity is generally 40,000 cm^3/cm^3-hr. The pelleted type catalyst samples are supported in small 12 cm^3 axial-flow converters. The monolithic samples have a 2.5 cm O.D. and they are 7.6 cm long.

In order to improve the overall effectiveness of a catalytic system during vehicle operation, it is necessary to test the aging behavior of the catalyst. Aging tests provide an estimate of catalyst deactivation in the presence of exhaust from automotive fuels. It is well-known that reduction as well as oxidation catalysts lose considerable activity during aging. To test the chemical stability and the activity as a function of aging time and of mileage, a catalyst aging test stand was developed. A schematic of the test stand is shown in Fig. 2.

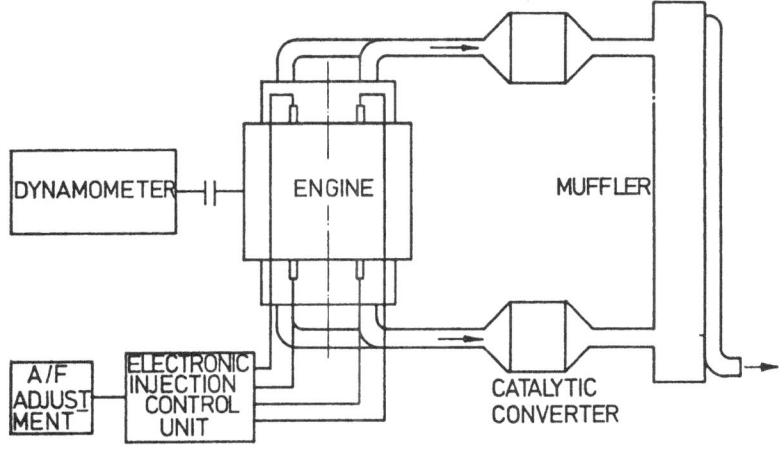

Fig. 2. Catalyst aging test stand.

This test stand consists of an engine operated on an engine-dynamometer with one converter at each side of the engine.

It can be seen from Fig. 2 that an electronic fuel injection control unit provides for stoichiometric as well as fuel-rich mixtures.

The aging tests are carried out in such a manner that road load conditions are simulated. The engine is controlled by a programmed electric motor which allows the engine to operate between 1000 and 3200 rpm under engine cruising conditions. The simulated average vehicle speed is about 71 km/hr.

Usually after 50 hours of operation (which is the equivalent of 7000 km), samples are removed for testing in the catalyst activity test apparatus in the laboratory. In some instances after activity evaluation, the samples are returned for further aging in the engine exhaust gas.

The aging tests are carried out with the engine exhaust from leaded and unleaded fuels. The sulphur content of the test fuels has to be known also because catalyst activity is affected by small amounts of sulphur in the exhaust gas. It was observed that near 400°C at a CO/O_2 ratio of 3.3, deactivation was very pronounced for sulphur contents greater than 0.01 wt.%.

Furthermore, small amounts of lead + phophorus have a pronounced effect on the deactivation of reduction and three-component catalysts, but this will be discussed in a later paper.

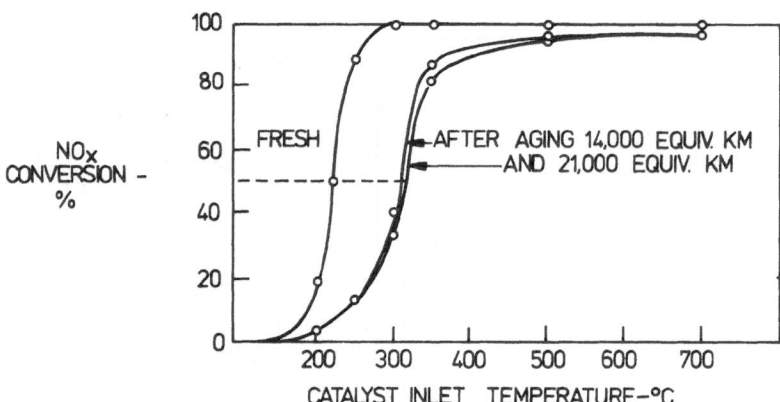

Fig. 3. Performance of a fresh NO_x catalyst and after aging
(SV = 40,000 h^{-1}; CO/O_2 = 4.3)

RESULTS

Fig. 3 was obtained as a typical result of measuring low temperature catalyst activity. This figure shows that the NO$_x$ reduction performance of this pelleted rhodium catalyst is promising. The ignition temperature of this catalyst lies near 220°C. After aging 21,000 equivalent km this catalyst lost much of its low temperature activity. The ignition temperature of the catalyst after the aging test is 100° higher (320°C).

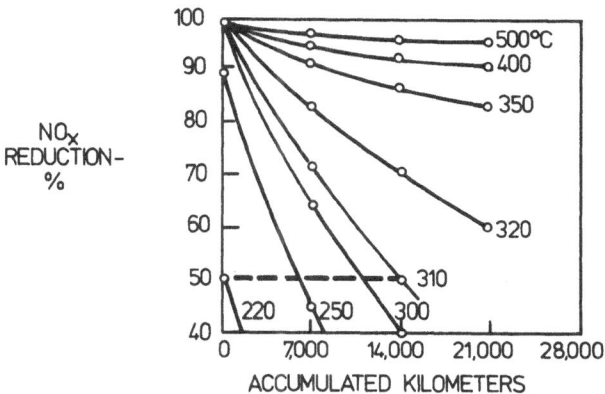

Fig. 4: Deactivation of NO$_x$ reduction activity as a function of accumulated kilometers and temperature

(SV = 40,000 h^{-1}; CO/O$_2$ = 4.3; unleaded fuel aging)

Fig. 4 shows this relationship in more detail. Besides the distinct moving of the ignition temperature with increasing kilometers of aging, it can be clearly seen that the aging influence is very stringent especially at low temperatures (less than 350°C). Furthermore, this supported rhodium catalyst has been found to be selective because NO$_x$ and CO were reduced simultaneously.

Fresh noble-metal NO$_x$ and three-component catalysts provide us with significant information. The catalytic performance of monolithic reduction and selective catalysts tested under reducing conditions is illustrated in Fig. 5. Some of these catalysts have promising conversion-efficiency. Catalyst U77 shows a low ignition temperature but poor high temperature performance and unstable catalytic activity even during the short time of testing. This is due to uncompleted recrystallization processes and to the fact that it is difficult to get specific active components (such as ruthenium) stabilized on the catalyst surface. Fig. 6 shows conversion-efficiency results which were obtained with fresh pelleted reduction and three-component catalysts. Our chemical activity studies indicate that there is no strong argument to decide on pelleted catalysts and to abandon monolithic catalysts or vice versa.

References p. 304.

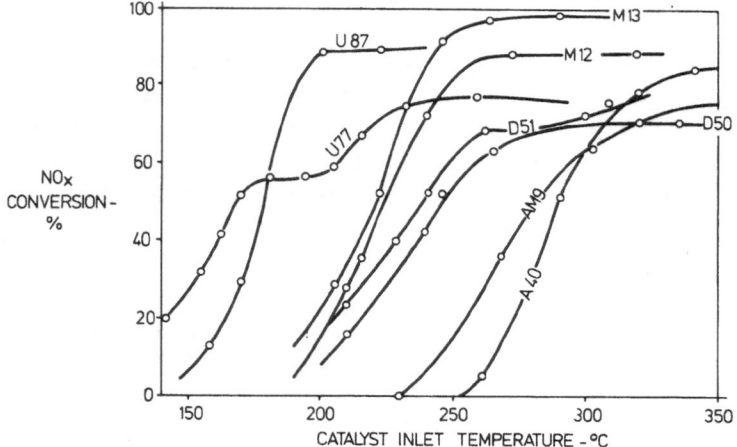

Fig. 5. Performance of fresh monolithic NO_x and three-component catalysts
(SV = 18,000 h-1; CO/O_2 = 5.0)

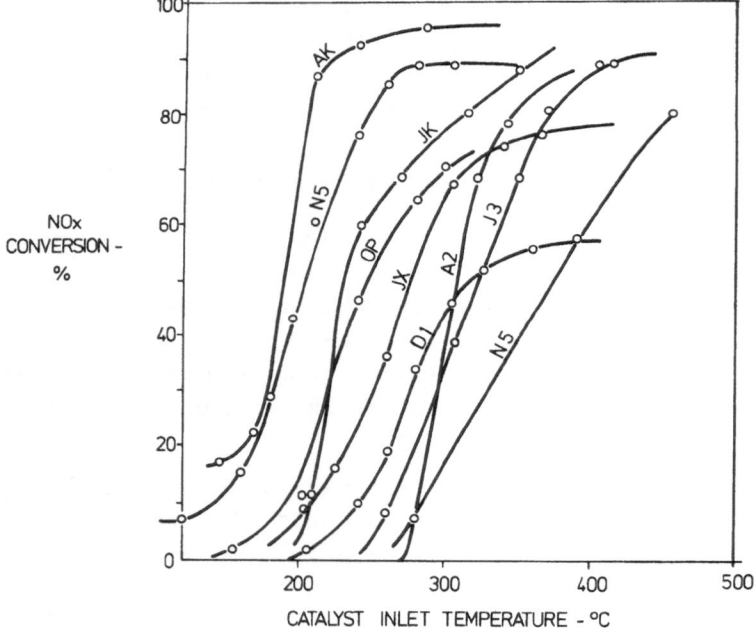

Fig. 6. Performance of fresh pelleted NO_x and three-component catalysts
(SV = 18,000 h-1; CO/O_2 = 5.0)

To date little information exists on the characteristics and performance of fresh
and aged three-component catalysts (1,2). The emission reduction behavior of a

recently developed monolithic three-component catalyst investigated under fresh and 50 hr aged testing conditions is shown in Fig. 7. It can be seen that if the O$_2$ concentration is allowed to increase (this means that the air-fuel ratio is allowed to go lean), NO$_x$ reduction efficiency drops considerably. On the other hand, if the O$_2$ concentration decreases (this means that air-fuel ratio becomes too rich), insufficient HC and CO conversion results. Hence, only over a very narrow range of air-fuel ratios are the concentrations of all three components significantly reduced. Volkswagen Research has examined nearly two dozen suitable catalysts, but most of them have an acceptable operation "window" only under fresh conditions.

The operation window is situated very near to the stoichiometric condition on the rich side for fresh three-component catalysts; see W$_F$ in Fig. 7. After 50 hrs aging, this window unfortunately gets narrower and it changes just a little to the lean side of stoichiometric*; hence, it is nearly impossible to synchronize accurately the optimal operation characteristics of the three-component catalyst with the oxygen-sensor characteristics, because control with the sensor is based on its step function change in voltage at stoichiometry.

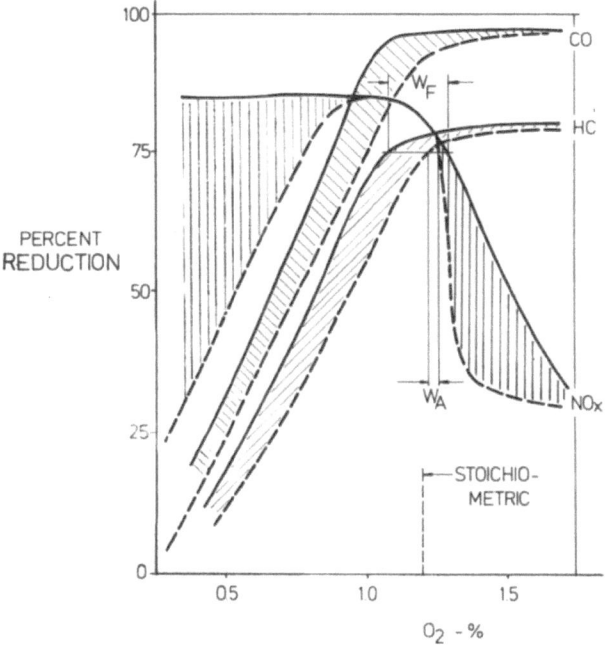

Fig. 7. Three-component catalyst characteristics tested under fresh conditions and after aging 7,000 equivalent kilometers (50 h)

(SV = 40,000 h^{-1}; temperature before catalyst = 380°C)

References p. 304.   *See W$_A$ in Fig. 7.*

The observed rapid conversion-efficiency degradation especially for NO_x has been found to be due to the unsatisfactory stability of selective noble metals such as ruthenium. After 7,000 km the ruthenium content was reduced down to 25% of its original value.

CONCLUSION

In order to use these catalysts in closed-loop electronic fuel injection control of internal combustion engines, it has been found that the monolithic and pelleted NO_x reduction and three-component catalysts still require a considerable amount of development effort. These catalysts should stay mechanically and chemically stable for the life of the car. Rapid progress has been made in this area but an excessive conversion-efficiency degradation together with a pronounced variation of the operation window with mileage for both pelleted and monolithic three-component catalysts has been found.

REFERENCES

1. *W. E. Bernhardt and E. Hoffmann: Methods for Fast Catalytic System Warm-Up During Vehicle Cold Starts, SAE Paper 720481, see also SAE Transactions 1972.*
2. *W. Buttgereit, D. Pundt and P. Oeser: "Engine Testing of Catalysts," SAE Paper 720482 (1972).*

DISCUSSION

J. W. Hightower *(Rice University)*

Which way did you say the sensor shift is?

Bernhardt

To the rich region.

Hightower

Therefore the window is shifting in the opposite direction to the way the sensor is changing, so you have problems?

Bernhardt

Yes. That's right.

APPROACHES TO CATALYTIC CONTROL
OF AUTOMOTIVE NO$_x$ EMISSIONS

R. L. KLIMISCH

General Motors Research Laboratories, Warren, Michigan

J. M. KOMARMY

General Motors AC Spark Plug Division, Flint, Michigan

ABSTRACT

This paper reviews the various concepts that have been suggested for controlling automotive nitrogen oxide emissions. Major emphasis is placed on the dual catalyst and the closed loop approach. The paper will also include a discussion of metallic vs. ceramic catalysts.

INTRODUCTION

There are two basic approaches for controlling automotive NO$_x$ emissions with catalysts — the dual catalyst approach (Figs. 1 and 2) and the single catalyst or closed loop approach (Fig. 3). The former approach uses one catalyst as an NO reduction catalyst and requires a strong reducing atmosphere (rich A/F or excess fuel) from the engine. The second catalyst in the dual catalyst system is an oxidation catalyst for the

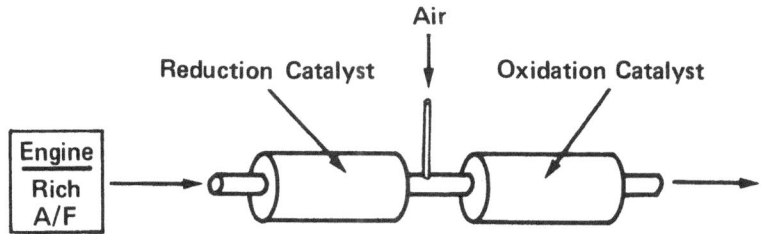

Fig. 1. Dual Catalyst Emission Control System.

removal of CO and hydrocarbon emissions. It is necessary to add air before the second catalyst to provide an oxidizing atmosphere.

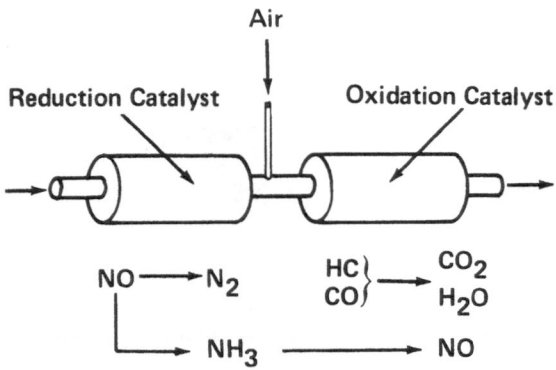

Fig. 2. Dual Catalyst Emission Control – The Ammonia Problem.

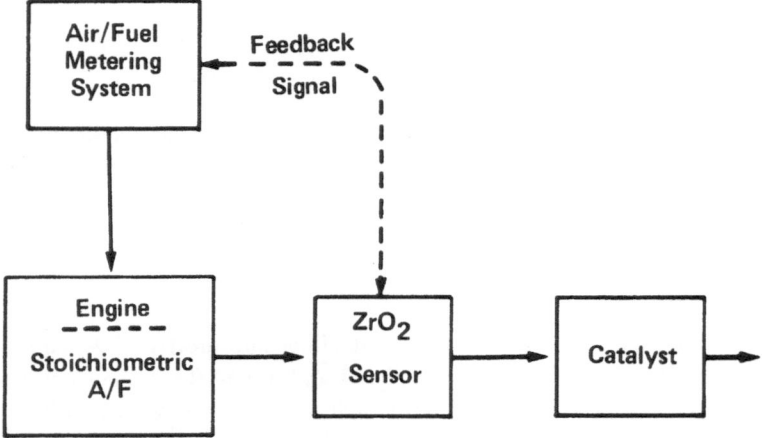

Fig. 3. Single Catalyst (Closed Loop) System for Automotive Emission Control.

The single catalyst or closed loop approach requires the engine to operate very near the stoichiometrically correct air-fuel ratio so that the single catalyst can remove CO and hydrocarbons as well as nitrogen oxides. Other suggested systems are generally combinations or modifications of these two approaches. Some of these will be discussed below. Both approaches, at least to some extent, have problems with ammonia formation, complexity, fuel economy, catalyst selection, and catalyst durability. This presentation will concentrate on the NO reduction aspects of these catalysts, but it is important to note that it is impossible to discuss these systems without considering CO and hydrocarbon control also. There are very strong interactions, and NO conversion cannot be considered separately.

DUAL CATALYST APPROACH

From a catalytic chemist's point of view, the most interesting aspect of this catalytic application is the ammonia formation problem. If ammonia is formed by the reduction catalyst, it is subsequently oxidized back to NO by the oxidation catalyst. Unfortunately, most catalysts tend to convert nitric oxide to ammonia rather than nitrogen. Only ruthenium appears to have the ability to convert NO directly to nitrogen and shows this selectivity at low temperature (<500°C), (1,2,3). Other catalyst systems including rhodium and nickel appear to convert NO to NH_3 and then decompose the ammonia to nitrogen afterward (1,4). Rhodium and nickel catalysts tend to show selectivity for nitrogen formation at higher temperature (>600°C), (4,5). This makes ruthenium the most desirable catalyst since high exhaust temperature is associated with loss in fuel economy. Ruthenium, however, has a serious durability problem in that it is readily converted to its volatile and toxic tetroxide upon exposure to oxidizing conditions (6). The dual catalyst approach requires the reduction catalyst to serve as an oxidation catalyst to control CO and hydrocarbons during the critical warm-up period so that the reduction catalyst is routinely exposed to oxidizing conditions (Fig. 2). This is another instance of the interaction between the various requirements. It is, of course, much easier to switch the air from place to place than it is to switch the corrosive exhaust mixture.

There have been a number of claims that ruthenium has been stabilized with respect to RuO_4 formation (7,8), but these catalysts have so far not shown the required durability performance for one reason or another. The very limited availability of ruthenium also is a major inhibition towards its use in this application. The availability situation for rhodium is worse than ruthenium; this, coupled with the metal's higher temperature requirements, represents a serious problem even though rhodium does not form a volatile oxide. The maximum quantities that could be used in each vehicle are very small for these two metals: approximately 0.3 g/car for Ru and 0.15 g/car for Rh. This results in a metal loading of less than 0.01 wt % on the catalyst which is usually considered to be too dilute to make a catalyst with the required activity and durability characteristics. These quanities, though very small, still would use up the world supply in this single application (9). It is expected that the selectivity of these catalysts for nitrogen formation will fall off with loading.

This leaves nickel containing catalysts for consideration. Two types of nickel catalysts have been considered: ceramic supported nickel catalysts and metal alloy supported nickel catalysts. The most widely used support for this type of catalyst is alumina, and it is the most advanced in terms of its development for this particular application. Nickel supported on alumina is plagued by the formation of nickel aluminate which destroys nickel's selectivity for nitrogen formation (6). It is generally

References pp. 313-314.

necessary to add platinum or palladium to this type of catalyst to provide the necessary activity for both NO reduction and for CO-HC oxidation (1). The activity and stability of nickel appears to be inferior on other supports.

There is much more information available on nickel metal alloy catalysts (10-14). This type of catalyst has been rather extensively investigated first at Esso (10-12) and more recently at Gould (14). An earlier presentation described these catalysts rather extensively (14). Apparently because of the lower active surface area and because of selectivity considerations, these catalysts require high temperatures for effective operation, which results in a substantial reduction in fuel economy (12). These catalysts are also subject to durability problems associated with irreversible oxidation resulting in exfoliation of catalyst material (12). In addition, running out of fuel causes catastrophic deterioration of the catalyst (12), partly because of the reactor configuration and partly because the whole catalyst structure is subject to highly exothermic oxidation.

One additional problem that should be mentioned with regard to nickel catalysts in this application is the possibility of forming highly toxic nickel carbonyl, $Ni(CO)_4$. Although early tests with the metallic nickel catalysts were not able to detect the presence of nickel carbonyl (10), because of the carbonyl's toxicity, this area needs to be thoroughly investigated.

So far as durability is concerned, none of the ceramic based NO_x catalysts have come close to the 50,000 mile requirement. Ruthenium catalysts have shown very good initial activity but have suffered a variety of durability problems. Some of these are associated with loss in CO and hydrocarbon oxidation ability. Carefully controlled tests of the nickel alloy catalysts have achieved more impressive results (14), but these catalysts are subject to some serious deficiencies as mentioned above.

It sometimes seems that platinum is a good catalyst for every application. We'd like to point out that platinum (as well as palladium) catalysts have not performed very well so far for NO_x reduction under rich conditions. Both the activity and selectivity of platinum have not been good enough. In addition, durability experience for both platinum and palladium has also been poor.

We have tested two basic configurations of the dual catalyst approach. These represent what we will call a conventional approach which uses an NO_x catalytic converter followed by an oxidation converter (Fig. 4). The second configuration, used in the TMECS (triple mode emission control system), involves a single monolith mounted in the exhaust manifold (Fig. 5) with the central portion the reducing catalyst and the outer portion the oxidation catalyst. Again, it should be emphasized that the quick warm-up requirement (for CO and HC control) imposes some rather severe design restrictions on the NO_x catalyst and converter.

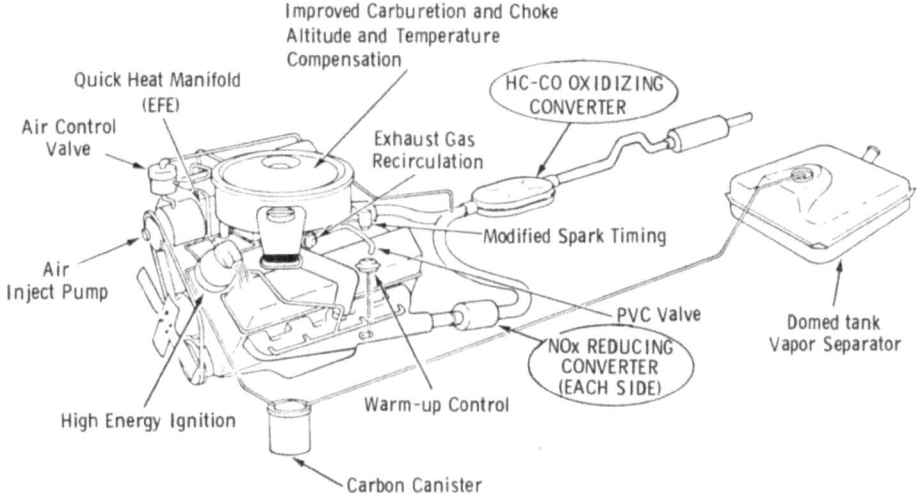

Fig. 4. Conventional Dual Catalyst System.

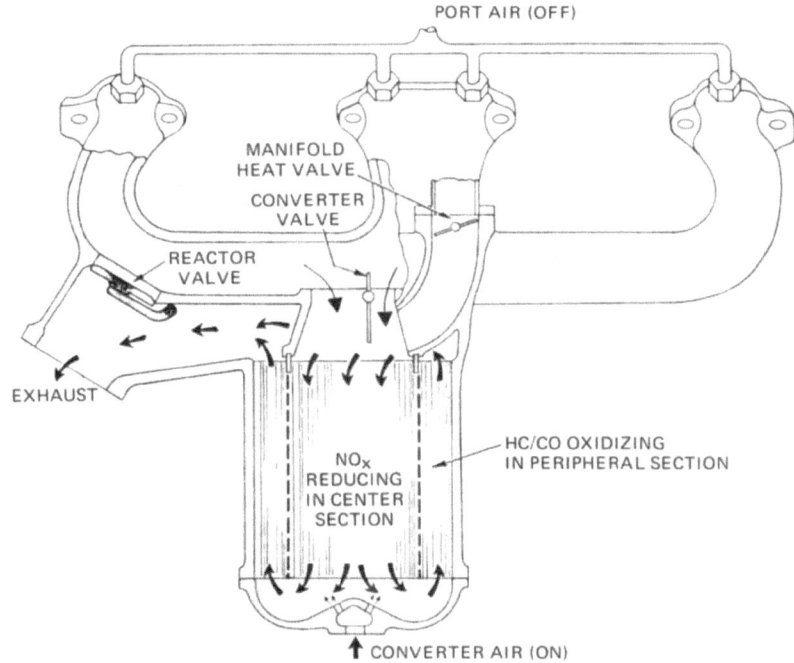

Fig. 5. T-MECS Dual Catalyst Converter.

SINGLE CATALYST APPROACH

The exhaust gas composition at the stoichiometric point should contain only CO_2, H_2O, and N_2 (15). However, because of flame quenching phenomena (16), the exhaust contains significant amounts of hydrocarbons. and because of kinetic effects, it also contains nonequilibrium quantities of CO, O_2, and NO (17). Table 1 shows typical exhaust composition at the stoichiometric point. Since the catalyst promotes the attainment of equilibrium and because the oxidizing and reducing agents are in balance stoichiometrically, a single catalyst can achieve reasonable success with fresh catalysts for removing all three pollutants if the exhaust composition is very close to the equivalence point, i.e., 14.7 pounds of air per pound of fuel for current gasolines.

A sensor and feedback system can be used to maintain precise control of the air-fuel ratio. The high temperature zirconium oxide solid electrolyte sensor has proved to be very sensitive to oxygen partial pressure near the stoichiometric point (18). The feedback system has been coupled both to carburetors and to electronic fuel injection systems. It should be mentioned that the stoichiometric air-fuel ratio is favorable for fuel economy and for vehicle performance.

TABLE 1

Approximate Exhaust Composition at Stoichiometric*

CO	0.3 − 1.0%
H_2	0.1 − 0.3%
O_2	0.2 − 0.5%
NO	0.05 − 0.15%
HC	0.03 − 0.08%
CO_2	ca. 12%
H_2O	ca. 13%

Calculated for a fuel with H/C = 2.0.

Catalysts that have been mentioned for this application include platinum, palladium, rhodium, and ruthenium (4,5,19). Our experience has shown that the first two have serious durability problems for this application while less is known about the durability characteristics of rhodium in this application. The durability of ruthenium here is highly suspect for reasons mentioned above. Of course, the limited availability of the two last mentioned metals must be considered in this situation also.

Base metals have also been mentioned for the closed loop application (5,19), and should not be overlooked since this type of catalyst has considerable oxygen storage capability which may be useful. However, the lower catalytic activity of base metals (6) coupled with their sensitivity to sulfur poisoning (20) will undoubtedly limit their effectiveness as closed loop catalysts.

Typical results for a fresh three component catalyst are shown in Fig. 6. Over 80% conversion for all three pollutants is achieved over a very narrow air-fuel ratio range, i.e., approximately one-tenth of an air-fuel ratio. The activity profile for the same catalyst after 500 hours of dynamometer aging is shown in Fig. 7. It is seen that the major activity loss is in NO$_x$ conversion while the CO and HC conversion of the catalyst is still high. These results are typical for the durability behavior of three component catalysts and suggest the processes represented by equation (1) occur before those represented by equation (2).

$$\left.\begin{array}{l} H_2 \\ CO \\ HC \end{array}\right\} + O_2 \longrightarrow CO_2 + H_2O \tag{1}$$

$$\left.\begin{array}{l} H_2 \\ CO \\ HC \end{array}\right\} + NO \longrightarrow CO_2 + H_2O + N_2 \tag{2}$$

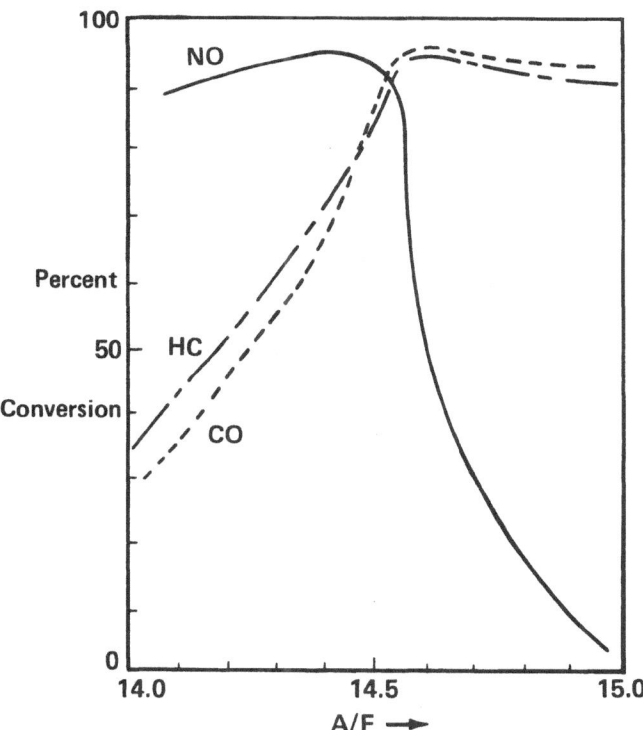

Fig. 6. Activity Profile for Fresh Catalyst: CO, NO, and HC Conversion.

References pp. 313-314.

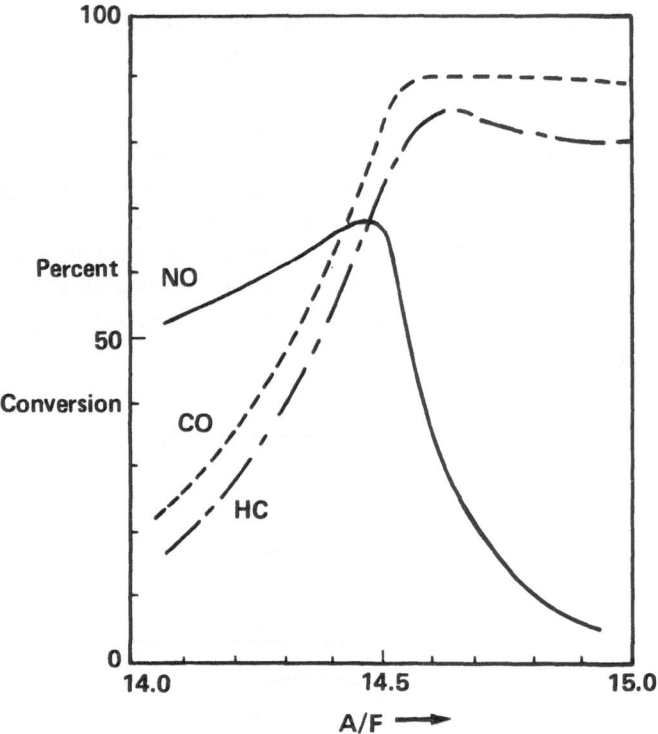

Fig. 7. Activity Profile for Catalyst After 500 hours Aging: CO, NO, and HC Conversion.

I. $\quad O_2 + \left.\begin{array}{l} H_2 \\ CO \\ HC \end{array}\right\} \longrightarrow H_2O + CO_2$

II. $\quad NO + \left.\begin{array}{l} H_2 \\ CO \\ HC \end{array}\right\} \longrightarrow N_2 + H_2O + CO_2$

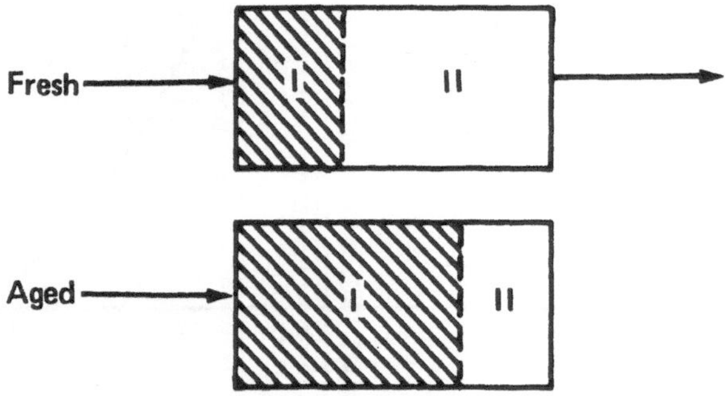

Fig. 8. Catalyst Aging in a Sequential Reactor.

The sequential nature of these processes (Fig. 8) is reasonable since one would expect oxygen to be a better oxidizing agent than NO except perhaps at low temperature (21). It may also explain the low ammonia formation if H$_2$ is removed before it can react with nitric oxide. The fact that NO reduction stops abruptly when there is excess oxygen also suggests that these processes are sequential (4).

This sequentiality has some fairly serious implications since NO$_x$ emissions are primarily associated with high flow conditions which tend to reduce the amount of catalyst available for NO removal. The aging processes (sintering and poisoning) also effectively reduce the amount of available catalyst. Thus, it is expected that NO conversion should fall off first. These considerations obviously have to be taken into account in sizing the automotive converter.

CONCLUSIONS

At the risk of appearing to be negative, there are still many problems associated with the application of catalysts to automotive NO$_x$ control. The most success in achieving the required emission performance has been obtained with metallic NO$_x$ catalysts, but only at the expense of a severe fuel economy loss as well as some serious uncertainties about the catalysts. Other approaches, particularly the single catalyst or closed loop approach, alleviate some of these problems but still have serious catalyst durability problems. Much progress has been made in nitrogen oxide catalysis; however, the development of a viable NO$_x$ catalyst for the automotive application remains as a great challenge for the catalytic scientist.

REFERENCES

1. R. L. Klimisch and K. C. Taylor, Envir. Sci. & Tech. 7, 127 (1973).
2. M. Shelef and H. S. Gandhi, Ind. Eng. Chem. Prod. Res. Dev. 11, 393 (1972).
3. T. P. Kobylinski and B. W. Taylor, J. Catal. 33, 376 (1974).
4. K. C. Taylor, "Simultaneous NO and CO Conversion over Rhodium," presented at the General Motors Research Symposium on "The Catalytic Chemistry of Nitrogen Oxides," October 8, 1974.
5. D. R. Ashmead, J. S. Campbell, P. Davies, and K. Farmery, "Nitrogen Oxide Removal Catalysts for Purification of Automobile Exhaust Gases," SAE Paper No. 740249, Detroit, 1974.
6. R. L. Klimisch and J. C. Schlatter, "The Control of Automotive Emissions by Catalysts," presented to the American Ceramic Society, Flint, Michigan, September 15, 1972.
7. T. P. Kobylinski, B. W. Taylor, and J. E. Young, "Stabilized Ruthenium Catalysts for NO$_x$ Reduction," SAE Paper No. 740250, Detroit, 1974.
8. M. Shelef and H. S. Gandhi, Platinum Metals Rev. 18, 2 (1974).
9. R. W. Ageton and J. P. Ryan, "Platinum Group Metals," Bulletin 650 (1970) Bureau of Mines, U. S. Department of the Interior.
10. L. S. Bernstein, K. K. Kearby, A. K. S. Raman, J. Vardi, and E. E. Wigg, "Application of Catalysts to Automotive NO$_x$ Control," SAE Paper No. 710014, 1971.
11. R. S. Lund, L. S. Bernstein, J. G. Hansel, and E. L. Holt, "Application of a Monel-Platinum Dual-Catalyst System to Automotive Emission Control," SAE Paper No. 720209, Detroit, 1972.

12. L. S. Bernstein, R. J. Lang, R. S. Lunt, G. S. Musser, and R. J. Fedor "Nickel-Copper Alloy NO_x Reduction Catalysts for Dual Catalyst Systems," SAE Paper No. 730567, Detroit, 1973.

13. G. H. Meguerian, F. W. Rakowsky, E. H. Hirschberg, C. R. Lang, and D. N. Schock, "NO_x Reduction Catalysts for Vehicle Emission Control," SAE Paper No. 720480, Detroit, 1972.

14. R. J. Fedor, "Application of Metallic Catalysts for Automotive NO_x Control," presented at the General Motors Research Symposium on "The Catalytic Chemistry of Nitrogen Oxides," October 8, 1974.

15. B. D'Alleva, "Procedure and Charts for Estimating Exhaust Gas Quantities and Compositions," GMR Report 372, 1960.

16. W. A. Daniel, SAE Transaction 79, 400 (1970).

17. H. K. Newhall, "Kinetics of Engine Generated Nitrogen Oxides and Carbon Monoxide," Twelfth International Symposium on Combustion, Poitiers, France, July 14-20, 1968.

18. W. J. Fleming, D. S. Howarth, and D. S. Eddy, "Sensor for On-Vehicle Detection of Engine Exhaust Gas Composition," SAE Paper No. 730575, Detroit, May 1973.

19. T. Ohara, "The Catalytic Reduction of NO_x over Supported Rhodium and Copper Nickel Catalysts," presented at General Motors Research Symposium on "The Catalytic Chemistry of Nitrogen Oxides," October 8, 1974.

20. J. E. Hunter, "Studies of Catalyst Degradation in Automotive Emission Control Systems," SAE Paper No. 720122, Detroit, 1972.

21. J. H. Jones, J. T. Kummer, K. Otto, M. Shelef, and E. E. Weaver, Envir. Sci. Technol. 5, 790 (1971).

DISCUSSION

W. K. Hall (University of Wisconsin – Milwaukee)

If your figures on the availability of ruthenium versus platinum today are compared with values of their natural abundances, I think you'll find the situation is much more favorable than you have indicated. Isn't it simply that the platinum has been mined because it has been needed for other purposes? If the huge automotive industry really wants to use ruthenium, my sources of information suggest that it can be made available.

Klimisch

No, that really is not the way we see it. It's one thing to ask for a large increase in platinum production since platinum is the major item that they go after in South Africa. One can simply open a new mine — if you've got a half billion dollars — and get a large increase in platinum production. But ruthenium and rhodium are basically by-products of platinum mining and it's more difficult to get a large increase in a by-product. Rhodium in particular is a very high-priced item and its demand is very high. The information that I've seen estimates world rhodium and ruthenium production at about 100,000 troy ounces per year.

Hall

The tables of natural abundance do not indicate a factor of 10 in the availability of ruthenium and platinum.

Klimisch

Well the ore in South Africa certainly is a factor of 10 . . . between those two.

J. Hightower *(Rice University)*

I think there have been some problems with some reports that I have seen that have misquoted ruthenium production by a factor of 10.

Klimisch

For 100 ounces of platinum in South Africa, I think you get 5 ounces of ruthenium, and I've seen estimates of 3 to 9 ounces of rhodium.

T. Kobylinski *(Gulf R & D)*

Two years ago, Hertha Skala of UOP showed us at the Gordon conference that for the 3-way catalyst to be really good it must contain rhodium. Today, Dr. Taylor more or less agrees with the same idea. My question is, do you feel that the rhodium catalyst is a viable 3-way material, is it feasible, or practical, because of the availability? And another question relates to our calculations which show that there is about ten times less rhodium than ruthenium.

Klimisch

No. I don't agree with you on your availability data. Certainly all of the statistics I've seen indicate that ruthenium and rhodium production are comparable. Rhodium is the most attractive 3-component material because it doesn't form a volatile oxide. But, of course, you look at the price and availability of rhodium and you wonder how you could ever consider using it in this application.

M. Shelef *(Ford Motor Company)*

I want to address myself to the question of ruthenium availability. There are considerable stock piles, because there was no use for ruthenium which is mined along with platinum. This might be the difference. I don't know what they are and where. There is another very big potential source but there is a hitch to it. As you know, ruthenium is almost on the peak of the distribution of the fission products. AEC at Hanford in Washington state has a lot of information available. Some of the isotopes can be separated and the half lives are relatively short and they can be decontaminated in a relatively short time. I don't know whether it would be acceptable for a customer to be riding on a radioactive source.

J. M. Komarmy *(GM AC Spark Plug Division)*

Mordecai, every time we try to find those stockpiles they become quite scarce and elusive. Where do you find them?

Klimisch

We think that Gulf has bought them all.

Harrison *(Matthey Bishop, Inc.)*

Since I'm in R&D, I hesitate to quote production figures from our sources. But, if it would be valuable to you I could have a statement sent to you to include in the minutes of this meeting.

Klimisch

We appreciate that. Thank you.

Editor's Note:

Sometime after the symposium the following statements were received from G. C. Harrison of Matthey Bishop:

Ruthenium

The availability of ruthenium is very much dependent upon the production of platinum. Rustenburg's reduced output of platinum will run at the rate of 900,000 troy ounces a year from this summer. Their capacity for production, if it is required, is already in excess of 1,200,000 ounces a year and will be expanded to 1,630,000 ounces by the end of 1977. For every 100 ounces of platinum refined, 10 ounces of ruthenium can be refined too. Thus, Rustenberg can produce at least 90,000 ounces of ruthenium in the next year if it is needed.

There are partially-refined stocks that could be brought forward quite quickly, but some positive idea of demand is necessary before the economics of recovering these can be judged.

Rhodium

The Western World production of rhodium during 1974 is projected at 120,000 troy ounces with South African mines being the major suppliers. Production could rise to slightly over 150,000 troy ounces in 1976 primarily as the result of increased platinum production in South Africa.

The USSR has traditionally supplied a nominal 30,000 troy ounces annually of rhodium to Western World purchasers.

It should be remembered that 1) present supplies of rhodium from all sources are seen to be used up in traditional industrial uses, 2) there is no assurance rhodium can or will be marketed by the USSR following the traditional pattern and 3) rhodium is produced as a by-product of platinum and will tend to track the production of platinum.

Dr. Harrison also mentioned that those contemplating use of rhodium should be aware of its limited availability.

R. L. Burwell, Jr. *(Northwestern University)*

Suppose, in some magic way I were to forbid you to use noble metals on the basis that the future might take a rather ill view of us for using up a large share of the noble metals for semi-cosmetic purposes. What would be achievable with base metals? Just a wild guess as to what kind of limits you could reach.

Klimisch

I don't want to answer it that way. One of the things that would have to be done is the sulfur level in the gasoline would have to be dropped. I think we could probably do fairly well with base metals if sulfur was reduced by an order of magnitude, or two. That's too easy an answer. What we've seen so far is that you've got lead and you find lead is the big problem. So you take the lead out of the gasoline and you find that in the absence of lead, sulfur is a problem. So if you take sulfur out, there might be something else that could get us. But we think we could do pretty well with lower sulfur levels.

E. Holt *(Exxon Research and Engineering Co.)*

May I just comment. After the sulfur comes out, you may find the hydrocarbons are the problem.

R. J. H. Voorhoeve *(Bell Laboratories)*

In respect to this last question, has anybody seriously looked at using base metal sulfides for NO_x reduction?

Klimisch

Yes, but I think inadvertently. Certainly, we've looked at base metal NO_x catalysts. Something like the copper-nickel that Dr. Ohara mentioned this morning is a reasonably effective NO_x catalyst but it does accumulate sulfur. I don't think those particular metals would form stable sulfides under these conditions. We've looked at other things that we haven't identified as sulfides but we haven't found any particularly striking activity with those kinds of materials.

S. Lynn *(University of California)*

I'd like to remind you that ferrous sulfide works very well as a catalyst for NO reduction as long as there is no oxygen present. The presence of oxygen led to sulfate formation which eliminated the catalytic activity.

Klimisch

There are some other base metals that also have problems with oxygen. For example, nickel on silica, or nickel unsupported, is pretty good, but as soon as you put oxygen in the system, it tends to poison the NO_x reduction activity.

PARTICIPANTS

I. A. Abu-Isa
General Motors Research
  Laboratories
General Motors Technical Center
Warren, Michigan 48090

D. E. Achey
AC Spark Plug Division
1300 North Dort Highway
Flint, Michigan 48556

G. J. K. Acres
Johnson Matthey & Company, Ltd.
Group Research Laboratories
South Way, Exhibition Grounds
Wembley, Middlesex HA9-OHW
England

W. G. Agnew
General Motors Research
  Laboratories
General Motors Technical Center
Warren, Michigan 48090

K. Aika
Tokyo Institute of Technology
Research Laboratory of Resources
  Utilization
O-Okayama, Megurio-Ku
Tokyo 152
Japan

H. Akamatsu
Toyota Motor Company, Ltd.
50 Poleto Avenue
Lyndshurst, New Jersey 07071

C. A. Amann
General Motors Research
  Laboratories
General Motors Technical Center
Warren, Michigan 48090

K. T. Antonius
General Motors Research
  Laboratories
General Motors Technical Center
Warren, Michigan 48090

D. R. Ashmead
Imperial Chemical Industries, Ltd.
Agricultural Division
P.O. Box 6
Billingham, Teesside TS23-1LB
England

G. J. Barnes
General Motors Research
  Laboratories
General Motors Technical Center
Warren, Michigan 48090

K. Baron
General Motors Research
  Laboratories
General Motors Technical Center
Warren, Michigan 48090

A. T. Bell
University of California
Department of Chemical
  Engineering
Berkeley, California 94720

J. D. Benson
General Motors Research
  Laboratories
General Motors Technical Center
Warren, Michigan 48090

W. E. Bernhardt
  Volkswagen AG
  Research & Development Division
  Research Department
  Combustion & Reaction Kinetics
  Wolfsburg D-3180
  Germany

F. Black
  Environmental Protection Agency
  National Environmental Research
    Center
  Research Triangle Park,
    North Carolina 27711

C. E. Bleil
  General Motors Research
    Laboratories
  General Motors Technical Center
  Warren, Michigan 48090

H. S. Bloch
  UOP, Inc.
  Ten UOP Plaza
  Algonquin & Mt. Prospect Roads
  Des Plaines, Illinois 60016

M. Boudart
  Stanford University
  Department of Chemical
    Engineering
  Stanford, California 94305

W. S. Briggs
  W. R. Grace & Company
  Davison Chemical Division
  Curtis Bay Works
  5500 Chemical Road
  Baltimore, Maryland 21226

J. M. Burkstrand
  General Motors Research
    Laboratories
  General Motors Technical Center
  Warren, Michigan 48090

R. L. Burwell, Jr.
  Northwestern University
  Department of Chemistry
  Evanston, Illinois 60201

J. B. Butt
  Northwestern University
  Department of Chemical
    Engineering
  Evanston, Illinois 60201

L. R. Buzan
  General Motors Research
    Laboratories
  General Motors Technical Center
  Warren, Michigan 48090

J. D. Caplan
  General Motors Research
    Laboratories
  General Motors Technical Center
  Warren, Michigan 48090

F. Caracciolo
  General Motors Research
    Laboratories
  General Motors Technical Center
  Warren, Michigan 48090

J. J. Carberry
  University of Notre Dame
  Department of Chemical
    Engineering
  Notre Dame, Indiana 46556

S. Carter
  GM Patent Section
  15-140 GM Building
  Detroit, Michigan 48202

J. A. Caton
  General Motors Research
    Laboratories
  General Motors Technical Center
  Warren, Michigan 48090

C. C. Chang
  General Motors Research
    Laboratories
  General Motors Technical Center
  Warren, Michigan 48090

J. M. Colucci
General Motors Research
   Laboratories
General Motors Technical Center
Warren, Michigan 48090

M. D. Cooper
General Motors Research
   Laboratories
General Motors Technical Center
Warren, Michigan 48090

G. M. Cornetti
Fiat Laboratories
Laboratori Centrali — D.C.R.
Dipartimento Chimica
Strada Torino, 50
Orbassano, Torino
Italy

W. D. Creps
General Motors Research
   Laboratories
General Motors Technical Center
Warren, Michigan 48090

A. L. Dent
Carnegie-Mellon University
Department of Chemical
   Engineering
Schenley Park
Pittsburgh, Pennsylvania 15213

T. W. DeWitt
National Science Foundation
Energy Related Research Office
1800 G Street, N.W.
Washington, D.C. 20550

D. S. Dickey
General Motors Research
   Laboratories
General Motors Technical Center
Warren, Michigan 48090

T. T. Dingo
General Motors Manufacturing Staff
General Motors Technical Center
Warren, Michigan 48090

F. G. Dwyer
Mobil Research &
   Development Corporation
Research Department
Paulsboro, New Jersey 08066

R. H. Ebel
Stamford Research Laboratories
Chemical Research Division
American Cyanamid Company
1937 West Main Street
Stamford, Connecticut 06904

P. H. Emmett
Department of Chemistry
Portland State University
P.O. Box 751
Portland, Oregon 07207

M. V. Ernest
W. R. Grace & Company
7379 Route 32
Columbia, Maryland 21044

E. Eusebi
General Motors Research
   Laboratories
General Motors Technical Center
Warren, Michigan 48090

L. J. Faix
Chevrolet Engineering Center
Engineering 1-217-A
Warren, Michigan 48090

R. J. Fedor
Gould, Inc.
Emission Controls Division
540 E. 105th Street
Cleveland, Ohio 44108

L. L. Fleck
General Motors Research
   Laboratories
General Motors Technical Center
Warren, Michigan 48090

W. J. Fleming
  General Motors Research
    Laboratories
  General Motors Technical Center
  Warren, Michigan 48090

W. D. France, Jr.
  General Motors Research
    Laboratories
  General Motors Technical Center
  Warren, Michigan 48090

Y. Fujitani
  Toyota Motor Company, Ltd.
  1 Toyota-Cho
  Toyota-Shi, Aichi-Ken
  Japan

N. E. Gallopoulos
  General Motors Research
    Laboratories
  General Motors Technical Center
  Warren, Michigan 48090

H. S. Gandhi
  Ford Motor Company
  P.O. Box 2053
  Scientific Research Staff
  Dearborn, Michigan 48121

K. D. Gardels
  General Motors Research
    Laboratories
  General Motors Technical Center
  Warren, Michigan 48090

R. A. Gast
  General Motors Research
    Laboratories
  General Motors Technical Center
  Warren, Michigan 48090

J. G. Gay
  General Motors Research
    Laboratories
  General Motors Technical Center
  Warren, Michigan 48090

N. A. Gjostein
  Ford Motor Company
  Scientific Research Staff
  Metallurgy Department
  Dearborn, Michigan 48020

J. L. Gland
  General Motors Research
    Laboratories
  General Motors Technical Center
  Warren, Michigan 48090

R. D. Gonzalez
  University of Rhode Island
  Department of Chemistry
  Kingston, Rhode Island 02881

W. L. Grube
  General Motors Research
    Laboratories
  General Motors Technical Center
  Warren, Michigan 48090

V. Haensel
  UOP, Inc.
  Ten UOP Plaza
  Algonquin & Mt. Prospect Roads
  Des Plaines, Illinois 60016

W. K. Hall
  University of Wisconsin
  Department of Chemistry
  Milwaukee, Wisconsin 53201

G. L. Haller
  Yale University
  Department of Engineering &
    Applied Science
  Nine Hillhouse Avenue
  New Haven, Connecticut 06520

R. S. Hansen
  Iowa State University
  Ames Laboratory, USAEC
  Ames, Iowa 50010

E. A. Hanysz
General Motors Research
Laboratories
General Motors Technical Center
Warren, Michigan 48090

H. F. Harnsberger
Chevron Research Company
576 Standard Avenue
Richmond, California 94802

G. Harrison
Matthey-Bishop, Inc.
Malvern, Pennsylvania 19355

W. E. Hauth
AC Spark Plug Division
1300 North Dort Highway
Flint, Michigan 48556

L. L. Hegedus
General Motors Research
Laboratories
General Motors Technical Center
Warren, Michigan 48090

K. Hellman
Environmental Protection Agency
Emission Control Technology
Division
Office of Air and Water Programs
Ann Arbor, Michigan 48105

J. Hensel
Degussa Wolfgang, FCPh
645 Hanau, Postfach 602
West Germany

R. J. Herrin
General Motors Research
Laboratories
General Motors Technical Center
Warren, Michigan 48090

J. M. Heuss
General Motors Research
Laboratories
General Motors Technical Center
Warren, Michigan 48090

J. W. Hightower
Rice University
Chemical Engineering Department
Houston, Texas 77001

D. Hill
Cadillac Motor Car Division
2860 Clark Avenue
Detroit, Michigan 48232

R. F. Hill
General Motors Research
Laboratories
General Motors Technical Center
Warren, Michigan 48090

R. N. Hollyer
General Motors Research
Laboratories
General Motors Technical Center
Warren, Michigan 48090

E. L. Holt
Exxon Research &
Engineering Company
P.O. Box 51
Linden, New Jersey 07036

J. C. Holzwarth
General Motors Research
Laboratories
General Motors Technical Center
Warren, Michigan 48090

D. S. Howarth
General Motors Research
Laboratories
General Motors Technical Center
Warren, Michigan 48090

T. Huls
Environmental Protection Agency
Emission Control Technology
Division
Office of Air and Water Programs
Ann Arbor, Michigan 48105

J. E. Hunter
General Motors Research
Laboratories
General Motors Technical Center
Warren, Michigan 48090

F. E. Jamerson
General Motors Research
Laboratories
General Motors Technical Center
Warren, Michigan 48090

W. C. Jones
American Motors Corporation
14250 Plymouth Road
Detroit, Michigan 48232

S. Katz
General Motors Research
Laboratories
General Motors Technical Center
Warren, Michigan 48090

J. R. Katzer
University of Delaware
Department of Chemical
Engineering
Newark, Delaware 19711

G. W. Keulks
University of Wisconsin
Department of Chemistry
Milwaukee, Wisconsin 53201

Michio Kimura
Japan Catalytic
International, Inc.
241 Walnut Road
Glen Cove, New York 11542

R. Klein
National Bureau of Standards
Washington, D.C. 20234

R. L. Klimisch
General Motors Research
Laboratories
General Motors Technical Center
Warren, Michigan 48090

T. P. Kobylinski
Gulf Research &
Development Company
P.O. Drawer 2038
Pittsburgh, Pennsylvania 15230

J. M. Komarmy
AC Spark Plug Division
1300 North Dort Highway
Flint, Michigan 48556

A. J. Kotwicki
General Motors Research
Laboratories
General Motors Technical Center
Warren, Michigan 48090

J. T. Kummer
Ford Motor Company
Scientific Laboratories
P.O. Box 2053
Dearborn, Michigan 48121

J. G. Larson
General Motors Research
Laboratories
General Motors Technical Center
Warren, Michigan 48090

A. Lawson
Ontario Research Foundation
Sheridan Park, Ontario
Canada

C. H. Lee
Gould, Inc.
540 E. 105th Street
Cleveland, Ohio 44108

L. L. Lewis
General Motors Research
Laboratories
General Motors Technical Center
Warren, Michigan 48090

K. H. Ludlum
Texaco Research Center
P.O. Box 509
Beacon, New York 12508

J. H. Lunsford
Texas A&M University
Department of Chemistry
College Station, Texas 77843

S. Lynn
University of California
Department of Chemical
Engineering
Berkeley, California 94720

D. P. McArthur
Union Research Center
P.O. Box 76
Brea, California 92621

W. Mannion
Engelhard Industries Division
Engelhard Minerals &
Chemicals Corporation
Menlo Park
Edison, New Jersey 08817

T. J. Mao
General Motors Research
Laboratories
General Motors Technical Center
Warren, Michigan 48090

C. Marks
General Motors Engineering Staff
General Motors Technical Center
Warren, Michigan 48090

S. W. Martens
General Motors Environmental
Activities Staff
General Motors Technical Center
Warren, Michigan 48090

R. J. McDonald
General Motors Research
Laboratories
General Motors Technical Center
Warren, Michigan 48090

M. L. McMillan
General Motors Research
Laboratories
General Motors Technical Center
Warren, Michigan 48090

G. H. Meguerian
Amoco Oil Company
Research & Development
Department
2500 New York Avenue
Whiting, Indiana 46394

D. Miles
General Motors Engineering Staff
General Motors Technical Center
Warren, Michigan 48090

E. J. Miller
General Motors Research
Laboratories
General Motors Technical Center
Warren, Michigan 48090

M. M. Mitchell, Jr.
Houdry Laboratories
Air Products and Chemicals, Inc.
P.O. Box 427
Marcus Hook, Pennsylvania 19061

J. R. Mondt
General Motors Research
Laboratories
General Motors Technical Center
Warren, Michigan 48090

J. J. Mooney
    Engelhard Industries Division
    Engelhard Minerals &
       Chemicals Corporation
    Menlo Park
    Edison, New Jersey 08817

J. Moran
    Environmental Protection Agency
    National Environmental Research
       Center
    Research Triangle Park,
       North Carolina 27711

W. R. Moser
    Exxon Research &
       Engineering Company
    P.O. Box 51
    Linden, New Jersey 07036

N. L. Muench
    General Motors Research
       Laboratories
    General Motors Technical Center
    Warren, Michigan 48090

P. C. Murray
    Chemical & Engineering News
    1155 Sixteenth Street, N.W.
    Washington, D.C. 20036

K. Nagano
    Isuzu Motors, Ltd.
    21415 Civic Center Drive
    Southfield, Michigan 48076

G. J. Nebel
    General Motors Research
       Laboratories
    General Motors Technical Center
    Warren, Michigan 48090

L. M. Niebylski
    Ethyl Corporation
    Research Laboratories
    1600 W. Eight Mile Road
    Ferndale, Michigan 48220

T. Ohara
    Nippon Shokubai Kagaku
       Kogyo Company, Ltd.
    Central Research Laboratory
    5-8 Nishlotabicho
    Osaka, Japan

T. Ono
    Japan Catalytic International, Inc.
    241 Walnut Road
    Glen Cove, New York 11542

K. Otto
    Ford Motor Company
    Scientific Laboratories
    Dearborn, Michigan 48121

A. Ozaki
    Tokyo Institute of Technology
    Research Laboratory of
       Resources Utilization
    O-Okayama, Meguru-Ku
    Tokyo 152, Japan

G. Parravano
    University of Michigan
    Department of Chemical
       Engineering
    Ann Arbor, Michigan 48104

D. Pashayan
    Environmental Protection Agency
    401 M Street, S.W.
    Washington, D.C. 20460

M. W. Pepper
    Exxon Research &
       Engineering Company
    P.O. Box 51
    Linden, New Jersey 07036

A. J. Perella
    Buick Motor Division
    902 E. Hamilton Avenue
    Flint, Michigan 48550

J. B. Peri
Amoco Oil Company
Research & Development
Department
P.O. Box 431
Whiting, Indiana 46394

M. S. Peters
University of Colorado
Department of Chemical
Engineering
Engineering Center AD-1-1
Boulder, Colorado 80302

E. E. Petersen
University of California
Department of Chemical
Engineering
Berkeley, California 94720

L. G. Pless
General Motors Research
Laboratories
General Motors Technical Center
Warren, Michigan 48090

N. M. Potter
General Motors Research
Laboratories
General Motors Technical Center
Warren, Michigan 48090

D. J. Pozniak
General Motors Research
Laboratories
General Motors Technical Center
Warren, Michigan 48090

R. A. Reck
General Motors Research
Laboratories
General Motors Technical Center
Warren, Michigan 48090

J. F. Rhodes
General Motors Research
Laboratories
General Motors Technical Center
Warren, Michigan 48090

G. H. Robinson
General Motors Research
Laboratories
General Motors Technical Center
Warren, Michigan 48090

T. Saito
Nissan Motor Company, Ltd.
P.O. Box 1606
Englewood Cliffs, New Jersey 07632

J. C. Schlatter
General Motors Research
Laboratories
General Motors Technical Center
Warren, Michigan 48090

H. Schwochert
Environmental Activities Staff
General Motors Technical Center
Warren, Michigan 48090

M. Shelef
Ford Motor Company
Scientific Research Staff
Dearborn, Michigan 48020

M. E. Sheridan
3M Company
3M Center — 201-2E
St. Paul, Minnesota 55101

H. Skala
UOP, Inc.
Ten UOP Plaza
Des Plaines, Illinois 60016

J. R. Smith
General Motors Research
Laboratories
General Motors Technical Center
Warren, Michigan 48090

J. Somers
Environmental Protection Agency
Emission Control Technology
Division
Office of Air and Water Programs
Ann Arbor, Michigan 48105

G. A. Somorjai
  University of California
  Department of Chemistry
  Berkeley, California 94720

R. A Spaulding
  General Motors Manufacturing Staff
  General Motors Technical Center
  Warren, Michigan 48090

J. A. Spearot
  General Motors Research
    Laboratories
  General Motors Technical Center
  Warren, Michigan 48090

T. Stephens
  Cadillac Motor Car Division
  2860 Clark Avenue
  Detroit, Michigan 48232

K. P. Stohl
  OxyCatalyst, Inc.
  East Biddle Street
  West Chester, Pennsylvania 19380

J. C. Summers
  General Motors Research
    Laboratories
  General Motors Technical Center
  Warren, Michigan 48090

K. C. Taylor
  General Motors Research
    Laboratories
  General Motors Technical Center
  Warren, Michigan 48090

D. M. Teague
  Chrysler Corporation
  P.O. Box 1118
  DIMS-418-19-40
  Detroit, Michigan 48231

R. F. Thomson
  General Motors Research
    Laboratories
  General Motors Technical Center
  Warren, Michigan 48090

J. A. Thorsen
  General Motors Overseas Operations
  9-135 General Motors Building
  Detroit, Michigan 48202

K. W. Thurston
  Oldsmobile Division
  920 Townsend Street
  Lansing, Michigan 48921

G. G. Tibbetts
  General Motors Research
    Laboratories
  General Motors Technical Center
  Warren, Michigan 48090

J. C. Tracy
  General Motors Research
    Laboratories
  General Motors Technical Center
  Warren, Michigan 48090

W. E. Tudor
  Pontiac Motor Division
  One Pontiac Plaza
  Pontiac, Michigan 48053

C. S. Tuesday
  General Motors Research
    Laboratories
  General Motors Technical Center
  Warren, Michigan 48090

R. J. Van Duyne
  AC Spark Plug Division
  1300 North Dort Highway
  Flint, Michigan 48556

P. T. Vickers
  General Motors Research
    Laboratories
  General Motors Technical Center
  Warren, Michigan 48090

R. J. H. Voorhoeve
  Bell Laboratories
  600 Mountain Avenue
  Murray Hill, New Jersey 07974

W. M. Wang
AC Spark Plug Division
1300 North Dort Highway
Flint, Michigan 48556

E. E. Weaver
Ford Motor Company
World Headquarters, Room 223
American Road
Dearborn, Michigan 48121

P. Weiss
General Motors Research
   Laboratories
General Motors Technical Center
Warren, Michigan 48090

E. F. Weller
General Motors Research
   Laboratories
General Motors Technical Center
Warren, Michigan 48090

D. W. Wendland
General Motors Research
   Laboratories
General Motors Technical Center
Warren, Michigan 48090

F. L. Williams
General Motors Research
   Laboratories
General Motors Technical Center
Warren, Michigan 48090

L. M. Wing
AC Spark Plug Division
1300 North Dort Highway
Flint, Michigan 48556

J. B. Wingo
Environmental Protection Agency
National Environmental Research
   Center
Research Triangle Park,
   North Carolina 27711

H. Wise
Stanford Research Institute
Menlo Park, California 94025

H. C. Yao
Ford Motor Company
Scientific Research Staff
Dearborn, Michigan 48020

K. C. Youn
Shell Development Company
P.O. Box 481
Houston, Texas 77001

J. S. Zwerner
Chevrolet Engineering Center
Engineering I-229
Warren, Michigan 48090

AUTHORS OF CITED REFERENCES

Numbers in parentheses indicate the numbers of the references when these are cited in the text without the names of the authors.

Aberdam, D., 230
Adams, C. R., 166, 243 (7, 10)
Adams, D. L., 29
Addison, C. C., 58
Ageton, R. W., 313
Ahmad, L. I., 230
Aika, K., 116 (1, 5, 6, 7, 8, 11, 12, 13, 14, 15)
Ablesova, K., 43
Aldag, A., 166
Alekseev, A. V., 58
Alexeyev, A., 90
Alkhazov, T. G., 243
Allen, M. G., 90
Amirnazmi, A., 89, 231
Anderson, H. C., 90 (43, 45, 46), 165, 166
Anderson, J. J., 43
Armijo, J. S., 58
Arnott, R. J., 231 (31, 32)
Ashmead, D. R., 313
Ashmore, P. G., 243

Bachman, P. W., 89
Baker, R. A., 43
Banks, E., 230
Barck, H., 90
Barnes, G. J., 127, 260
Bartok, W., 259
Bauerle, G. L., 212 (6, 7)

Beck, W., 43
Bedford, G., 166
Bell, E. W., 212
Benard, J., 165 (21, 22, 23, 26)
Benesi, H. A., 166
Benson, J. E., 89, 166, 231
Bernhardt, W. E., 304
Bernstein, L. S., 212, 290 (1, 3), 313 (10, 11), 314
Berthier, Y., 165
Bettman, M., 58
Bishop, R. J., 16
Blumenthal, J. L., 43
Blyholder, G., 90, 165
Bodenstein, M., 90
Bond, G. C., 231
Bonzel, H. P., 29, 166
Boreskov, G. K., 90 (48, 49), 116, 231
Bouchet, G., 230
Boudart, M., 89, 104 (2, 7, 9), 116, 166, 231
Braphy, J. H., 128
Bridges, J. M., 58
Brill, R., 104
Brunauer, S., 104, 212
Buttgereit, W., 304

Callahan, J. L., 243
Campau, R. M., 165
Campbell, J. S., 313

Prichard, C. R., 90
Princiotta, F. R., 259
Pundt, D., 304
Pusateri, R. J., 166

Raccah, P. M., 231
Rahman, M. M., 230
Rakowsky, F. W., 165, 314
Raman, A. K. S., 313
Randall, J. J., 230
Ray, J. D., 89
Redhead, P. A., 43
Remeika, J. P., 89, 230 (6, 8, 9, 11)
Rewick, R. T), 243
Rhodin, T. N., 166
Richter, E. L., 104
Riesz, C. H., 43
Rifkin, E. B., 279 (8, 10)
Riley, R. F., 16
Rinker, R. G., 89
Robbins, R. C., 260
Robertson, W. D., 166
Robinson, E., 260
Roev, L., 90
Romeo, P. L., 166
Rose, R., 128
Ross, A., 279 (8, 9)
Roth, J. F., 43, 89
Ruch, E., 104
Russell, H., 89
Ryan, J. P., 313
Rybalko, V. F., 166
Rymer, G. T., 58

Sachtler, W. M. H., 90, 231, 243 (7, 8)
Sakaida, R. R., 89
Saleh, J., 165 (18, 19, 20)
Sancier, K. M., 243 (11, 22)
Schachner, H., 90
Schachner, M., 58
Schaefer, H., 116
Schiavello, M., 58
Schlatter, J. C., 313

Schluze, V., 43
Schmidt, K. H., 43
Schmidt, L. D., 166
Schock, D. N., 165, 314
Schoenherr, R., 279
Schrieffer, T. R., 29
Schowalter, W. R., 243
Schuit, G. C. A., 243 (7, 18)
Scott, W. R., 230
Service, G. R., 212
Shvachko, V. I., 166
Sheets, R. W., 165
Shelef, M., 16, 29 (3, 4), 43, 58 (2, 3, 4,
5, 6, 11), 89 (6, 23, 27), 90, 128 (8, 15,
16, 17, 18), 165 (6, 7, 8, 14), 166 (38,
59), 186, 212, 231 (19, 30), 243, 260,
279, 313 (2, 8), 314
Shirane, G., 230
Shulz, G., 116
Siepmann, R., 89
Simpson, H. D., 166
Sin-chou, Hu., 165
Sis, L. B., 231
Slaughter, J. I., 89
Sloane, R. J., 90
Smith, C. S., 166
Smith, J. M., 166
Solbakken, A., 90
Sorensen, L. L., 212
Sorenson, S. C., 231
Sourirajan, S., 43, 89
Steele, D. R., 90 (43, 45), 165
Stefan, A., 165
Sternling, C. V., 90
Stone, F. S., 58
Street, J. C., 279
Szostak, R. J., 165

Taarit, Y. Ben, 16 (6, 10)
Tagami, M., 212
Takezawa, N., 116
Tamaru, K., 116
Tanaka, K., 116

SUBJECT INDEX